Student Solutions Manual

to accompany

Calculus: Single Variable

Seventh Edition

Deborah Hughes-Hallett

University of Arizona

Andrew M. Gleason

Harvard University

William G. McCallum

University of Arizona

et al.

Prepared by

Elliot J. Marks

WILEY

To order books or for customer service please, call 1-800-CALL WILEY (225-5945).

This material is based upon work supported by the National Science Foundation under Grant No. DUE-9352905. Opinions expressed are those of the authors and not necessarily those of the Foundation.

ISBN-13 978-1-119-37899-0

Printed in the United States of America

V10003535_081018

CONTENTS

CONTENTS

CHAPTER ONE

Solutions for Section 1.1

Exercises

1. Since t represents the number of years since 2010, we see that $f(5)$ represents the population of the city in 2015. In 2015, the city's population was 7 million.

5. The slope is $(3 - 2)/(2 - 0) = 1/2$. So the equation of the line is $y = (1/2)x + 2$.

9. Rewriting the equation as

$$y = -\frac{12}{7}x + \frac{2}{7}$$

shows that the line has slope $-12/7$ and vertical intercept $2/7$.

13. (a) is (V), because slope is negative, vertical intercept is 0
    (b) is (VI), because slope and vertical intercept are both positive
    (c) is (I), because slope is negative, vertical intercept is positive
    (d) is (IV), because slope is positive, vertical intercept is negative
    (e) is (III), because slope and vertical intercept are both negative
    (f) is (II), because slope is positive, vertical intercept is 0

17. $y = 5x - 3$. Since the slope of this line is 5, we want a line with slope $-\frac{1}{5}$ passing through the point $(2, 1)$. The equation is $(y - 1) = -\frac{1}{5}(x - 2)$, or $y = -\frac{1}{5}x + \frac{7}{5}$.

21. Since x goes from 1 to 5 and y goes from 1 to 6, the domain is $1 \le x \le 5$ and the range is $1 \le y \le 6$.

25. The domain is all x-values, as the denominator is never zero. The range is $0 < y \le \frac{1}{2}$.

29. For some constant k, we have $S = kh^2$.

Problems

33. (a) When the car is 5 years old, it is worth $6000.
    (b) Since the value of the car decreases as the car gets older, this is a decreasing function. A possible graph is in Figure 1.1:

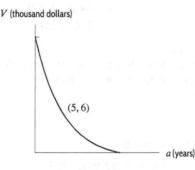

Figure 1.1

    (c) The vertical intercept is the value of V when $a = 0$, or the value of the car when it is new. The horizontal intercept is the value of a when $V = 0$, or the age of the car when it is worth nothing.

37. The year 1949 was $2012 - 1949 = 63$ years before 2012 so 1949 corresponds to $t = 63$. Similarly, we see that the year 2000 corresponds to $t = 12$. Thus, an expression that represents the statement is:

$$f(63) = f(12).$$

41. See Figure 1.2.

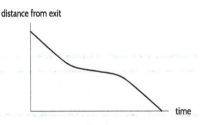

distance from exit

time

Figure 1.2

45. See Figure 1.3.

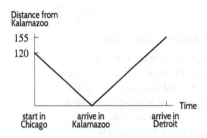

Distance from
Kalamazoo

155

120

Time

start in
Chicago

arrive in
Kalamazoo

arrive in
Detroit

Figure 1.3

49. The difference quotient $\Delta Q / \Delta r$ equals the slope of the line and represents the increase in the quantity of grass per millimeter of rainfall. We see from the graph that the slope of the line for 1939 is larger than the slope of the line for 1997. Thus, each additional 1 mm of rainfall in 1939 led to a larger increase in the quantity of grass than in 1997.

53. (a) We find the slope m and intercept b in the linear equation $C = b + mw$. To find the slope m, we use

$$m = \frac{\Delta C}{\Delta w} = \frac{12.32 - 8}{68 - 32} = 0.12 \text{ dollars per gallon.}$$

We substitute to find b:

$$C = b + mw$$
$$8 = b + (0.12)(32)$$
$$b = 4.16 \text{ dollars.}$$

The linear formula is $C = 4.16 + 0.12w$.

(b) The slope is 0.12 dollars per gallon. Each additional gallon of waste collected costs 12 cents.

(c) The intercept is $4.16. The flat monthly fee to subscribe to the waste collection service is $4.16. This is the amount charged even if there is no waste.

57. (a) (i) $f(2001) = 272$
           (ii) $f(2014) = 525$

(b) The average yearly increase is the rate of change.

$$\text{Yearly increase } = \frac{f(2014) - f(2001)}{2014 - 2001} = \frac{525 - 272}{13} = 19.46 \text{ billionaires per year.}$$

(c) Since we assume the rate of increase remains constant, we use a linear function with slope 19.46 billionaires per year. The equation is

$$f(t) = b + 19.46t$$

where $f(2001) = 272$, so

$$272 = b + 19.46(2001)$$
$$b = -38{,}667.5.$$

Thus, $f(t) = 19.46t - 38{,}667.5$.

61. (a) The scale of the graph makes it impossible to read values of $g(4210)$ and $g(4209)$ accurately enough to evaluate their difference. But that difference equals the slope of the line, which we can estimate. Using the points $(3000, 70)$ and $(5000, 50)$ on the line, we have

$$\text{Slope} = \frac{50 - 70}{5000 - 3000} = -0.01.$$

So

$$\Delta y = \text{Slope} \times \Delta x$$
$$= (-0.01)(4210 - 4209) = -0.01.$$

Thus $g(4210) - g(4209) = -0.01$.

(b) We have

$$\Delta y = \text{Slope} \times \Delta x$$
$$= (-0.01)(3760 - 3740) = -0.2.$$

Thus $g(3760) - g(3740) = -0.2$.

65. (a) Since 2008 corresponds to $t = 0$, the average annual sea level in Aberdeen in 2008 was 7.094 meters.

(b) Looking at the table, we see that the average annual sea level was 7.019 twenty five years before 2008, or in the year 1983. Similar reasoning shows that the average sea level was 6.957 meters 125 years before 2008, or in 1883.

(c) Because 125 years before 2008 the year was 1883, we see that the sea level value corresponding to the year 1883 is 6.957 (this is the sea level value corresponding to $t = 125$). Similar reasoning yields the table:

Year	1883	1908	1933	1958	1983	2008
S	6.957	6.938	6.965	6.992	7.019	7.094

69. (a) If the bakery owner decreases the price, the customers want to buy more. Thus, the slope of $d(q)$ is negative. If the owner increases the price, she is make more cakes. Thus the slope of $s(q)$ is positive.

(b) To determine whether an ordered pair (q, p) is a solution to the inequality, we substitute the values of q and p into the inequality and see whether the resulting statement is true. Substituting the two ordered pairs gives

$(60, 18)$: $18 \leq 20 - 60/20$, or $18 \leq 17$.   This is false, so $(60, 18)$ is not a solution.

$(120, 12)$: $12 \leq 20 - 120/20$, or $12 \leq 14$.   This is true, so $(120, 12)$ is a solution.

The pair $(60, 18)$ is not a solution to the inequality $p \leq 20 - q/20$. This means that the price $18 is higher than the unit price at which customers would be willing to buy a total of 60 cakes. So customers are not willing to buy 60 cakes at $18. The pair $(120, 12)$ is a solution, meaning that $12 is not more than the price at which customers would be willing to buy 120 cakes. Thus customers are willing to buy a total of 120 (and more) cakes at $12. Each solution (q, p) represents a quantity of cakes q that customers would be willing to buy at the unit price p.

(c) In order to be a solution to both of the given inequalities, a point (q, p) must lie on or below the line $p = 20 - q/20$ and on or above the line $p = 11 + q/40$. Thus, the solution set of the given system of inequalities is the region shaded in Figure 1.4.

A point (q, p) in this region represents a quantity q of cakes that customers would be willing to buy, and that the bakery-owner would be willing to make and sell, at the price p.

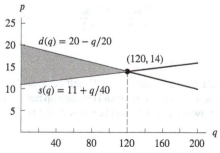

Figure 1.4: Possible cake sales at different prices and quantities

(d) To find the rightmost point of this region, we need to find the intersection point of the lines $p = 20 - q/20$ and $p = 11 + q/40$. At this point, p is equal to both $20 - q/20$ and $11 + q/40$, so these two expressions are equal to each other:

$$20 - \frac{q}{20} = 11 + \frac{q}{40}$$
$$9 = \frac{q}{20} + \frac{q}{40}$$
$$9 = \frac{3q}{40}$$
$$q = 120.$$

Therefore, $q = 120$ is the maximum number of cakes that can be sold at a price at which customers are willing to buy them all, and the owner of the bakery is willing to make them all. The price at this point is $p = 20 - 120/20 = 14$ dollars. (In economics, this price is called the *equilibrium price*, since at this point there is no incentive for the owner of the bakery to raise or lower the price of the cakes.)

Strengthen Your Understanding

73. The line $x = 3$ is vertical, so it is not a function of x: it fails the vertical line test. The line $y = 3$ is horizontal, hence a linear function of x with slope 0.

77. One possible answer is $f(x) = 2x + 3$.

81. True. Solving for y on the second equation, we see that the second linear function has the same equation as the first:

$$x = -y + 1$$
$$x - 1 = -y$$
$$-x + 1 = y.$$

85. (b) and (c). For $g(x) = \sqrt{x}$, the domain and range are all nonnegative numbers, and for $h(x) = x^3$, the domain and range are all real numbers.

Solutions for Section 1.2

Exercises

1. The graph shows a concave up function.

5. Initial quantity = 5; growth rate = 0.07 = 7%.

9. Since $e^{0.25t} = \left(e^{0.25}\right)^t \approx (1.2840)^t$, we have $P = 15(1.2840)^t$. This is exponential growth since 0.25 is positive. We can also see that this is growth because $1.2840 > 1$.

13. (a) Let $Q = Q_0 a^t$. Then $Q_0 a^5 = 75.94$ and $Q_0 a^7 = 170.86$. So

$$\frac{Q_0 a^7}{Q_0 a^5} = \frac{170.86}{75.94} = 2.25 = a^2.$$

So $a = 1.5$.
(b) Since $a = 1.5$, the growth rate is $r = 0.5 = 50\%$.

17. The function is increasing and concave up between D and E, and between H and I. It is increasing and concave down between A and B, and between E and F. It is decreasing and concave up between C and D, and between G and H. Finally, it is decreasing and concave down between B and C, and between F and G

Problems

21. Table D is the only table that could represent an exponential function of x. This is because, in Table D, the ratio of y values is the same for all equally spaced x values. Thus, the y values in the table have a constant percent rate of decrease:

$$\frac{9}{18} = \frac{4.5}{9} = \frac{2.25}{4.5} = 0.5.$$

Table A represents a constant function of x, so it cannot represent an exponential function. In Table B, the ratio between y values corresponding to equally spaced x values is not the same. In Table C, y decreases and then increases as x increases. So neither Table B nor Table C can represent exponential functions.

25. (a) This is a linear function, corresponding to $g(x)$, whose rate of decrease is constant, 0.6.
  (b) This graph is concave down, so it corresponds to a function whose rate of decrease is increasing, like $h(x)$. (The rates are $-0.2, -0.3, -0.4, -0.5, -0.6$.)
  (c) This graph is concave up, so it corresponds to a function whose rate of decrease is decreasing, like $f(x)$. (The rates are $-10, -9, -8, -7, -6$.)

29. We look for an equation of the form $y = y_0 a^x$ since the graph looks exponential. The points $(1, 6)$ and $(2, 18)$ are on the graph, so

$$6 = y_0 a^1 \quad \text{and} \quad 18 = y_0 a^2$$

Therefore $a = \frac{y_0 a^2}{y_0 a} = \frac{18}{6} = 3$, and so $6 = y_0 a = y_0 \cdot 3$; thus, $y_0 = 2$. Hence $y = 2(3^x)$.

33. (a) See Figure 1.5.
  (b) No. The points in the plot do not lie even approximately on a straight line, so a linear model is a poor choice.
  (c) No. The plot shows that H is a decreasing function of Y that might be leveling off to an asymptote at $H = 0$. These are features of an exponential decay function, but their presence does not show that H is an exponential function of Y. We can do a more precise check by dividing each value of H by the previous year's H.

$$\frac{\text{Houses in 2011}}{\text{Houses in 2010}} = \frac{13 \text{ million}}{18.3 \text{ million}} = 0.710$$

$$\frac{\text{Houses in 2012}}{\text{Houses in 2011}} = \frac{7.8 \text{ million}}{13 \text{ million}} = 0.600$$

$$\frac{\text{Houses in 2013}}{\text{Houses in 2012}} = \frac{3.9 \text{ million}}{7.8 \text{ million}} = 0.500$$

$$\frac{\text{Houses in 2014}}{\text{Houses in 2013}} = \frac{1 \text{ million}}{3.9 \text{ million}} = 0.256$$

$$\frac{\text{Houses in 2015}}{\text{Houses in 2014}} = \frac{0.5 \text{ million}}{1 \text{ million}} = 0.500.$$

If H were an exponential function of Y, the ratios would be approximately constant. Since they are not approximately constant, we see that an exponential model is a poor choice.

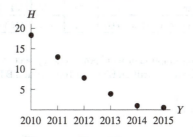

Figure 1.5

37. If production continues to decay exponentially at a continuous rate of 9.1% per year, the production $f(t)$ at time t years after 2008 is

$$f(t) = 84e^{-0.091t}.$$

In 2025, production is predicted to be

$$f(17) = 84e^{-0.091(17)} = 17.882 \text{ million barrels per day.}$$

41. (a) Since y is the number of US colonies, we need a growth factor smaller than 1 in order to describe a colony loss. A loss of 18.9% per year means a yearly growth factor equal to

$$1 - 0.189 = 0.811.$$

An annual percentage loss smaller than 18.9% would still be sustainable, and would result in a growth factor larger than 0.811 but still smaller than 1. Hence, II is the only function that could describe an economically sustainable colony loss trend.

(b) Function I says the number of US bee colonies is growing by 18.9% a year. Function II says bee colony loss is 10.9% per year. Function III indicates a bee colony loss of 20.2% a year.

45. (a) After 50 years, the amount of money is

$$P = 2P_0.$$

After 100 years, the amount of money is

$$P = 2(2P_0) = 4P_0.$$

After 150 years, the amount of money is

$$P = 2(4P_0) = 8P_0.$$

(b) The amount of money in the account doubles every 50 years. Thus in t years, the balance doubles $t/50$ times, so

$$P = P_0 2^{t/50}.$$

49. (a) The 1.3 represents the number of food bank users, in thousands, when $t = 0$, that is in 2006.

(b) The continuous growth rate is 0.81, or 81%, per year.

(c) Since $e^{0.81t} = (e^{0.81})^t = 2.248^t$, we see that the annual growth factor is 2.248. Thus, the annual growth rate is $2.248 - 1 = 1.248 = 124.8\%$ per year.

(d) Since the annual growth rate is over 100% and the annual growth factor is 2.248, the number of users more than doubles each year. The doubling time is less than one year.

53. (a) From the figure we can read-off the approximate percent growth for each year over the previous year:

Year	2009	2010	2011	2012	2013	2014
% growth over previous yr	32	28	26	15	20	8

Since the annual growth factor from 2009 to 2010 was $1 + 0.28 = 1.28$ and $721(1 + 0.28) = 922.88$, the US consumed approximately 923 trillion BTUs of wind power energy in 2010. Since the annual growth factor from 2010 to 2011 was $1 + 0.26 = 1.26$ and $922.88(1 + 0.26) = 1162.829$, the US consumed about 1163 trillion BTUs of wind power energy in 2010.

(b) Completing the table of annual consumption of wind power and plotting the data gives Figure 1.6.

Year	2009	2010	2011	2012	2013	2014
Consumption of wind power (trillion BTU)	721	923	1163	1337	1605	1733

(c) The largest increase in the US consumption of wind power energy occurred in 2013. In this year the US consumption of wind power energy rose by about 268 trillion BTUs to 1605 trillion BTUs, up from 1337 trillion BTUs in 2012.

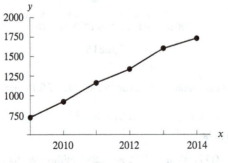

Figure 1.6

Strengthen Your Understanding

57. The function $y = e^{-0.25x}$ is decreasing but its graph is concave up.

61. One possible answer is $q = 2.2(0.97)^t$.

65. True. Using algebra and rules of exponents, we can rewrite $f(x)$ so that it has the form $P_0 e^{kx}$:

$$f(x) = e^{2x}/(2e^{5x})$$
$$= \frac{1}{2}e^{2x-5x}$$
$$= \frac{1}{2}e^{-3x}.$$

69. True. Suppose $y = Ab^x$ and we start at the point (x_1, y_1), so $y_1 = Ab^{x_1}$. Then increasing x_1 by 1 gives $x_1 + 1$, so the new y-value, y_2, is given by

$$y_2 = Ab^{x_1+1} = Ab^{x_1}b = (Ab^{x_1})b,$$

so

$$y_2 = by_1.$$

Thus, y has increased by a factor of b, so $b = 3$, and the function is $y = A3^x$.

However, if x_1 is increased by 2, giving $x_1 + 2$, then the new y-value, y_3, is given by

$$y_3 = A3^{x_1+2} = A3^{x_1}3^2 = 9A3^{x_1} = 9y_1.$$

Thus, y has increased by a factor of 9.

Solutions for Section 1.3

Exercises

1.

(a)
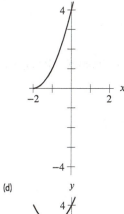

(b)
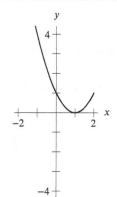

(c)
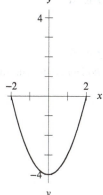

(d)

(e)

(f)
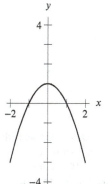

5. This graph is the graph of $m(t)$ shifted to the right by one unit. See Figure 1.7.

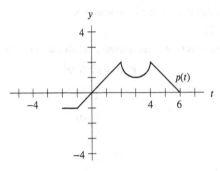

Figure 1.7

9. (a) $f(g(1)) = f(1 + 1) = f(2) = 2^2 = 4$
   (b) $g(f(1)) = g(1^2) = g(1) = 1 + 1 = 2$
   (c) $f(g(x)) = f(x + 1) = (x + 1)^2$
   (d) $g(f(x)) = g(x^2) = x^2 + 1$
   (e) $f(t)g(t) = t^2(t + 1)$

13. (a) $f(t + 1) = (t + 1)^2 + 1 = t^2 + 2t + 1 + 1 = t^2 + 2t + 2.$
   (b) $f(t^2 + 1) = (t^2 + 1)^2 + 1 = t^4 + 2t^2 + 1 + 1 = t^4 + 2t^2 + 2.$
   (c) $f(2) = 2^2 + 1 = 5.$
   (d) $2f(t) = 2(t^2 + 1) = 2t^2 + 2.$
   (e) $(f(t))^2 + 1 = (t^2 + 1)^2 + 1 = t^4 + 2t^2 + 1 + 1 = t^4 + 2t^2 + 2.$

17. $m(z) - m(z - h) = z^2 - (z - h)^2 = 2zh - h^2.$

21. Since

$$f(-x) = (-x)^4 - (-x)^2 + 3 = x^4 - x^2 + 3 = f(x),$$

we see f is even

25. Since

$$f(-x) = (-x)((-x)^2 - 1) = -x(x^2 - 1) = -f(x),$$

we see f is odd

29. Since a horizontal line cuts the graph of $f(x) = x^2 + 3x + 2$ two times, f is not invertible. See Figure 1.8.

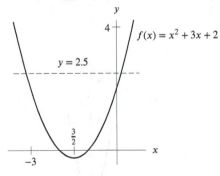

Figure 1.8

33. (a) $f(10,000)$ represents the value of C corresponding to $A = 10,000$, or in other words the cost of building a 10,000 square-foot store.
   (b) $f^{-1}(20,000)$ represents the value of A corresponding to $C = 20,000$, or the area in square feet of a store which would cost $20,000 to build.

Problems

37. This looks like a shift of the graph $y = -x^2$. The graph is shifted to the left 1 unit and up 3 units, so a possible formula is $y = -(x+1)^2 + 3$.

41. We have approximately $u(10) = 13$ and $v(13) = 60$ so $v(u(10)) = 60$.

45. $f(f(1)) \approx f(-0.4) \approx -0.9$.

49. $f(x) = x^3$, $g(x) = x + 1$.

53. The tree has $B = y - 1$ branches on average and each branch has $n = 2B^2 - B = 2(y-1)^2 - (y-1)$ leaves on average. Therefore

$$\text{Average number of leaves } = Bn = (y-1)(2(y-1)^2 - (y-1)) = 2(y-1)^3 - (y-1)^2.$$

57. (a) We find $f^{-1}(2)$ by finding the x value corresponding to $f(x) = 2$. Looking at the graph, we see that $f^{-1}(2) = -1$.

   (b) We construct the graph of $f^{-1}(x)$ by reflecting the graph of $f(x)$ over the line $y = x$. The graphs of $f^{-1}(x)$ and $f(x)$ are shown together in Figure 1.9.

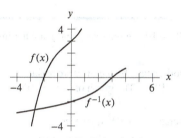

Figure 1.9

61. Not invertible, since it costs the same to mail a 50-gram letter as it does to mail a 51-gram letter.

65. The combined carbon footprint, in kilograms of GHGs, of a water bottle that has traveled 150 km from its production source and one that has remained at the source.

69. The time elapsed is: $f^{-1}(30)$ min.

73. (a) The graph shows that $f(15)$ is approximately 48. So, the place to find find 15 million-year-old rock is about 48 meters below the Atlantic sea floor.

   (b) Since f is increasing (not decreasing, since the depth axis is reversed!), f is invertible. To confirm, notice that the graph of f is cut by a horizontal line at most once.

   (c) Look at where the horizontal line through 120 intersects the graph of f and read downward: $f^{-1}(120)$ is about 35. In practical terms, this means that at a depth of 120 meters down, the rock is 35 million years old.

   (d) First, we standardize the graph of f so that time and depth are increasing from left to right and bottom to top. Points (t, d) on the graph of f correspond to points (d, t) on the graph of f^{-1}. We can graph f^{-1} by taking points from the original graph of f, reversing their coordinates, and connecting them. This amounts to interchanging the t and d axes, thereby reflecting the graph of f about the line bisecting the 90° angle at the origin. Figure 1.10 is the graph of f^{-1}. (Note that we cannot find the graph of f^{-1} by flipping the graph of f about the line $t = d$ in because t and d have different scales in this instance.)

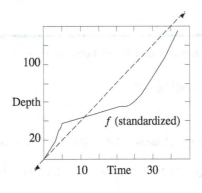
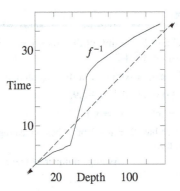

Figure 1.10: Graph of f, reflected to give that of f^{-1}

Strengthen Your Understanding

77. Since $f(g(x)) = 3(-3x - 5) + 5 = -9x - 10$, we see that f and g are not inverse functions.

81. One possibility is $f(x) = x^2 + 2$.

85. True. The graph of $y = 10^x$ is moved horizontally by h units if we replace x by $x - h$ for some number h. Writing $100 = 10^2$, we have $f(x) = 100(10^x) = 10^2 \cdot 10^x = 10^{x+2}$. The graph of $f(x) = 10^{x+2}$ is the graph of $g(x) = 10^x$ shifted two units to the left.

89. False. For $x < 0$, as x increases, x^2 decreases, so e^{-x^2} increases.

93. True. A linear function can be written as $y = b + mx$. Hence, writing $f(x) = c + nx$ and $g(x) = d + px$ and using algebra, we see that:

$$f(g(x)) = f(d + px) = c + n(d + px) = c + nd + npx.$$

So, setting $b = c + nd$ and $m = np$, we see that $f(g(x))$ is of the form $b + mx$. Similarly, we can check $g(f(x))$ is of this form.

97. This is impossible. As x increases, $g(x)$ decreases. As $g(x)$ decreases, so does $f(g(x))$ because f is increasing (an increasing function increases as its variable increases, so it decreases as its variable decreases).

Solutions for Section 1.4

Exercises

1. Using the identity $e^{\ln x} = x$, we have $e^{\ln(1/2)} = \frac{1}{2}$.

5. Using the rules for ln, we have

$$\ln\left(\frac{1}{e}\right) + \ln AB = \ln 1 - \ln e + \ln A + \ln B$$
$$= 0 - 1 + \ln A + \ln B$$
$$= -1 + \ln A + \ln B.$$

9. Isolating the exponential term

$$20 = 50(1.04)^x$$
$$\frac{20}{50} = (1.04)^x.$$

Taking logs of both sides

$$\log \frac{2}{5} = \log(1.04)^x$$

$$\log \frac{2}{5} = x \log(1.04)$$

$$x = \frac{\log(2/5)}{\log(1.04)} = -23.4.$$

13. To solve for x, we first divide both sides by 600 and then take the natural logarithm of both sides.

$$\frac{50}{600} = e^{-0.4x}$$

$$\ln(50/600) = -0.4x$$

$$x = \frac{\ln(50/600)}{-0.4} \approx 6.212.$$

17. Using the rules for ln, we have

$$2x - 1 = x^2$$

$$x^2 - 2x + 1 = 0$$

$$(x - 1)^2 = 0$$

$$x = 1.$$

21. Taking logs of both sides yields

$$nt = \frac{\log \left(\frac{Q}{Q_0} \right)}{\log a}.$$

Hence

$$t = \frac{\log \left(\frac{Q}{Q_0} \right)}{n \log a} = \frac{\log Q - \log Q_0}{n \log a}.$$

25. Since we want $(1.5)^t = e^{kt} = (e^k)^t$, so $1.5 = e^k$, and $k = \ln 1.5 = 0.4055$. Thus, $P = 15e^{0.4055t}$. Since 0.4055 is positive, this is exponential growth.

29. If $p(t) = (1.04)^t$, then, for p^{-1} the inverse of p, we should have

$$(1.04)^{p^{-1}(t)} = t,$$

$$p^{-1}(t) \log(1.04) = \log t,$$

$$p^{-1}(t) = \frac{\log t}{\log(1.04)} \approx 58.708 \log t.$$

Problems

33. At the doubling time, $t = 10$, we have $p = 2p_0$. Thus

$$p_0 e^{10k} = 2p_0$$

$$e^{10k} = 2$$

$$10k = \ln 2$$

$$k = \frac{1}{10} \ln 2 = 0.0693.$$

The function $p = p_0 e^{0.0693t}$ has doubling time equal to 10.

37. (a) Assuming the US population grows exponentially, we have population $P(t) = 281.4e^{kt}$ at time t years after 2000. Using the 2013 population, we have

$$316.1 = 281.4e^{13k}$$

$$k = \frac{\ln(316.1/281.4)}{13} = 0.00894.$$

We want to find the time t in which

$$350 = 281.4e^{0.00894t}$$
$$t = \frac{\ln(350/281.4)}{0.00926} = 24.4 \text{ years.}$$

This model predicts the population to go over 350 million 24.4 years after 2000, in the year 2024.

(b) Evaluate $P = 281.4e^{0.00894t}$ for $t = 20$ to find $P = 336.49$ million people.

41. (a) Since the percent increase in deaths during a year is constant for constant increase in pollution, the number of deaths per year is an exponential function of the quantity of pollution. If Q_0 is the number of deaths per year without pollution, then the number of deaths per year, Q, when the quantity of pollution is x micrograms per cubic meter of air is

$$Q = Q_0(1.0033)^x.$$

(b) We want to find the value of x making $Q = 2Q_0$, that is,

$$Q_0(1.0033)^x = 2Q_0.$$

Dividing by Q_0 and then taking natural logs yields

$$\ln\left((1.0033)^x\right) = x \ln 1.0033 = \ln 2,$$

so

$$x = \frac{\ln 2}{\ln 1.0033} = 210.391.$$

When there are 210.391 micrograms of pollutants per cubic meter of air, respiratory deaths per year are double what they would be in the absence of air pollution.

45. (a) The initial dose is 10 mg.

(b) Since $0.82 = 1 - 0.18$, the decay rate is 0.18, so 18% leaves the body each hour.

(c) When $t = 6$, we have $A = 10(0.82)^6 = 3.04$. The amount in the body after 6 hours is 3.04 mg.

(d) We want to find the value of t when $A = 1$. Using logarithms:

$$1 = 10(0.82)^t$$
$$0.1 = (0.82)^t$$
$$\ln(0.1) = t \ln(0.82)$$
$$t = 11.60 \text{ hours.}$$

After 11.60 hours, the amount is 1 mg.

49. (a) $B(t) = B_0 e^{0.067t}$

(b) $P(t) = P_0 e^{0.033t}$

(c) If the initial price is \$50, then

$$B(t) = 50e^{0.067t}$$
$$P(t) = 50e^{0.033t}.$$

We want the value of t such that

$$B(t) = 2P(t)$$
$$50e^{0.067t} = 2 \cdot 50e^{0.033t}$$
$$\frac{e^{0.067t}}{e^{0.033t}} = e^{0.034t} = 2$$
$$t = \frac{\ln 2}{0.034} = 20.387 \text{ years .}$$

Thus, when $t = 20.387$ the price of the textbook was predicted to be double what it would have been had the price risen by inflation only. This occurred in the year 2000.

53. Let $P(t)$ be the world population in billions t years after 2010.

(a) Assuming exponential growth, we have

$$P(t) = 6.9e^{kt}.$$

In 2050, we have $t = 40$ and we expect the population then to be 9 billion, so

$$9 = 6.9e^{k \cdot 40}.$$

Solving for k, we have

$$e^{k \cdot 40} = \frac{9}{6.9}$$

$$k = \frac{1}{40} \ln\left(\frac{9}{6.9}\right) = 0.00664 = 0.664\% \text{ per year.}$$

(b) The "Day of 8 Billion" should occur when

$$8 = 6.9e^{0.00664t}.$$

Solving for t gives

$$e^{0.00664t} = \frac{8}{6.9}$$

$$t = \frac{\ln(8/6.9)}{0.00664} = 22.277 \text{ years.}$$

So the "Day of 8 Billion" should be 22.277 years after the end of 2010. This is 22 years and $0.227 \cdot 365 \approx 101$ days; so 101 days into 2032. That is, April 11, 2032.

57. We assume exponential decay and solve for k using the half-life:

$$e^{-k(5730)} = 0.5 \quad \text{so} \quad k = 1.21 \cdot 10^{-4}.$$

Now find t, the age of the painting:

$$e^{-1.21 \cdot 10^{-4}t} = 0.995, \quad \text{so} \quad t = \frac{\ln 0.995}{-1.21 \cdot 10^{-4}} = 41.43 \text{ years.}$$

Since Vermeer died in 1675, the painting is a fake.

61. The function e^x has a vertical intercept of 1, so must be A. The function $\ln x$ has an x-intercept of 1, so must be D. The graphs of x^2 and $x^{1/2}$ go through the origin. The graph of $x^{1/2}$ is concave down so it corresponds to graph C and the graph of x^2 is concave up so it corresponds to graph B.

65. (a) The y-intercept of $g(x) = \ln(ax + 2)$ is $g(0) = \ln 2$. Thus increasing a does not effect the y-intercept.

(b) The x-intercept of $g(x) = \ln(ax + 2)$ is where $g(x) = 0$. Since this occurs where $ax + 2 = 1$, or $x = -1/a$, increasing a moves the x-intercept toward the origin. (The intercept is to the left of the origin if $a > 0$ and to the right if $a < 0$.)

Strengthen Your Understanding

69. The function $-\log|x|$ is even, since $|-x| = |x|$, which means $-\log|-x| = -\log|x|$.

73. One of the properties of logarithms states that the logarithm of a product is the sum of the logarithms, that is

$$\ln(AB) = \ln A + \ln B.$$

There is no property of logarithms that states the logarithm of a sum is the sum of logarithms. We can check by taking $A = B = 1$. Then the statement gives us the incorrect statement:

$$\ln 2 = \ln 1 + \ln 1 = 0.$$

77. One possibility is $f(x) = \ln(x - 3)$.

81. False, since $ax + b = 0$ if $x = -b/a$. Thus $y = \ln(ax + b)$ has a vertical asymptote at $x = -b/a$.

Solutions for Section 1.5

Exercises

1. See Figure 1.11.

$$\sin\left(\frac{3\pi}{2}\right) = -1 \quad \text{is negative.}$$

$$\cos\left(\frac{3\pi}{2}\right) = 0$$

$$\tan\left(\frac{3\pi}{2}\right) \quad \text{is undefined.}$$

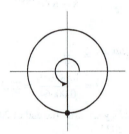

Figure 1.11

5. See Figure 1.12.

$$\sin\left(\frac{\pi}{6}\right) \text{ is positive.}$$

$$\cos\left(\frac{\pi}{6}\right) \text{ is positive.}$$

$$\tan\left(\frac{\pi}{6}\right) \text{ is positive.}$$

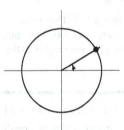

Figure 1.12

9. $-1 \text{ radian} \cdot \dfrac{180°}{\pi \text{ radians}} = -\left(\dfrac{180°}{\pi}\right) \approx -60°$. See Figure 1.13.

$$\sin(-1) \quad \text{is negative}$$

$$\cos(-1) \quad \text{is positive}$$

$$\tan(-1) \quad \text{is negative.}$$

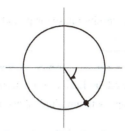

Figure 1.13

13. The period is $2\pi/\pi = 2$, since when t increases from 0 to 2, the value of πt increases from 0 to 2π. The amplitude is 0.1, since the function oscillates between 1.9 and 2.1.

17. This graph is an inverted cosine curve with amplitude 8 and period 20π, so it is given by $f(x) = -8\cos\left(\frac{x}{10}\right)$.

21. This can be represented by a sine function of amplitude 3 and period 18. Thus,

$$f(x) = 3\sin\left(\frac{\pi}{9}x\right).$$

25.

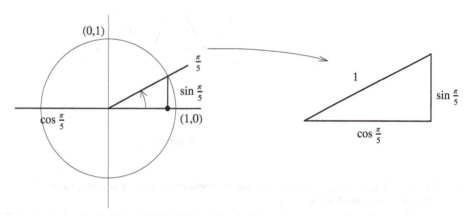

By the Pythagorean Theorem, $(\cos\frac{\pi}{5})^2 + (\sin\frac{\pi}{5})^2 = 1^2$;

so $(\sin\frac{\pi}{5})^2 = 1 - (\cos\frac{\pi}{5})^2$, and $\sin\frac{\pi}{5} = \sqrt{1 - (\cos\frac{\pi}{5})^2} = \sqrt{1 - (0.809)^2} \approx 0.588$.

We take the positive square root since by the picture we know that $\sin\frac{\pi}{5}$ is positive.

29. Since $\tan\theta = \sin\theta/\cos\theta$, we need to find $\sin\frac{\pi}{5}$ first. We can use the Pythagorean identity $\sin^2\theta + \cos^2\theta = 1$ to find

$$\sin\frac{\pi}{5} = \pm\sqrt{1 - 0.809^2} = \pm 0.588.$$

Since the angle $\theta = \frac{\pi}{5}$ lies in the first quadrant, $\sin\frac{\pi}{5} = 0.588$. Thus,

$$\tan\frac{\pi}{5} = \frac{0.588}{0.809} = 0.727.$$

33. We first isolate $(2x + 1)$ and then use inverse tangent:

$$1 = 8\tan(2x + 1) - 3$$
$$4 = 8\tan(2x + 1)$$
$$0.5 = \tan(2x + 1)$$
$$\arctan(0.5) = 2x + 1$$
$$x = \frac{\arctan(0.5) - 1}{2} = -0.268.$$

There are infinitely many other possible solutions since the tangent is periodic.

37. 200 revolutions per minute is $\frac{1}{200}$ minutes per revolution, so the period is $\frac{1}{200}$ minutes, or 0.3 seconds.

Problems

41. All three functions have amplitude 2 and period 2π. Each is just a horizontal shift of the others.

    (a) We see that $h(t)$ looks like a cosine graph with amplitude 2 and period 2π, but shifted right by $\pi/2$. Thus, $h(t) = 2\cos(t - \pi/2)$.

    (b) We see that $f(t)$ has and amplitude of 2 and a period of 2π. Furthermore, since $f(0) = 2$, we know that $f(t) = 2\cos t$.

    (c) We see that $g(t)$ looks like a cosine graph with amplitude 2 and period 2π, but shifted left by $\pi/2$. Thus, $g(t) = 2\cos(t + \pi/2)$.

45. The graph is a stretched cosine function shifted vertically, so it starts at its maximum when $x = 0$.

    We begin by finding the midline, which is k, since the sine function is shifted up by k. The amplitude is k, so the sinusoidal graph oscillates between a minimum of $k - k = 0$ and a maximum of $k + k = 2k$.

    The period is 2π. See Figure 1.14.

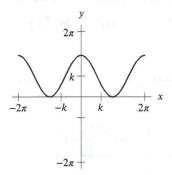

Figure 1.14

49. The graph is a vertically stretched sine function horizontally compressed by a factor of $2\pi/k$. The function takes on the value of its midline, 0, when $x = k$.

    The amplitude is k, so the sinusoidal graph oscillates between a minimum of $-k$ and a maximum of k.

    Since k is positive, the function is increasing at $x = 0$. The period is $2\pi/(2\pi/k) = k$. See Figure 1.15.

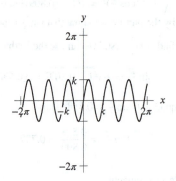

Figure 1.15

53. Since the maximum value of the function appears at about $t = 7.5$, we see that the graph is obtained by horizontally shifting the graph of $A\cos(Bt) + C$ to the right by 7.5 units. Thus $h = 7.5$ months.

57. Since the graph increases across the midline at about $t = 1$, we see that it is obtained by horizontally shifting the graph $A\sin(Bt) + C$ to the right by 1 unit. Thus $h = 1$ month.

61. (a) See Figure 1.16.

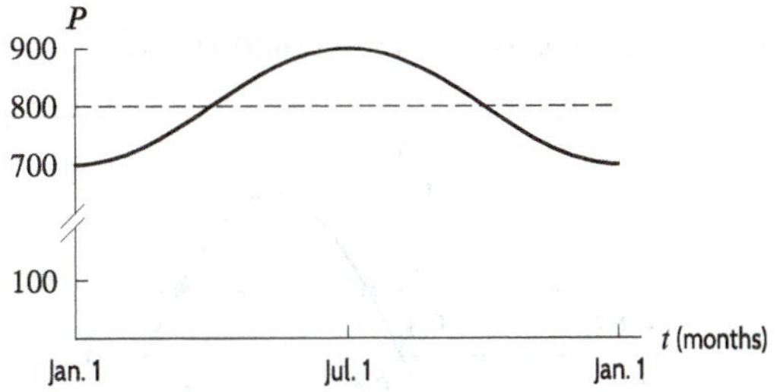

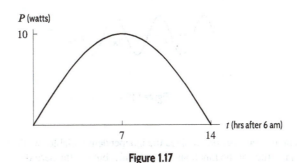

Figure 1.16

(b) Average value of population $= \frac{700+900}{2} = 800$, amplitude $= \frac{900-700}{2} = 100$, and period = 12 months, so $B = 2\pi/12 = \pi/6$. Since the population is at its minimum when $t = 0$, we use a negative cosine:

$$P = 800 - 100\cos\left(\frac{\pi t}{6}\right).$$

65. (a) When the time is t hours after 6 am, the solar panel outputs $f(t) = P(\theta(t))$ watts. So,

$$f(t) = 10\sin\left(\frac{\pi}{14}t\right)$$

where $0 \le t \le 14$ is the number of hours after 6 am.
   (b) The graph of $f(t)$ is in Figure 1.17:

P (watts)

10

7   14   t (hrs after 6 am)

Figure 1.17

(c) The power output is greatest when $\sin(\pi t/14) = 1$. Since $0 \le \pi t/14 \le \pi$, the only point in the domain of f at which $\sin(\pi t/14) = 1$ is when $\pi t/14 = \pi/2$. Therefore, the power output is greatest when $t = 7$, that is, at 1 pm. The output at this time will be $f(7) = 10$ watts.
   (d) On a typical winter day, there are 9 hours of sun instead of the 14 hours of sun. So, if t is the number of hours since 8 am, the angle between a solar panel and the sun is

$$\phi = \frac{14}{9}\theta = \frac{\pi}{9}t \quad \text{where } 0 \le t \le 9.$$

The solar panel outputs $g(t) = P(\phi(t))$ watts:

$$g(t) = 10\sin\left(\frac{\pi}{9}t\right)$$

where $0 \le t \le 9$ is the number of hours after 8 am.

69. (a) $D =$ the average depth of the water.
   (b) $A =$ the amplitude $= 15/2 = 7.5$.
   (c) Period = 12.4 hours. Thus $(B)(12.4) = 2\pi$ so $B = 2\pi/12.4 \approx 0.507$.
   (d) C is the time of a high tide.

73. (a) Two solutions: 0.4 and 2.7. See Figure 1.18.
   (b) arcsin(0.4) is the first solution approximated above; the second is an approximation to $\pi - $ arcsin(0.4).
   (c) By symmetry, there are two solutions: -0.4 and -2.7.

(d) $-0.4 \approx -\arcsin(0.4)$ and $-2.7 \approx -(\pi - \arcsin(0.4)) = \arcsin(0.4) - \pi$.

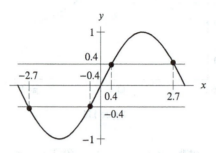

Figure 1.18

Strengthen Your Understanding

77. The functions $f(x) = \sin x$ and $g(x) = \cos x$ both have period 2π. The product, fg, of these two functions is periodic, but has a smaller period, π. See Figure 1.19.

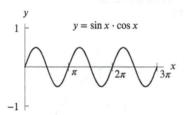

Figure 1.19

81. True. All functions in this family have period 2π since the independent variable, x, (the input) is not scaled by a constant. The constant a varies the amplitude of the functions in the family, but not the period.

85. True. The period is $2\pi/(200\pi) = 1/100$ seconds. Thus, the function executes 100 cycles in 1 second.

89. False. When $\pi/2 < x < 3\pi/2$, we have $\cos |x| = \cos x < 0$ but $|\cos x| > 0$.

93. False. A counterexample is given by $f(x) = \sin x$, which has period 2π, and $g(x) = x^2$. The graph of $f(g(x)) = \sin(x^2)$ in Figure 1.20 is not periodic with period 2π.

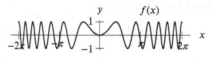

Figure 1.20

97. True. Since $\sin(t + 2\pi) = \sin t$, we have

$$f(t + 2\pi) = \sin^{-1}(\sin(t + 2\pi)) = \sin^{-1}(\sin t) = f(t).$$

To see that this is the smallest interval on which we have a complete cycle, notice that since $f(t) = 0$ only for $t = n\pi$, $n = 0, \pm 1, \pm 2, \ldots$, the only possible period shorter than 2π would be π. But $f(\pi/2)) = \pi/2$ and $f(\pi/2 + \pi) = -\pi/2$. Thus, π is not a period.

Solutions for Section 1.6

Exercises

1. As $x \to \infty$, $y \to \infty$.
   As $x \to -\infty$, $y \to -\infty$.

5. As $x \to \pm\infty$, the lower-degree terms of $f(x)$ become insignificant, and $f(x)$ becomes approximated by the highest degree term of $5x^4$. Thus, as $x \to \pm\infty$, we see that $f(x) \to +\infty$.

9. As $x \to \pm\infty$, we see that $3x^{-4}$ gets closer and closer to 0, so $f(x) \to 0$ as $x \to \pm\infty$.

13. An exponential growth function always dominates a power function as $x \to \infty$, so 1.05^x is larger.

17. (I) (a) Minimum degree is 3 because graph turns around twice.
       (b) Leading coefficient is negative because $y \to -\infty$ as $x \to \infty$.
    (II) (a) Minimum degree is 4 because graph turns around three times.
       (b) Leading coefficient is positive because $y \to \infty$ as $x \to \infty$.
    (III) (a) Minimum degree is 4 because graph turns around three times.
       (b) Leading coefficient is negative because $y \to -\infty$ as $x \to \infty$.
    (IV) (a) Minimum degree is 5 because graph turns around four times.
       (b) Leading coefficient is negative because $y \to -\infty$ as $x \to \infty$.
    (V) (a) Minimum degree is 5 because graph turns around four times.
       (b) Leading coefficient is positive because $y \to \infty$ as $x \to \infty$.

21. $f(x) = kx(x + 3)(x - 4) = k(x^3 - x^2 - 12x)$, where $k < 0$. ($k \approx -\frac{2}{9}$ if the horizontal and vertical scales are equal; otherwise one can't tell how large k is.)

25. There is only one function, r, who can be put in the form $y = Ax^2 + Bx + C$:

$$r(x) = -x + b - \sqrt{cx^4} = -\sqrt{c}\, x^2 - x + b, \quad \text{where } A = -\sqrt{c}, \text{ since } c \text{ is a constant.}$$

Thus, r is the only quadratic function.

29. Since b and c are constants, so is b/c^a. In addition, since a is a positive integer, $h(x)$, which can be rewritten as

$$h(x) = b\left(\frac{x}{c}\right)^a = \frac{b}{c^a}x^a,$$

is a power function. It is also a polynomial, since a polynomial is a sum of power functions with positive integer exponents, and it is a rational function, since any polynomial can be viewed as a rational function with denominator 1. However, $h(x)$ is not an exponential function.
   Thus, $h(x)$ falls into families (II), (III) and (IV).

33. IV, since expanding gives

$$y = x^3 - x.$$

37. V, since the denominator has exactly one zero, $x = 0$.

41. Since the graph of this function is only defined for positive values of x, this is the graph of a logarithmic function.

Problems

45. Consider the end behavior of the graph; that is, as $x \to +\infty$ and $x \to -\infty$. The ends of a degree 5 polynomial are in Quadrants I and III if the leading coefficient is positive or in Quadrants II and IV if the leading coefficient is negative. Thus, there must be at least one root. Since the degree is 5, there can be no more than 5 roots. Thus, there may be 1, 2, 3, 4, or 5 roots. Graphs showing these five possibilities are shown in Figure 1.21.

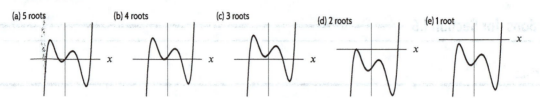

Figure 1.21

49. Let us represent the height by h. Since the volume is V, we have

$$x^2 h = V.$$

Solving for h gives

$$h = \frac{V}{x^2}.$$

The graph is in Figure 1.22. We are assuming V is a positive constant.

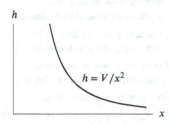

Figure 1.22

53. (a) Since the rate R varies directly with the fourth power of the radius r, we have the formula

$$R = kr^4$$

where k is a constant.

(b) Given $R = 400$ for $r = 3$, we can determine the constant k.

$$400 = k(3)^4$$
$$400 = k(81)$$
$$k = \frac{400}{81} \approx 4.93827.$$

So the formula is

$$R = 4.93827r^4$$

(c) Evaluating the formula above at $r = 5$ yields

$$R = 4.93827(5)^4 = 3086.42 \frac{\text{cm}^3}{\text{sec}}.$$

57. Yes. The graph of this power function is increasing and concave down for positive mass M. See Figure 1.23. This function could represent the length of a plant species that stretches relatively less per unit mass increase as the plant gets larger.

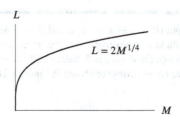

Figure 1.23

61. No. The graph of this rational function is decreasing for positive mass M. See Figure 1.24. A plant growing according to this function would actually shrink in length (rather than stretch) as its mass increases.

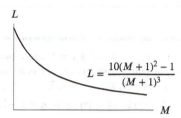

$$L = \frac{10(M+1)^2 - 1}{(M+1)^3}$$

Figure 1.24

65. (a) A polynomial has the same end behavior as its leading term, so this polynomial behaves as $-5x^4$ globally. Thus we have:
$$f(x) \to -\infty \text{ as } x \to -\infty, \quad \text{and} \quad f(x) \to -\infty \text{ as } x \to +\infty.$$
  (b) Polynomials behave globally as their leading term, so this rational function behaves globally as $(3x^2)/(2x^2)$, or $3/2$. Thus we have:
$$f(x) \to 3/2 \text{ as } x \to -\infty, \quad \text{and} \quad f(x) \to 3/2 \text{ as } x \to +\infty.$$
  (c) We see from a graph of $y = e^x$ that
$$f(x) \to 0 \text{ as } x \to -\infty, \quad \text{and} \quad f(x) \to +\infty \text{ as } x \to +\infty.$$

69. $h(t)$ cannot be of the form ct^2 or kt^3 since $h(0.0) = 2.04$. Therefore $h(t)$ must be the exponential, and we see that the ratio of successive values of h is approximately 1.5. Therefore $h(t) = 2.04(1.5)^t$. If $g(t) = ct^2$, then $c = 3$ since $g(1.0) = 3.00$. However, $g(2.0) = 24.00 \neq 3 \cdot 2^2$. Therefore $g(t) = kt^3$, and using $g(1.0) = 3.00$, we obtain $g(t) = 3t^3$. Thus $f(t) = ct^2$, and since $f(2.0) = 4.40$, we have $f(t) = 1.1t^2$.

73. Since $z = \ln x$, we have $x = e^z$. Hence
$$y = 100x^{-0.2} = 100(e^z)^{-0.2} = 100e^{-0.2z}.$$

Strengthen Your Understanding

77. As seen from the graphs of the functions in Figure 1.25, this inequality only holds for $x > 1$. The functions cross at the point $(1, 1)$ and the inequality is reversed for $0 < x < 1$.

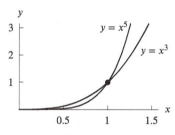

Figure 1.25

81. A possible function is
$$f(x) = \frac{3x}{x - 10}.$$

85. The function $f(x) = \frac{x - 1}{x - 2}$ has $y = 1$ as the horizontal asymptote and $x = 2$ as the vertical asymptote. These lines cross at the point $(2, 1)$. Other answers are possible.

89. True. If we factor and simplify the rational expression, we see that
$$f(x) = \frac{5x(x + 1)(x - 1)}{x(x + 1)} = 5x - 5, \text{ for } x \neq -1, 0.$$

Hence, over $x > 0$ the graph of the f is the same at the graph of the linear function $y = 5x - 5$.

Solutions for Section 1.7

Exercises

1. (a) The function f has jumps in the graph at $x = -1$ and at $x = 1$. Therefore, f is not continuous at $x = -1$ and at $x = 1$.
    (b) The domain of f is $-3 \leq x \leq 3$, and by part (a), we know that f is not continuous at $x = -1$ and at $x = 1$. Therefore, f is continuous on $-3 < x < -1$, $-1 < x < 1$, and $1 < x < 3$.

5. (a) Since the graph has a hole at $x = -2$, $f(x)$ is not continuous at $x = -2$. Since the graph has an asymptote at $x = 3$, $f(x)$ is not continuous at $x = 3$.
    (b) As x approaches -2 from either side, the values of $f(x)$ get closer and closer to -3, so the limit appears to be about -3. As x approaches 3 from either side, the values of $f(x)$ approach ∞. Thus $\lim_{x \to 3} f(x)$ does not exist.

9. (a) Substituting x-values into $f(x)$ gives:

x	-0.1	-0.01	0.01	0.1
$f(x)$	2.592	2.955	3.045	3.499

    (b) Since the values of $f(x)$ appear to get closer and closer to 3 as x gets closer and closer to 0, we estimate

    $$\lim_{x \to 0} \frac{e^{3x} - 1}{x} = 3.$$

    Notice that $f(x) = \dfrac{e^{3x} - 1}{x}$ is undefined at $x = 0$.

13. Yes, because $2x - 5$ is positive for $3 \leq x \leq 4$.

17. (a) At $x = 1$, on the line $y = x$, we have $y = 1$. At $x = 1$, on the parabola $y = x^2$, we have $y = 1$. Thus, $f(x)$ is continuous. See Figure 1.26.
    (b) At $x = 3$, on the line $y = x$, we have $y = 3$. At $x = 3$, on the parabola $y = x^2$, we have $y = 9$. Thus, $g(x)$ is not continuous. See Figure 1.27.

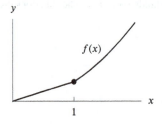

Figure 1.26

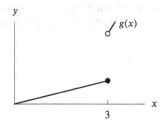

Figure 1.27

21. We have that $f(0) = -1 < 0$ and $f(1) = 1 - \cos 1 > 0$ and that f is continuous. Thus, by the Intermediate Value Theorem applied to $k = 0$, there is a number c in $[0, 1]$ such that $f(c) = k = 0$.

25. Evaluating $\dfrac{x^2 + 4x - 5}{x - 1}$ at $x = 1$ gives us $0/0$, so we see if we can rewrite the function using algebra. We have

    $$\frac{x^2 + 4x - 5}{x - 1} = \frac{(x + 5)(x - 1)}{x - 1}.$$

    Since $x \neq 1$ in the limit, we can cancel the common factor $x - 1$ to see

    $$\lim_{x \to 1} \frac{x^2 + 4x - 5}{x - 1} = \lim_{x \to 1} \frac{(x + 5)(x - 1)}{x - 1} = \lim_{x \to 1}(x + 5) = 6.$$

29. Since $kx + 10$ is continuous at $x = 5$, the limit can be found by substituting $x = 5$. Thus,

$$k(5) + 10 = 20$$
$$k = 2.$$

33. For any value of k, the function is continuous at every point except $x = 1$. We choose k to make the function continuous at $x = 1$.

Since $x + 3$ takes the value $1 + 3 = 4$ at $x = 1$, we choose k so that the graph of kx goes through the point $(1, 4)$. Thus $k = 4$.

Problems

37. (a) The function f has a jump in the graph at $t = 5$ so it is not continuous. The discontinuity at $t = 5$ occurs because the speed of the rock suddenly drops to zero when it hits the ground.

(b) The stone starts at the top of the cliff and its height decreases over time until $t = 5$ when it hits the ground. For $5 \leq t \leq 7$, we have $h = 0$ since the stone is on the ground. Therefore Figure 1.28 is a possible graph. The function g is continuous because there is no sudden change in the height of the stone.

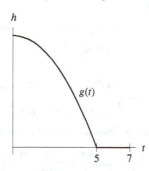

Figure 1.28

41. The break in the graph at $x = 0$ suggests that $\lim_{x \to 0} \dfrac{|x|}{x}$ does not exist. See Figure 1.29.

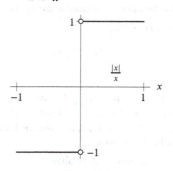

Figure 1.29

45. A graph of $y = \dfrac{e^h - 1}{h}$ in a window such as $-0.5 \leq h \leq 0.5$ and $0 \leq y \leq 3$ appears to indicate $y \to 1$ as $h \to 0$. We estimate that

$$\lim_{h \to 0} \frac{e^h - 1}{h} = 1.$$

49. At $x = \pi$, the curve $y = k \cos x$ has $y = k \cos \pi = -k$. At $x = \pi$, the line $y = 12 - x$ has $y = 12 - \pi$. If $h(x)$ is continuous, we need

$$-k = 12 - \pi$$
$$k = \pi - 12.$$

53. For any value of k, the function is continuous at every point except $x = 1$. We choose k to make the function continuous at $x = 1$.

Since $\sin(kx)$ takes the value $\sin k$ at $x = 1$, we choose k so that the graph of $0.5x$ goes through the point $(1, \sin k)$. This gives

$$\sin k = 0.5$$
$$k = \sin^{-1} 0.5 = \frac{\pi}{6}.$$

Other solutions are possible.

57. The drug first increases linearly for half a second, at the end of which time there is 0.6 ml in the body. Thus, for $0 \le t \le 0.5$, the function is linear with slope $0.6/0.5 = 1.2$:

$$Q = 1.2t \quad \text{for} \quad 0 \le t \le 0.5.$$

At $t = 0.5$, we have $Q = 0.6$. For $t > 0.5$, the quantity decays exponentially at a continuous rate of 0.002, so Q has the form

$$Q = Ae^{-0.002t} \qquad 0.5 < t.$$

We choose A so that $Q = 0.6$ when $t = 0.5$:

$$0.6 = Ae^{-0.002(0.5)} = Ae^{-0.001}$$
$$A = 0.6e^{0.001}.$$

Thus

$$Q = \begin{cases} 1.2t & 0 \le t \le 0.5 \\ 0.6e^{0.001}e^{-.002t} & 0.5 < t. \end{cases}$$

61. At $t = 3$ and $t = -3$, the function is not defined. Thus, $q(t)$ is not continuous at $t = \pm 3$. We have

$$\lim_{t \to 3} \frac{-t^3 + 9t}{t^2 - 9} = \lim_{t \to 3} \frac{-t(t-3)(t+3)}{(t-3)(t+3)} = \lim_{t \to 3} -t = -3.$$

Similarly, we have

$$\lim_{t \to -3} \frac{-t^3 + 9t}{t^2 - 9} = \lim_{t \to -3} \frac{-t(t-3)(t+3)}{(t-3)(t+3)} = \lim_{t \to -3} -t = 3.$$

So if we let $q(3) = -3$ and $q(-3) = 3$, the function is continuous everywhere.

65. Since $2x/x = 2$ for $x \ne 0$, we have $\lim_{x \to 0} f(x) = 2$, so

$$\lim_{x \to 0} f(x) \ne f(0) = 3.$$

Thus, $f(x)$ is not continuous at $x = 0$.

69. (a) Since $f(x)$ is not continuous at $x = 1$, it does not satisfy the conditions of the Intermediate Value Theorem.
   (b) We see that $f(0) = e^0 = 1$ and $f(2) = 4 + (2-1)^2 = 5$. Since e^x is increasing between $x = 0$ and $x = 1$, and since $4 + (x-1)^2$ is increasing between $x = 1$ and $x = 2$, any value of k between $e^1 = e$ and $4 + (1-1)^2 = 4$, such as $k = 3$, is a value such that $f(x) = k$ has no solution.

73. (a) Since $\sin(n\pi) = 0$ for $n = 1, 2, 3, \ldots$ the sequence of x-values

$$\frac{1}{\pi}, \frac{1}{2\pi}, \frac{1}{3\pi}, \ldots$$

works. These x-values $\to 0$ and are zeroes of $f(x)$.
   (b) Since $\sin(n\pi/2) = 1$ for $n = 1, 5, 9 \ldots$ the sequence of x-values

$$\frac{2}{\pi}, \frac{2}{5\pi}, \frac{2}{9\pi}, \ldots$$

works.
   (c) Since $\sin(n\pi)/2 = -1$ for $n = 3, 7, 11, \ldots$ the sequence of x-values

$$\frac{2}{3\pi}, \frac{2}{7\pi}, \frac{2}{11\pi} \ldots$$

works.

(d) Any two of these sequences of x-values show that if the limit were to exist, then it would have to have two (different) values: 0 and 1, or 0 and −1, or 1 and −1. Hence, the limit can not exist.

Strengthen Your Understanding

77. We want a function which has a value at every point but where the graph has a break at $x = 15$. One possibility is

$$f(x) = \begin{cases} 1 & x \geq 15 \\ -1 & x < 15 \end{cases}$$

81. False. For example, let $f(x) = \begin{cases} 1 & x \leq 3 \\ 2 & x > 3 \end{cases}$, then $f(x)$ is defined at $x = 3$ but it is not continuous at $x = 3$. (Other examples are possible.)

Solutions for Section 1.8

Exercises

1. (a) As x approaches -1 from the right, $f(x)$ approaches 1. So $\lim\limits_{x \to -1^+} f(x) = 1$.

(b) As x approaches 0 from the left, $f(x)$ approaches 0. So $\lim\limits_{x \to 0^-} f(x) = 0$.

(c) As x approaches 0 from either side, $f(x)$ approaches 0, regardless of the fact that $f(0) = 2$. So $\lim\limits_{x \to 0} f(x) = 0$.

(d) As x approaches 1 from the left, $f(x)$ approaches 1. So $\lim\limits_{x \to 1^-} f(x) = 1$.

(e) As x approaches 1 from the left, $f(x)$ approaches 1. However, as x approaches 1 from the right, $f(x)$ approaches 2. Since there is no single value that $f(x)$ approaches as x approaches 1 from both sides, $\lim\limits_{x \to 1} f(x)$ does not exist.

(f) As x approaches 2 from the left, $f(x)$ approaches 1. So $\lim\limits_{x \to 2^-} f(x) = 1$.

5. (a) As x gets larger and larger, the values of $f(x)$ oscillate, but the oscillations get smaller and smaller around 3. Therefore, we estimate that $\lim\limits_{x \to \infty} f(x) = 3$

(b) As x gets larger and larger in magnitude but negative, the values of $f(x)$ get larger and larger in magnitude but negative, so the $f(x)$ does not approach a real number, so the limit does not exist. Alternatively, we can say $\lim\limits_{x \to -\infty} f(x) = -\infty$.

9. From the graphs of f and g, we estimate $\lim\limits_{x \to 1^-} f(x) = 3$, $\lim\limits_{x \to 1^-} g(x) = 5$, $\lim\limits_{x \to 1^+} f(x) = 4$, $\lim\limits_{x \to 1^+} g(x) = 1$.

(a) Using Theorem 1.2 we have $\lim\limits_{x \to 1^-} (f(x) + g(x)) = 3 + 5 = 8$.

(b) Using Theorem 1.2 we have $\lim\limits_{x \to 1^+} (f(x) + 2g(x)) = \lim\limits_{x \to 1^+} f(x) + 2 \lim\limits_{x \to 1^+} g(x) = 4 + 2(1) = 6$

(c) Using Theorem 1.2 we have $\lim\limits_{x \to 1^-} (f(x)g(x)) = (\lim\limits_{x \to 1^-} f(x))(\lim\limits_{x \to 1^-} g(x)) = (3)(5) = 15$

(d) Using Theorem 1.2 we have $\lim\limits_{x \to 1^+} (f(x)/g(x)) = \left(\lim\limits_{x \to 1^+} f(x)\right) / \left(\lim\limits_{x \to 1^+} g(x)\right) = 4/1 = 4$

13. We see that $f(x)$ approaches a y-value of 3 on the left and goes to $-\infty$ on the right. One possible graph is shown in Figure 1.30. Other answers are possible.

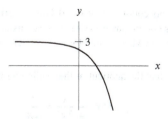

Figure 1.30

17. We have $\lim_{x \to -\infty} x^2 = \infty$ because x^2 increases and becomes arbitrarily large and positive as x becomes arbitrarily large in magnitude and negative.

21. We have $\lim_{x \to \infty} e^{-x} = \lim_{x \to \infty} \frac{1}{e^x} = 0$ because e^x increases and becomes arbitrarily large and positive as x becomes arbitrarily large and positive, so $1/e^x$ gets closer and closer to 0 as x becomes arbitrarily large and positive.

25. We have $\lim_{x \to \infty} x^{-2} = \lim_{x \to \infty} \frac{1}{x^2} = 0$ because x^2 increases and becomes arbitrarily large and positive as x becomes arbitrarily large and positive, so $1/x^2$ gets closer and closer to 0 as x becomes arbitrarily large and positive.

29. We see that $\lim_{x \to -\infty} f(x) = -\infty$ and $\lim_{x \to \infty} f(x) = -\infty$.

33. As $x \to \pm\infty$, we know that x^{-3} gets closer and closer to zero, so we have $\lim_{x \to -\infty} f(x) = 0$ and $\lim_{x \to \infty} f(x) = 0$.

37. $f(x) = \dfrac{|x - 2|}{x} = \begin{cases} \dfrac{x - 2}{x}, & x > 2 \\[2mm] -\dfrac{x - 2}{x}, & x < 2 \end{cases}$

Figure 1.31 confirms that $\lim_{x \to 2^+} f(x) = \lim_{x \to 2^-} f(x) = \lim_{x \to 2} f(x) = 0$.

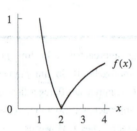

Figure 1.31

Problems

41. (a)

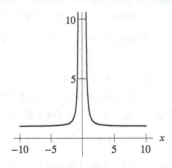

Figure 1.32

(b) The graph of $f(x)$ suggests that it is not continuous at $x = 0$. However, the function $f(x)$ is the composition of a rational function $1/(x^2 + 0.0001)$, which is continuous everywhere since its denominator is never zero, with the exponential function e^x, which is continuous everywhere. Therefore $f(x)$ is continuous everywhere, so must be continuous at $x = 0$.

45. As $x \to \infty$, we know that $f(x)$ behaves like the quotient of the leading terms of its numerator and denominator. Since

$$f(x) \to \frac{x}{2x^2} = \frac{1}{2x},$$

we have $\lim_{x \to \infty} f(x) = 0$ because $1/(2x) \to 0$ as $x \to \infty$.

Alternatively, we could divide numerator and denominator by x^2 and apply the properties of limits:

$$f(x) = \frac{x-5}{5+2x^2} = \frac{1/x - 5/x^2}{5/x^2 + 2},$$

so

$$\lim_{x\to\infty} f(x) = \lim_{x\to\infty} \frac{1/x - 5/x^2}{5/x^2 + 2} = \frac{\lim_{x\to\infty}\left((1/x) - (5/x^2)\right)}{\lim_{x\to\infty}\left((5/x^2) + 2\right)} = \frac{\lim_{x\to\infty}(1/x) - \lim_{x\to\infty}(5/x^2)}{\lim_{x\to\infty}(5/x^2) + \lim_{x\to\infty}2} = \frac{0}{2} = 0$$

since $\lim_{x\to\infty} a/x^n = 0$ for any nonzero number a and positive n.

49. As $x \to \infty$, we know that $f(x)$ behaves like the quotient of the leading terms of its numerator and denominator. Since

$$f(x) \to \frac{x^4}{2x^5} = \frac{1}{2x},$$

we have $\lim_{x\to\infty} f(x) = 0$ because $1/(2x) \to 0$ as $x \to \infty$.

Alternatively, we could divide numerator and denominator by x^5 and apply the properties of limits:

$$f(x) = \frac{x^4 + 3x}{x^4 + 2x^5} = \frac{1/x + 3/x^4}{1/x + 2},$$

so

$$\lim_{x\to\infty} f(x) = \frac{\lim_{x\to\infty}\left(1/x + 3/x^4\right)}{\lim_{x\to\infty}\left(1/x + 2\right)} = \frac{\lim_{x\to\infty}(1/x) + \lim_{x\to\infty}(3/x^4)}{\lim_{x\to\infty}(1/x) + \lim_{x\to\infty}2} = \frac{0}{2} = 0$$

since $\lim_{x\to\infty} a/x^n = 0$ for any nonzero number a and positive n.

53. (a) Figure 1.33 shows a possible graph of $f(x)$, yours may be different.

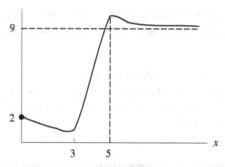

Figure 1.33

(b) In order for f to approach the horizontal asymptote at 9 from above it is necessary that f eventually become concave up. It is therefore not possible for f to be concave down for all $x > 6$.

57. (a) (i) As t approaches 1 from the left, Q approaches 50. Therefore, $\lim_{t\to 1^-} f(t) = 50$. This means that the amount of the drug in the patient's body is about 50 mg right before the second dose is taken.

(ii) As t approaches 1 from the right, Q approaches 150. Therefore, $\lim_{t\to 1^+} f(t) = 150$. This means that the amount of drug in the patient's body is about 150 mg right after the second dose is taken.

(b) The function f is discontinuous at $t = 1, 2, 3,$ and 4 days. These points of discontinuity occur at the times when the drug is taken because the amount of the drug in the body suddenly increases at these times.

61. (a) We have

$$\frac{1}{x-5} - \frac{10}{x^2 - 5} = \frac{x+5}{(x+5)(x-5)} - \frac{10}{(x+5)(x-5)}$$

$$= \frac{x+5-10}{(x+5)(x-5)}$$

$$= \frac{x-5}{(x+5)(x-5)}.$$

(b) We have

$$\lim_{x \to 5} \left(\frac{1}{x - 5} - \frac{10}{x^2 - 25} \right) = \lim_{x \to 5} \frac{x - 5}{(x + 5)(x - 5)}$$

$$= \lim_{x \to 5} \frac{1}{x + 5} \qquad \text{Canceling } x - 5 \text{ since } x \neq 5 \text{ in the limit}$$

$$= \frac{1}{5 + 5} = \frac{1}{10} \qquad \text{Substituting } x = 5 \text{ since } 1/(x + 5) \text{ is continuous at } x = 5.$$

(c) Property 2 of Theorem 1.2 states

$$\lim_{x \to c} (f(x) + g(x)) = \lim_{x \to c} f(x) + \lim_{x \to c} g(x)$$

provided the limits on the right hand side exist. In this case, neither $\lim\limits_{x \to 5} \dfrac{1}{x - 5}$ nor $\lim\limits_{x \to 5} -\dfrac{10}{x^2 - 25}$ exist, so we cannot invoke the property.

Strengthen Your Understanding

65. Though $P(x)$ and $Q(x)$ are both continuous for all x, it is possible for $Q(x)$ to be equal to zero for some x. For any such value of x, where $Q(x) = 0$, we see that $P(x)/Q(x)$ is undefined, and thus not continuous. For example,

$$\frac{P(x)}{Q(x)} = \frac{x}{x - 1}$$

is not defined or continuous at $x = 1$.

69. True, by Property 3 of limits in Theorem 1.2, since $\lim\limits_{x \to 3} x = 3$.

73. True. Suppose instead that $\lim\limits_{x \to 3} g(x)$ does not exist but $\lim\limits_{x \to 3} (f(x)g(x))$ did exist. Since $\lim\limits_{x \to 3} f(x)$ exists and is not zero, then $\lim\limits_{x \to 3} ((f(x)g(x))/f(x))$ exists, by Property 4 of limits in Theorem 1.2. Furthermore, $f(x) \neq 0$ for all x in some interval about 3, so $(f(x)g(x))/f(x) = g(x)$ for all x in that interval. Thus $\lim\limits_{x \to 3} g(x)$ exists. This contradicts our assumption that $\lim\limits_{x \to 3} g(x)$ does not exist.

77. False, since we don't know if $f(x)$ or $(f(x))^2$ is defined at $x = 7$.

81. False, since as $x \to 0^+$ the function $1/x \to \infty$. Also, as $x \to \infty$ the function $\sin(1/x)$ oscillates faster and faster between $+1$ and -1. Therefore the product of the two does not tend to either ∞ or $-\infty$ as $x \to 0$ on either side, but instead oscillates between large positive and large negative values.

Additional Problems (online only)

85. We have $(\sin 3x)/x = 3(\sin 3x)/(3x)$. If we let $u = 3x$, then $u \to 0$ as $x \to 0$, so

$$\lim_{x \to 0} \frac{\sin 3x}{3x} = \lim_{u \to 0} \frac{\sin u}{u} = 1.$$

Thus

$$\lim_{x \to 0} \frac{\sin 3x}{x} = 3 \lim_{x \to 0} \frac{\sin 3x}{3x} = 3.$$

Solutions for Section 1.9

Exercises

1. Canceling, we have

$$\lim_{x \to 0} \frac{3x^2}{x^2} = \lim_{x \to 0} 3 = 3.$$

5. Factoring the numerator and the denominator, we obtain

$$\frac{t^4 + t^2}{2t^3 - 9t^2} = \frac{t^2(t^2 + 1)}{t^2(2t - 9)} = \frac{t^2 + 1}{2t - 9}$$

provided $t \neq 0$. Thus

$$\lim_{t \to 0} \frac{t^4 + t^2}{2t^3 - 9t^2} = \lim_{t \to 0} \frac{t^2 + 1}{2t - 9} = -\frac{1}{9}.$$

9. Factoring the numerator and denominator and simplifying, we get

$$\frac{x^2 + 2x - 3}{x^2 - 3x + 2} = \frac{(x - 1)(x + 3)}{(x - 1)(x - 2)} = \frac{x + 3}{x - 2}$$

provided $x \neq 1$. Thus

$$\lim_{x \to 1} \frac{x^2 + 2x - 3}{x^2 - 3x + 2} = \lim_{x \to 1} \frac{x + 3}{x - 2} = -4.$$

13. Expanding the numerator, we get

$$(3 + h)^2 - 9 = 9 + 6h + h^2 - 9 = 6h + h^2.$$

Next, factoring and simplifying, we get

$$\frac{(3 + h)^2 - 9}{h} = \frac{h(6 + h)}{h} = 6 + h$$

provided $h \neq 0$. Thus

$$\lim_{h \to 0} \frac{(3 + h)^2 - 9}{h} = \lim_{h \to 0} (6 + h) = 6.$$

17. Simplifying, we obtain

$$\frac{1/t - 1/3}{t - 3} = \frac{\frac{3 - t}{3t}}{t - 3} = \frac{3 - t}{3t} \cdot \frac{1}{t - 3} = \frac{-(t - 3)}{3t(t - 3)} = \frac{-1}{3t}$$

provided $t \neq 3$. Thus

$$\lim_{t \to 3} \frac{1/t - 1/3}{t - 3} = \lim_{t \to 3} \frac{-1}{3t} = -\frac{1}{9}.$$

21. There are two ways to do this. One way is to multiply the numerator and denominator by $\sqrt{9 + h} + 3$ and then simplify:

$$\frac{\left(\sqrt{9 + h} - 3\right)\left(\sqrt{9 + h} + 3\right)}{h\left(\sqrt{9 + h} + 3\right)} = \frac{9 + h - 9}{h\left(\sqrt{9 + h} + 3\right)} = \frac{h}{h\left(\sqrt{9 + h} + 3\right)} = \frac{1}{\sqrt{9 + h} + 3}$$

provided $h \neq 0$. Thus

$$\lim_{h \to 0} \frac{\sqrt{9 + h} - 3}{h} = \lim_{h \to 0} \frac{1}{\sqrt{9 + h} + 3} = \frac{1}{6}.$$

A second way is to let $u = \sqrt{9 + h}$ so $h = u^2 - 9$. Since $u \to 3$ as $h \to 0$, we can rewrite

$$\lim_{h \to 0} \frac{\sqrt{9 + h} - 3}{h} = \lim_{u \to 3} \frac{u - 3}{u^2 - 9}.$$

Factoring $u^2 - 9 = (u - 3)(u + 3)$ and canceling the $u - 3$ factors since $u \neq 3$ in the limit, we get

$$\lim_{u \to 3} \frac{u - 3}{u^2 - 9} = \lim_{u \to 3} \frac{1}{u + 3} = \frac{1}{6}.$$

25. (a) We have $f(1) = 2(1)^3 + (1)^2 - 3(1) = 0$ and $g(1) = (1)^2 + 3(1) - 4 = 0$, so the limit is of the form $0/0$.

(b) Factoring, we have

$$\lim_{x \to 1} \frac{2x^3 + x^2 - 3x}{x^2 + 3x - 4} = \lim_{x \to 1} \frac{x(2x^2 + x - 3)}{(x - 1)(x + 4)}$$

$$= \lim_{x \to 1} \frac{x(2x + 3)(x - 1)}{(x - 1)(x + 4)}$$

$$= \lim_{x \to 1} \frac{x(2x + 3)}{(x + 4)}$$

$$= \lim_{x \to 1} \frac{(1)(2(1) + 3)}{(1 + 4)} = \frac{5}{5} = 1.$$

So $\lim_{x \to 1} \dfrac{2x^3 + x^2 - 3x}{x^2 + 3x - 4} = 1$.

29. Dividing both numerator and denominator by e^t, we get

$$\lim_{t \to \infty} \frac{4e^t + 3e^{-t}}{5e^t + 2e} = \lim_{t \to \infty} \frac{4 + \frac{3}{e^{2t}}}{5 + \frac{2e}{e^t}} = \frac{4}{5}.$$

33. We have,

$$\lim_{t \to \infty} (4e^t)(7e^{-t}) = \lim_{t \to \infty} \frac{28e^t}{e^t} = 28.$$

37. Given that $\dfrac{4x^2 - 5}{x^2} \le f(x) \le \dfrac{4x^6 + 3}{x^6}$, we see that

$$\lim_{x \to \infty} \frac{4x^2 - 5}{x^2} \le \lim_{x \to \infty} f(x) \le \lim_{x \to \infty} \frac{4x^6 + 3}{x^6}$$

$$\lim_{x \to \infty} \frac{4x^2}{x^2} - \lim_{x \to \infty} \frac{5}{x^2} \le \lim_{x \to \infty} f(x) \le \lim_{x \to \infty} \frac{4x^6}{x^6} + \lim_{x \to \infty} \frac{3}{x^6}$$

(using the sum property since each limit exists)

$$4 - 0 \le \lim_{x \to \infty} f(x) \le 4 + 0$$

$$4 \le \lim_{x \to \infty} f(x) \le 4.$$

Since $4 \le \lim_{x \to \infty} f(x) \le 4$, we know that $\lim_{x \to \infty} f(x) = 4$.

Problems

41. Because the denominator equals 0 when $x = 5$, so must the numerator. So $25 - 5k + 5 = 0$ and the only possible value of k is 6.

45. For the numerator, $\lim_{x \to -\infty} (e^{2x} - 5) = -5$. If $k > 0$, $\lim_{x \to -\infty} (e^{kx} + 3) = 3$, so the quotient has a limit of $-5/3$. If $k = 0$, $\lim_{x \to -\infty} (e^{kx} + 3) = 4$, so the quotient has limit of $-5/4$. If $k < 0$, the limit of the quotient is given by $\lim_{x \to -\infty} (e^{2x} - 5)/(e^{kx} + 3) = 0$.

49. In the denominator, we have $\lim_{x \to -\infty} 3^{2x} + 4 = 4$. In the numerator, if $k < 0$, we have $\lim_{x \to -\infty} 3^{kx} + 6 = \infty$, so the quotient has a limit of ∞. If $k = 0$, we have $\lim_{x \to -\infty} 3^{kx} + 6 = 7$, so the quotient has a limit of $7/4$. If $k > 0$, we have $\lim_{x \to -\infty} 3^{kx} + 6 = 6$, so the quotient has a limit of $6/4$.

53. Let $t = \sqrt[3]{x}$ so that $x = t^3$. Since $t \to 1$ as $x \to 1$, we have

$$\lim_{x \to 1} \frac{\sqrt[3]{x} - 1}{x - 1} = \lim_{t \to 1} \frac{t - 1}{t^3 - 1}$$

$$= \lim_{t \to 1} \frac{t - 1}{(t - 1)(t^2 + t + 1)}$$

$$= \lim_{t \to 1} \frac{1}{t^2 + t + 1}$$

$$= \frac{1}{3}.$$

57. For $x \geq 0$, since $0 < e^{-x} \leq 1$, we have $x < x + e^{-x} < x + 1$. Therefore, $\dfrac{1}{x+1} < \dfrac{1}{x+e^{-x}} < \dfrac{1}{x}$. Therefore, since $\lim\limits_{x \to \infty} \dfrac{1}{x+1} = 0 = \lim\limits_{x \to \infty} \dfrac{1}{x}$, we have

$$\lim_{x \to \infty} \frac{1}{x+1} \leq \lim_{x \to \infty} \frac{1}{x+e^{-x}} \leq \lim_{x \to \infty} \frac{1}{x}$$

$$0 \leq \lim_{x \to \infty} \frac{1}{x+e^{-x}} \leq 0$$

Thus, $\lim\limits_{x \to \infty} \dfrac{1}{x+e^{-x}} = 0$.

61. We know that $-1 \leq \cos x \leq 1$, and hence that $0 \leq 2\cos^2 x \leq 2$. Therefore, $x \leq x + 2\cos^2 x \leq x + 2$, so for $x > 0$,

$$\frac{1}{x+2} \leq \frac{1}{x+2\cos^2 x} \leq \frac{1}{x}.$$

Since $\lim\limits_{x \to \infty} \dfrac{1}{x+2} = 0$ and $\lim\limits_{x \to \infty} \dfrac{1}{x} = 0$, we have $\lim\limits_{x \to \infty} \dfrac{1}{x+2\cos^2 x} = 0$ by the Squeeze Theorem.

65. There are many possible choices for $g(x)$. One possibility is

$$g(x) = \left(\frac{x-1}{x-1}\right) \cdot f(x) = \left(\frac{x-1}{x-1}\right) \ln x = \frac{x\ln x - \ln x}{x-1}.$$

Observe that $f(x) = g(x)$ wherever $f(x)$ is defined except for $x = 1$ since $g(1)$ is undefined. This means the graphs of $f(x)$ and $g(x)$ will be the same except at $x = 1$ where the graph of $g(x)$ has a hole.

Since the graphs of $f(x)$ and $g(x)$ are identical except at $x = 1$, the hole will have the same coordinates as the function $f(x)$ at $x = 1$, so the coordinates of the hole will be $(1, 0)$.

69. We have

$$\lim_{x \to 1} f(x) = \lim_{x \to 1} \frac{(x-1)^2}{(x-1)^2}$$

$$= \lim_{x \to 1}(1) \qquad \text{Cancelling } (x-1)^2 \text{ since } x \neq 1$$

$$= 1.$$

73. The Squeeze Theorem can be applied when $c = \infty$ or $c = -\infty$ since $\lim\limits_{x \to \infty} -\dfrac{1}{x} = 0 = \lim\limits_{x \to \infty} \dfrac{1}{x}$ and $\lim\limits_{x \to -\infty} -\dfrac{1}{x} = 0 = \lim\limits_{x \to -\infty} \dfrac{1}{x}$. In both cases $L = 0$. The Squeeze Theorem cannot be applied to other cases since for any other c,

$$\lim_{x \to c} -\frac{1}{x} = -\frac{1}{c} \neq \frac{1}{c} = \lim_{x \to c} \frac{1}{x}.$$

Strengthen Your Understanding

77. True, by the Squeeze Theorem, using the constant function $b(x) = 0$.

81. False, for example if $f(x) = 1/x$, and $g(x) = 1/x^2$ then $\lim\limits_{x \to 0} \dfrac{f(x)}{g(x)} = \lim\limits_{x \to 0} x = 0$ but the limits of $f(x)$ and $g(x)$ do not exist, since $\lim\limits_{x \to 0^+} \dfrac{1}{x} = \infty$ (and $\lim\limits_{x \to 0^-} \dfrac{1}{x} = -\infty$) and $\lim\limits_{x \to 0} \dfrac{1}{x^2} = \infty$.

Solutions for Section 1.10

Exercises

1. For each δ we determine whether the graph of $f(x)$ on the interval $3 - \delta < x < 3 + \delta$ stays in the shaded region $L - \epsilon < y < L + \epsilon$ in the figure. Notice that as δ gets smaller, the interval gets shorter.

   (a) No for $\delta = 1$. The graph of $f(x)$ on the interval $2 < x < 4$ extends outside the shaded region.

(b) No for $\delta = 0.75$. The graph of $f(x)$ on the interval $2.25 < x < 3.75$ extends outside the shaded region.
(c) Yes for $\delta = 0.5$. The graph of $f(x)$ on the interval $2.5 < x < 3.5$ fits inside the shaded region.
(d) Yes for $\delta = 0.25$. The graph of $f(x)$ on the interval $2.75 < x < 3.25$ fits inside the shaded region.
(e) Yes for $\delta = 0.1$. The graph of $f(x)$ on the interval $2.9 < x < 3.1$ fits inside the shaded region.

5. Since $|4x - 12| = 4|x - 3|$, to get $|4x - 12| < 0.1$ we require that $|x - 3| < 0.1/4 = 0.025$. Thus we take $\delta = 0.025$.

Problems

9. The statement

$$\lim_{h \to a} g(h) = K$$

means that we can make the value of $g(h)$ as close to K as we want by choosing h sufficiently close to, but not equal to, a. In symbols, for any $\epsilon > 0$, there is a $\delta > 0$ such that

$$|g(h) - K| < \epsilon \quad \text{for all } 0 < |h - a| < \delta.$$

13. We want to find $\delta > 0$ such that if $-\delta < x - 1 < \delta$, then $|\ln x| < 0.1$. Observe that

$$|\ln x| < 0.1$$
$$-0.1 < \ln x < 0.1$$
$$e^{-0.1} < x < e^{0.1}$$
$$e^{-0.1} - 1 < x - 1 < e^{0.1} - 1$$

and that $1 - e^{-0.1} = 0.095$ and $e^{0.1} - 1 = 0.105$. We therefore choose $\delta = 1 - e^{-0.1}$ so that $-\delta < x - 1 < \delta$ implies

$$e^{-0.1} - 1 < x - 1 < 1 - e^{-0.1} < e^{0.1} - 1,$$

which, in turn, implies that $|\ln x| < 0.1$ (see Figure 1.34).

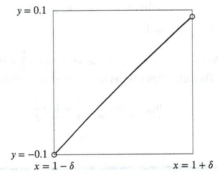

Figure 1.34: Graph of $\ln x$ with
$1 - \delta < x < 1 + \delta$, where $\delta = 1 - e^{-0.1}$.

17. From Table 1.1, it appears the limit is 0. This is confirmed by Figure 1.35. An appropriate window is $-0.015 < x < 0.015$, $-0.01 < y < 0.01$.

Table 1.1

x	$f(x)$		x	$f(x)$
0.1	0.0666		-0.0001	-0.0001
0.01	0.0067		-0.001	-0.0007
0.001	0.0007		-0.01	-0.0067
0.0001	0.0001		-0.1	-0.0666

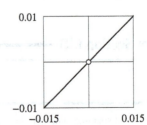

Figure 1.35

21. For any $\epsilon > 0$, we must show that there exists $\delta > 0$ such that, if $|x| < \delta$ with $x \neq 0$, then $|(-2x + 3) - 3| < \epsilon$. Observe that

$$|(-2x + 3) - 3| = |-2x| = 2|x|.$$

Choose $\delta = \epsilon/2$. If $|x| < \delta$ with $x \neq 0$, we have

$$|(-2x + 3) - 3| = 2|x| < 2\delta = \epsilon,$$

proving that $\lim_{x \to 0} (-2x + 3) = 3$.

25. For any $\epsilon > 0$, we must show that there exists $\delta > 0$ such that, if $|x + 3| < \delta$ with $x \neq -3$, then

$$\left| \frac{x^2 + 2x - 3}{x + 3} + 4 \right| < \epsilon.$$

Observe that, if $x \neq -3$, then

$$\left| \frac{x^2 + 2x - 3}{x + 3} + 4 \right| = \left| \frac{(x + 3)(x - 1)}{x + 3} + 4 \right| = |x + 3|.$$

Choose $\delta = \epsilon$. If $|x + 3| < \delta$ with $x \neq -3$, we have

$$\left| \frac{x^2 + 2x - 3}{x + 3} + 4 \right| = |x + 3| < \delta = \epsilon,$$

proving that $\lim_{x \to -3} \dfrac{x^2 + 2x - 3}{x + 3} = -4$.

29. The only change is that, instead of considering all x near c, we only consider x near to and less than c. Thus the phrase "$|x - c| < \delta$" must be replaced by "$c - \delta < x < c$." Thus, we define

$$\lim_{x \to c^-} f(x) = L$$

to mean that for any $\epsilon > 0$ (as small as we want), there is a $\delta > 0$ (sufficiently small) such that if $c - \delta < x < c$, then $|f(x) - L| < \epsilon$.

33. (a) We need to show that for any given $\epsilon > 0$, there is a $\delta > 0$ so that $|x - c| < \delta$ implies $|f(x)g(x)| < \epsilon$. If $\epsilon > 0$ is given, choose δ_1 so that when $|x - c| < \delta_1$, we have $|f(x)| < \sqrt{\epsilon}$. This can be done since $\lim_{x \to 0} f(x) = 0$. Similarly, choose δ_2 so that when $|x - c| < \delta_2$, we have $|g(x)| < \sqrt{\epsilon}$. Then, if we take δ to be the smaller of δ_1 and δ_2, we'll have that $|x - c| < \delta$ implies both $|f(x)| < \sqrt{\epsilon}$ and $|g(x)| < \sqrt{\epsilon}$. So when $|x - c| < \delta$, we have $|f(x)g(x)| = |f(x)| \, |g(x)| < \sqrt{\epsilon} \cdot \sqrt{\epsilon} = \epsilon$. Thus $\lim_{x \to c} f(x)g(x) = 0$.

(b) $\big(f(x) - L_1\big)\big(g(x) - L_2\big) + L_1 g(x) + L_2 f(x) - L_1 L_2$
$= f(x)g(x) - L_1 g(x) - L_2 f(x) + L_1 L_2 + L_1 g(x) + L_2 f(x) - L_1 L_2 = f(x)g(x).$

(c) $\lim_{x \to c} \big(f(x) - L_1\big) = \lim_{x \to c} f(x) - \lim_{x \to c} L_1 = L_1 - L_1 = 0$, using the second limit property. Similarly, $\lim_{x \to c} \big(g(x) - L_2\big) = 0$.

(d) Since $\lim_{x \to c} \big(f(x) - L_1\big) = \lim_{x \to c} \big(g(x) - L_2\big) = 0$, we have that $\lim_{x \to c} \big(f(x) - L_1\big)\big(g(x) - L_2\big) = 0$ by part (a).

(e) From part (b), we have

$$\lim_{x \to c} f(x)g(x) = \lim_{x \to c} \Big(\big(f(x) - L_1\big)\big(g(x) - L_2\big) + L_1 g(x) + L_2 f(x) - L_1 L_2 \Big)$$

$$= \lim_{x \to c} \big(f(x) - L_1\big)\big(g(x) - L_2\big) + \lim_{x \to c} L_1 g(x) + \lim_{x \to c} L_2 f(x) + \lim_{x \to c}(-L_1 L_2)$$

(using limit property 2)

$$= 0 + L_1 \lim_{x \to c} g(x) + L_2 \lim_{x \to c} f(x) - L_1 L_2$$

(using limit property 1 and part (d))

$$= L_1 L_2 + L_2 L_1 - L_1 L_2 = L_1 L_2.$$

Strengthen Your Understanding

37. False. The definition of a limit guarantees that, for any positive ϵ, there is a δ. This statement, which guarantees an ϵ for a specific $\delta = 10^{-3}$, is not equivalent to $\lim_{x \to c} f(x) = L$. For example, consider a function with a vertical asymptote within 10^{-3} of 0, such as $c = 0$, $L = 0$, $f(x) = x/(x - 10^{-4})$.

Solutions for Chapter 1 Review

Exercises

1. The line of slope m through the point (x_0, y_0) has equation

$$y - y_0 = m(x - x_0),$$

so the line we want is

$$y - 0 = 2(x - 5)$$
$$y = 2x - 10.$$

5. A circle with center (h, k) and radius r has equation $(x - h)^2 + (y - k)^2 = r^2$. Thus $h = -1$, $k = 2$, and $r = 3$, giving

$$(x + 1)^2 + (y - 2)^2 = 9.$$

Solving for y, and taking the positive square root gives the top half, so

$$(y - 2)^2 = 9 - (x + 1)^2$$
$$y = 2 + \sqrt{9 - (x + 1)^2}.$$

See Figure 1.36.

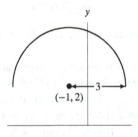

Figure 1.36: Graph of $y = 2 + \sqrt{9 - (x + 1)^2}$

9. See Figure 1.37.

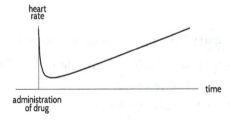

Figure 1.37

13. (a) Since $m = f(A)$, we see that $f(100)$ represents the value of m when $A = 100$. Thus $f(100)$ is the minimum annual gross income needed (in thousands) to take out a 30-year mortgage loan of $100,000 at an interest rate of 6%.
 (b) Since $m = f(A)$, we have $A = f^{-1}(m)$. We see that $f^{-1}(75)$ represents the value of A when $m = 75$, or the size of a mortgage loan that could be obtained on an income of $75,000.

17. Collecting similar factors yields $\left(\frac{1.04}{1.03}\right)^t = \frac{12.01}{5.02}$. Solving for t yields

$$t = \frac{\log\left(\frac{12.01}{5.02}\right)}{\log\left(\frac{1.04}{1.03}\right)} = 90.283.$$

21. $f(x) = x^3, \quad g(x) = \ln x.$

25. (a) Since $f(x)$ is an odd polynomial with a positive leading coefficient, it follows that $f(x) \to \infty$ as $x \to \infty$ and $f(x) \to -\infty$ as $x \to -\infty$.

(b) Since $f(x)$ is an even polynomial with negative leading coefficient, it follows that $f(x) \to -\infty$ as $x \to \pm\infty$.

(c) As $x \to \pm\infty$, $x^4 \to \infty$, so $x^{-4} = 1/x^4 \to 0$.

(d) As $x \to \pm\infty$, the lower-degree terms of $f(x)$ become insignificant, and $f(x)$ becomes approximated by the highest degree terms in its numerator and denominator. So as $x \to \pm\infty$, $f(x) \to 6$.

29. Starting with the general exponential equation $y = Ae^{kx}$, we first find that for $(0, 1)$ to be on the graph, we must have $A = 1$. Then to make $(3, 4)$ lie on the graph, we require

$$4 = e^{3k}$$
$$\ln 4 = 3k$$
$$k = \frac{\ln 4}{3} \approx 0.4621.$$

Thus the equation is

$$y = e^{0.4621x}.$$

Alternatively, we can use the form $y = a^x$, in which case we find $y = (1.5874)^x$.

33. $z = 1 - \cos\theta$

37. This looks like a fourth degree polynomial with roots at -5 and -1 and a double root at 3. The leading coefficient is negative, and so a possible formula is

$$y = -(x + 5)(x + 1)(x - 3)^2.$$

41. This graph has period 5, amplitude 1 and no vertical shift or horizontal shift from $\sin x$, so it is given by

$$f(x) = \sin\left(\frac{2\pi}{5}x\right).$$

45. $f(x) = \dfrac{x^3|2x - 6|}{x - 3} = \begin{cases} \dfrac{x^3(2x - 6)}{x - 3} = 2x^3, & x > 3 \\ \dfrac{x^3(-2x + 6)}{x - 3} = -2x^3, & x < 3 \end{cases}$

Figure 1.38 confirms that $\lim\limits_{x \to 3^+} f(x) = 54$ while $\lim\limits_{x \to 3^-} f(x) = -54$; thus $\lim\limits_{x \to 3} f(x)$ does not exist.

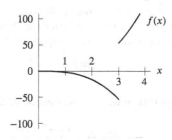

Figure 1.38

49. Yes, because the denominator is never zero.

53. For $-1 \le \theta \le 1$, $-1 \le y \le 1$, the graph of $y = \dfrac{\cos \theta - 1}{\theta}$ is shown in Figure 1.39. The graph suggests that

$$\lim_{\theta \to 0} \frac{\cos \theta - 1}{\theta} = 0.$$

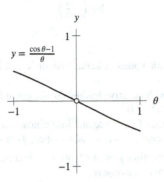

Figure 1.39

57. (a) Substituting x-values into $f(x)$ gives:

x	0.9	0.99	1.01	1.1
$f(x)$	1.054	1.005	0.995	0.953

(b) Since the values of $f(x)$ appear to get closer and closer to 1 as x gets closer and closer to 1, we estimate

$$\lim_{x \to 1} \frac{\ln x}{x - 1} = 1.$$

Notice that $f(x) = \dfrac{\ln x}{x - 1}$ is undefined at $x = 1$.

61. For large values of x, $-5x^2 + 2$ behaves like $-5x^2$ and $3x^2$ behaves like $3x^2$, so $(-5x^2 + 2)/(3x^2)$ behaves like $-5x^2/3x^2$. Thus

$$\lim_{x \to \infty} \frac{-5x^2 + 2}{3x^2} = \lim_{x \to \infty} \frac{-5x^2}{3x^2} = -\frac{5}{3}.$$

65. (a) For $x < 2$, we have $f(x) = 3x - 5$, so as x approaches 2 from the left, we have

$$\lim_{x \to 2^-} f(x) = 3(2) - 5 = 1.$$

(b) For $x > 2$, we have $f(x) = 3 - 2x$, so as x approaches 2 from the right, we have

$$\lim_{x \to 2^+} f(x) = 3 - 2(2) = -1.$$

(c) The one-sided limits $\lim_{x \to 2^-} f(x)$ and $\lim_{x \to 2^+} f(x)$ do not agree. Therefore, the limit $\lim_{x \to 2} f(x)$ does not exist, since $f(x)$ does not approach a single value as x approaches 2 from both sides.

(d) If we are looking at the behavior of $f(x)$ close to $x = 0$, then we are looking at values of $f(x)$ for $x \le 2$. So for values of x near zero, $f(x) = 3x - 5$. As x approaches zero, $f(x)$ approaches $3(0) - 5 = -5$. Therefore,

$$\lim_{x \to 0} f(x) = -5.$$

69. Evaluating $\dfrac{x^2 - 5x + 6}{x - 2}$ at $x = 2$ gives us $0/0$, so we see if we can rewrite the function using algebra. We have

$$\frac{x^2 - 5x + 6}{x - 2} = \frac{(x - 3)(x - 2)}{x - 2}.$$

Since $x \ne 2$ in the limit, we can cancel the common factor $x - 2$ to see

$$\lim_{x \to 2} \frac{x^2 - 5x + 6}{x - 2} = \lim_{x \to 2} \frac{(x - 3)(x - 2)}{x - 2} = \lim_{x \to 2}(x - 3) = -1.$$

Problems

73. (a) More fertilizer increases the yield until about 40 lbs.; then it is too much and ruins crops, lowering yield.

(b) The vertical intercept is at $Y = 200$. If there is no fertilizer, then the yield is 200 bushels.

(c) The horizontal intercept is at $a = 80$. If you use 80 lbs. of fertilizer, then you will grow no apples at all.

(d) The range is the set of values of Y attainable over the domain $0 \leq a \leq 80$. Looking at the graph, we can see that Y goes as high as 550 and as low as 0. So the range is $0 \leq Y \leq 550$.

(e) Looking at the graph, we can see that Y is decreasing at $a = 60$.

(f) Looking at the graph, we can see that Y is concave down everywhere, so it is certainly concave down at $a = 40$.

77. (a)   (i) If the atoms are moved farther apart, then $r > a$ so, from the graph, F is negative, indicating an attractive force, which pulls the atoms back together.

(ii) If the atoms are moved closer together, then $r < a$ so, from the graph, F is positive, indicating an attractive force, which pushes the atoms apart again.

(b) At $r = a$, the force is zero. The answer to part (a)(i) tells us that if the atoms are pulled apart slightly, so $r > a$, the force tends to pull them back together; the answer to part (a)(ii) tells us that if the atoms are pushed together, so $r < a$, the force tends to push them back apart. Thus, $r = a$ is a stable equilibrium.

81. (a) The US consumption of hydroelectric power increased by at least 20% in 2011 and decreased by at least 10% in 2012, relative to each corresponding previous year. .

(b) False. In 2012 hydroelectric power consumption increased only by 22.1% over consumption in 2011.

(c) True. From 2011 to 2012 consumption decreased by 15.2%, which means $x(1 - 0.152)$ units of hydroelectric power were consumed in 2012 if x had been consumed in 2011. Similarly,

$$(x(1 - 0.152)(1 - 0.025)$$

units of hydroelectric power were consumed in 2013 if x had been consumed in 2012, and

$$(x(1 - 0.152)(1 - 0.025)(1 - 0.036)$$

units of hydroelectric power were consumed in 2014 if x had been consumed in 2013. Since

$$x(1 - 0.152)(1 - 0.025)(1 - 0.036) = x(0.797) = x(1 - 0.203),$$

the percent growth in hydroelectric power consumption was -20.3%, in 2014 relative to consumption in 2011. This amounts to about 20% decrease in hydroelectric power consumption from 2011 to 2014.

85. Given the doubling time of 2 hours, $200 = 100e^{k(2)}$, we can solve for the growth rate k using the equation:

$$2P_0 = P_0 e^{2k}$$
$$\ln 2 = 2k$$
$$k = \frac{\ln 2}{2}.$$

Using the growth rate, we wish to solve for the time t in the formula

$$P = 100e^{\frac{\ln 2}{2}t}$$

where $P = 3{,}200$, so

$$3{,}200 = 100e^{\frac{\ln 2}{2}t}$$
$$t = 10 \text{ hours.}$$

89. (a) We know the decay follows the equation

$$P = P_0 e^{-kt},$$

and that 10% of the pollution is removed after 5 hours (meaning that 90% is left). Therefore,

$$0.90P_0 = P_0 e^{-5k}$$
$$k = -\frac{1}{5}\ln(0.90).$$

Thus, after 10 hours:

$$P = P_0 e^{-10((-0.2)\ln 0.90)}$$
$$P = P_0(0.9)^2 = 0.81P_0$$

so 81% of the original amount is left.

(b) We want to solve for the time when $P = 0.50P_0$:

$$0.50P_0 = P_0 e^{t((0.2)\ln 0.90)}$$
$$0.50 = e^{\ln(0.90^{0.2t})}$$
$$0.50 = 0.90^{0.2t}$$
$$t = \frac{5\ln(0.50)}{\ln(0.90)} \approx 32.9 \text{ hours.}$$

(c)

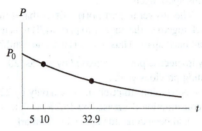

(d) When highly polluted air is filtered, there is more pollutant per liter of air to remove. If a fixed amount of air is cleaned every day, there is a higher amount of pollutant removed earlier in the process.

93. The US voltage has a maximum value of 156 volts and has a period of $1/60$ of a second, so it executes 60 cycles a second. The European voltage has a higher maximum of 339 volts, and a slightly longer period of $1/50$ seconds, so it oscillates at 50 cycles per second.

97. (a) See Figure 1.40.

(b) The graph is made of straight line segments, rising from the x-axis at the origin to height a at $x = 1$, b at $x = 2$, and c at $x = 3$ and then returning to the x-axis at $x = 4$. See Figure 1.41.

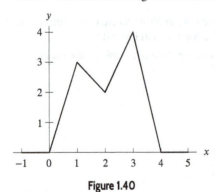

Figure 1.40

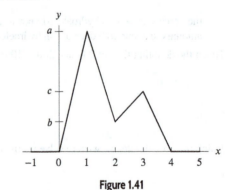

Figure 1.41

101. For any values of k, the function is continuous on any interval that does not contain $x = 2$.

Since $5x^3 - 10x^2 = 5x^2(x - 2)$, we can cancel $(x - 2)$ provided $x \neq 2$, giving

$$f(x) = \frac{5x^3 - 10x^2}{x - 2} = 5x^2 \qquad x \neq 2.$$

Thus, if we pick $k = 5(2)^2 = 20$, the function is continuous.

105. (a) To find $\lim_{x \to 4^+} f(x)$, we observe how $f(x)$ behaves as x approaches 4 from the right. If $x > 4$, then $f(x) = 9 - x$; so as x approaches 4 from the right, $f(x)$ approaches $9 - 4 = 5$. So $\lim_{x \to 4^+} f(x) = 5$.

(b) When $x < 4$, we have $f(x) = 2x - 3$; so as x approaches 4 from the left, $f(x)$ approaches $2(4) - 3 = 5$. So $\lim_{x \to 4^-} f(x) = 5$.

(c) From (a) and (b), we know $\lim_{x \to 4^+} f(x) = \lim_{x \to 4^-} f(x) = 5$, so $f(x)$ approaches 5 as x approaches 4 from either direction. Therefore, $\lim_{x \to 4} f(x) = 5$.

109. This statement is false. Since the graph of $g(t)$ has horizontal asymptotes at both $y = -1$ and at $y = 2$, the graph must approach one of these lines as t approaches ∞ and the other as t approaches $-\infty$. So $\lim\limits_{t \to -\infty} g(t)$ must exist, and must be equal to -1 or 2.

113. (a) The values of $f(x)$ as x gets closer to 5 are getting closer to 3.5, which suggests $\lim\limits_{x \to 5} f(x) = 3.5$.

(b) $f(x)$ is continuous at $x = 5$ if the values of $f(x)$ approach $f(5)$ as x approaches 5. The values approach 3.5, and $f(5) = 3.5$, so the function is continuous.

117.

$$
\begin{aligned}
\lim\limits_{x \to 5} \frac{x^2 + 4x + 3}{x + 2} &= \frac{\lim\limits_{x \to 5}(x^2 + 4x + 3)}{\lim\limits_{x \to 5}(x + 2)} & \text{Property 4} \\[2mm]
&= \frac{\lim\limits_{x \to 5}(x^2) + \lim\limits_{x \to 5}(4x) + \lim\limits_{x \to 5}(3)}{\lim\limits_{x \to 5}(x) + \lim\limits_{x \to 5}(2)} & \text{Property 2} \\[2mm]
&= \frac{(\lim\limits_{x \to 5} x)^2 + 4\lim\limits_{x \to 5}(x) + \lim\limits_{x \to 5}(3)}{\lim\limits_{x \to 5}(x) + \lim\limits_{x \to 5}(2)} & \text{Properties 1 and 3} \\[2mm]
&= \frac{5^2 + 4(5) + 3}{5 + 2} & \text{Properties 5 and 6} \\[2mm]
&= \frac{48}{7}.
\end{aligned}
$$

121. We have

$$
\lim\limits_{x \to 4} 2x + 3 = \left(\lim\limits_{x \to 4} 2\right)\left(\lim\limits_{x \to 4} x\right) + \left(\lim\limits_{x \to 4} 3\right) = 2 \cdot 4 + 3 = 11.
$$

125. The limit appears to be 1; a graph and table of values is shown below.

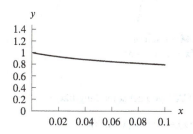

x	x^x
0.1	0.7943
0.01	0.9550
0.001	0.9931
0.0001	0.9990
0.00001	0.9999

129. From Table 1.2, it appears the limit is -1. This is confirmed by Figure 1.42. An appropriate window is $-0.099 < x < 0.099$, $-1.01 < y < -0.99$.

Table 1.2

x	$f(x)$	x	$f(x)$
0.1	−0.99	−0.0001	−0.99999999
0.01	−0.9999	−0.001	−0.999999
0.001	−0.999999	−0.01	−0.9999
0.0001	−0.99999999	−0.1	−0.99

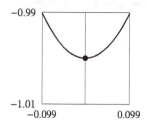

Figure 1.42

133. From Table 1.3, it appears the limit is 3. This is confirmed by Figure 1.43. An appropriate window is $-0.047 < x < 0.047$, $2.99 < y < 3.01$.

Table 1.3

x	$f(x)$
0.1	2.9552
0.01	2.9996
0.001	3.0000
0.0001	3.0000

x	$f(x)$
−0.0001	3.0000
−0.001	3.0000
−0.01	2.9996
−0.1	2.9552

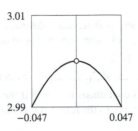

Figure 1.43

CAS Challenge Problems

137. (a) A CAS gives $f(x) = -(x-1)^2(x-3)^3$.

(b) For large $|x|$, the graph of $f(x)$ looks like the graph of $y = -x^5$, so $f(x) \to \infty$ as $x \to -\infty$ and $f(x) \to -\infty$ as $x \to \infty$. The answer to part (a) shows that f has a double root at $x = 1$, so near $x = 1$, the graph of f looks like a parabola touching the x-axis at $x = 1$. Similarly, f has a triple root at $x = 3$. Near $x = 3$, the graph of f looks like the graph of $y = x^3$, flipped over the x-axis and shifted to the right by 3, so that the "seat" is at $x = 3$. See Figure 1.44.

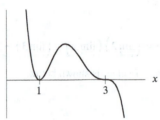

Figure 1.44: Graph of
$f(x) = -(x-1)^2(x-3)^3$

141. Using the trigonometric expansion capabilities of your CAS, you get something like

$$\sin(5x) = 5\cos^4(x)\sin(x) - 10\cos^2(x)\sin^3(x) + \sin^5(x).$$

Answers may vary. To get rid of the powers of cosine, use the identity $\cos^2(x) = 1 - \sin^2(x)$. This gives

$$\sin(5x) = 5\sin(x)\left(1 - \sin^2(x)\right)^2 - 10\sin^3(x)\left(1 - \sin^2(x)\right) + \sin^5(x).$$

Finally, using the CAS to simplify,

$$\sin(5x) = 5\sin(x) - 20\sin^3(x) + 16\sin^5(x).$$

CHAPTER TWO

Solutions for Section 2.1

Exercises

1. For t between 2 and 5, we have

$$\text{Average velocity} = \frac{\Delta s}{\Delta t} = \frac{400 - 135}{5 - 2} = \frac{265}{3} \text{ km/hr.}$$

The average velocity on this part of the trip was 265/3 km/hr.

5. The average velocity over a time period is the change in position divided by the change in time. Since the function $s(t)$ gives the distance of the particle from a point, we read off the graph that $s(1) = 2$ and $s(3) = 6$. Thus,

$$\text{Average velocity} = \frac{\Delta s(t)}{\Delta t} = \frac{s(3) - s(1)}{3 - 1} = \frac{6 - 2}{2} = 2 \text{ meters/sec.}$$

9. (a) Let $s = f(t)$.
   (i) We wish to find the average velocity between $t = 1$ and $t = 1.1$. We have

$$\text{Average velocity} = \frac{f(1.1) - f(1)}{1.1 - 1} = \frac{3.63 - 3}{0.1} = 6.3 \text{ m/sec.}$$

   (ii) We have

$$\text{Average velocity} = \frac{f(1.01) - f(1)}{1.01 - 1} = \frac{3.0603 - 3}{0.01} = 6.03 \text{ m/sec.}$$

   (iii) We have

$$\text{Average velocity} = \frac{f(1.001) - f(1)}{1.001 - 1} = \frac{3.006003 - 3}{0.001} = 6.003 \text{ m/sec.}$$

(b) We see in part (a) that as we choose a smaller and smaller interval around $t = 1$ the average velocity appears to be getting closer and closer to 6, so we estimate the instantaneous velocity at $t = 1$ to be 6 m/sec.

13. See Figure 2.1.

Figure 2.1

Problems

17. Using $h = 0.1, 0.01, 0.001$, we see

$$\frac{7^{0.1} - 1}{0.1} = 2.148$$

$$\frac{7^{0.01} - 1}{0.01} = 1.965$$

$$\frac{7^{0.001} - 1}{0.001} = 1.948$$

$$\frac{7^{0.0001} - 1}{0.0001} = 1.946.$$

This suggests that $\lim\limits_{h \to 0} \dfrac{7^h - 1}{h} \approx 1.9$.

21.

Slope	-3	-1	0	1/2	1	2
Point	F	C	E	A	B	D

25.

$$\left(\begin{array}{c} \text{Average velocity} \\ 0 < t < 0.2 \end{array} \right) = \frac{s(0.2) - s(0)}{0.2 - 0} = \frac{0.5}{0.2} = 2.5 \text{ ft/sec.}$$

$$\left(\begin{array}{c} \text{Average velocity} \\ 0.2 < t < 0.4 \end{array} \right) = \frac{s(0.4) - s(0.2)}{0.4 - 0.2} = \frac{1.3}{0.2} = 6.5 \text{ ft/sec.}$$

A reasonable estimate of the velocity at $t = 0.2$ is the average: $\frac{1}{2}(6.5 + 2.5) = 4.5$ ft/sec.

29. (a) When $t = 0$, the ball is on the bridge and its height is $f(0) = 36$, so the bridge is 36 feet above the ground.
 (b) After 1 second, the ball's height is $f(1) = -16 + 50 + 36 = 70$ feet, so it traveled $70 - 36 = 34$ feet in 1 second, and its average velocity was 34 ft/sec.
 (c) At $t = 1.001$, the ball's height is $f(1.001) = 70.017984$ feet, and its velocity about $\frac{70.017984 - 70}{1.001 - 1} = 17.984 \approx 18$ ft/sec.
 (d) We complete the square:

$$\begin{aligned} f(t) &= -16t^2 + 50t + 36 \\ &= -16\left(t^2 - \frac{25}{8}t\right) + 36 \\ &= -16\left(t^2 - \frac{25}{8}t + \frac{625}{256}\right) + 36 + 16\left(\frac{625}{256}\right) \\ &= -16(t - \tfrac{25}{16})^2 + \tfrac{1201}{16} \end{aligned}$$

so the graph of f is a downward parabola with vertex at the point $(25/16, 1201/16) = (1.6, 75.1)$. We see from Figure 2.2 that the ball reaches a maximum height of about 75 feet. The velocity of the ball is zero when it is at the peak, since the tangent is horizontal there.
 (e) The ball reaches its maximum height when $t = \frac{25}{16} = 1.6$.

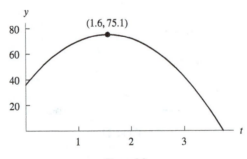

Figure 2.2

33. $\displaystyle \lim_{h \to 0} \frac{(3 + h)^2 - (3 - h)^2}{2h} = \lim_{h \to 0} \frac{9 + 6h + h^2 - 9 + 6h - h^2}{2h} = \lim_{h \to 0} \frac{12h}{2h} = \lim_{h \to 0} 6 = 6.$

Strengthen Your Understanding

37. $f(t) = t^2$. The slope of the graph of $y = f(t)$ is negative for $t < 0$ and positive for $t > 0$.
 Many other answers are possible.

41. True. During a short enough time interval the car can not change its velocity very much, and so it velocity will be nearly constant. It will be nearly equal to the average velocity over the interval.

Solutions for Section 2.2

Exercises

1. The derivative, $f'(2)$, is the rate of change of x^3 at $x = 2$. Notice that each time x changes by 0.001 in the table, the value of x^3 changes by 0.012. Therefore, we estimate

$$f'(2) = \frac{\text{Rate of change}}{\text{of } f \text{ at } x = 2} \approx \frac{0.012}{0.001} = 12.$$

The function values in the table look exactly linear because they have been rounded. For example, the exact value of x^3 when $x = 2.001$ is 8.012006001, not 8.012. Thus, the table can tell us only that the derivative is approximately 12. Example 5 on page 104 shows how to compute the derivative of $f(x)$ exactly.

5. (a)

Table 2.1

x	1	1.5	2	2.5	3
$\log x$	0	0.18	0.30	0.40	0.48

(b) The average rate of change of $f(x) = \log x$ between $x = 1$ and $x = 3$ is

$$\frac{f(3) - f(1)}{3 - 1} = \frac{\log 3 - \log 1}{3 - 1} \approx \frac{0.48 - 0}{2} = 0.24$$

(c) First we find the average rates of change of $f(x) = \log x$ between $x = 1.5$ and $x = 2$, and between $x = 2$ and $x = 2.5$.

$$\frac{\log 2 - \log 1.5}{2 - 1.5} = \frac{0.30 - 0.18}{0.5} \approx 0.24$$

$$\frac{\log 2.5 - \log 2}{2.5 - 2} = \frac{0.40 - 0.30}{0.5} \approx 0.20$$

Now we approximate the instantaneous rate of change at $x = 2$ by finding the average of the above rates, i.e.

$$\left(\begin{array}{c} \text{the instantaneous rate of change} \\ \text{of } f(x) = \log x \text{ at } x = 2 \end{array} \right) \approx \frac{0.24 + 0.20}{2} = 0.22.$$

9. We estimate $f'(2)$ using the average rate of change formula on a small interval around 2. We use the interval $x = 2$ to $x = 2.001$. (Any small interval around 2 gives a reasonable answer.) We have

$$f'(2) \approx \frac{f(2.001) - f(2)}{2.001 - 2} = \frac{3^{2.001} - 3^2}{2.001 - 2} = \frac{9.00989 - 9}{0.001} = 9.89.$$

13. One possible choice of points is shown below.

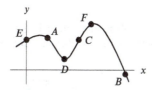

17. On February 1st, 2015, there were almost 14 million square kilometers of Arctic Sea covered by ice.

Problems

21. (a) Since f is increasing, $f(4) > f(3)$.
(b) From Figure 2.3, it appears that $f(2) - f(1) > f(3) - f(2)$.

(c) The quantity $\dfrac{f(2) - f(1)}{2 - 1}$ represents the slope of the secant line connecting the points on the graph at $x = 1$ and

$x = 2$. This is greater than the slope of the secant line connecting the points at $x = 1$ and $x = 3$ which is $\dfrac{f(3) - f(1)}{3 - 1}$.

(d) The function is steeper at $x = 1$ than at $x = 4$ so $f'(1) > f'(4)$.

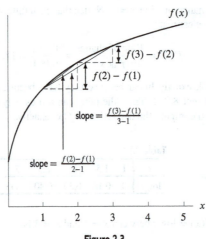

Figure 2.3

25. See Figure 2.4.

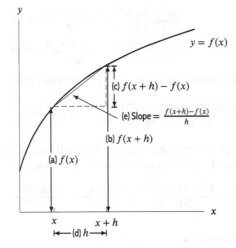

Figure 2.4

29. Figure 2.5 shows the graph of an odd function. We see that since g is symmetric about the origin, its tangent line at $x = -4$ is just the tangent line at $x = 4$ flipped about the origin, so they have the same slope. Thus, $g'(-4) = 5$.

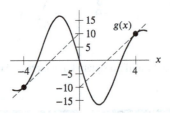

Figure 2.5

33. We want $f'(2)$. The exact answer is

$$f'(2) = \lim_{h \to 0} \frac{f(2+h) - f(2)}{h} = \lim_{h \to 0} \frac{(2+h)^{2+h} - 4}{h},$$

but we can approximate this. If $h = 0.001$, then

$$\frac{(2.001)^{2.001} - 4}{0.001} \approx 6.779$$

and if $h = 0.0001$ then

$$\frac{(2.0001)^{2.0001} - 4}{0.0001} \approx 6.773,$$

so $f'(2) \approx 6.77$.

37. The quantity $f(0)$ represents the population on October 17, 2006, so $f(0) = 300$ million.
The quantity $f'(0)$ represents the rate of change of the population (in millions per year). Since

$$\frac{1 \text{ person}}{11 \text{ seconds}} = \frac{1/10^6 \text{ million people}}{11/(60 \cdot 60 \cdot 24 \cdot 365) \text{ years}} = 2.867 \text{ million people/year},$$

so we have $f'(0) = 2.867$.

41. We have

$$p(400 + h) - p(400) \approx p'(400)h.$$

Taking $h = -2$ we have

$$\text{Change in } y = p(398) - p(400) \approx (2)(-2) = -4.$$

45. $\lim\limits_{h \to 0} \dfrac{(-3+h)^2 - 9}{h} = \lim\limits_{h \to 0} \dfrac{9 - 6h + h^2 - 9}{h} = \lim\limits_{h \to 0} \dfrac{h(-6+h)}{h} = \lim\limits_{h \to 0} -6 + h = -6.$

49. $\sqrt{4+h} - 2 = \dfrac{(\sqrt{4+h} - 2)(\sqrt{4+h} + 2)}{\sqrt{4+h} + 2} = \dfrac{4 + h - 4}{\sqrt{4+h} + 2} = \dfrac{h}{\sqrt{4+h} + 2}.$

Therefore $\lim\limits_{h \to 0} \dfrac{\sqrt{4+h} - 2}{h} = \lim\limits_{h \to 0} \dfrac{1}{\sqrt{4+h} + 2} = \dfrac{1}{4}$

53. The difference quotient for $f(x) = \sqrt{x}$ is given by

$$\frac{f(1+h) - f(1)}{h} = \frac{\sqrt{1+h} - 1}{h}$$

Table 2.2 shows the difference quotient evaluated at several small values of h.

Table 2.2

h	0.1	0.01	0.001	0.0001
$\dfrac{f(1+h) - f(1)}{h}$	0.4881	0.4988	0.4999	0.49999

Therefore, we estimate that $f'(1) = 1/2$. The exact calculation of the derivative is

$$
\begin{aligned}
f'(1) &= \lim_{h \to 0} \frac{\sqrt{1+h} - 1}{h} \\
&= \lim_{h \to 0} \frac{\sqrt{1+h} - 1}{h} \cdot \frac{\sqrt{1+h} + 1}{\sqrt{1+h} + 1} \\
&= \lim_{h \to 0} \frac{1 + h - 1}{h(\sqrt{1+h} + 1)} \\
&= \lim_{h \to 0} \frac{1}{\sqrt{1+h} + 1} \\
&= \frac{1}{2}.
\end{aligned}
$$

Thus the exact value of $f'(1) = 1/2$.

57. Using the definition of the derivative, we have

$$
\begin{aligned}
f'(-2) &= \lim_{h \to 0} \frac{f(-2 + h) - f(-2)}{h} \\
&= \lim_{h \to 0} \frac{(-2 + h)^3 - (-2)^3}{h} \\
&= \lim_{h \to 0} \frac{(-8 + 12h - 6h^2 + h^3) - (-8)}{h} \\
&= \lim_{h \to 0} \frac{12h - 6h^2 + h^3}{h} \\
&= \lim_{h \to 0} \frac{h(12 - 6h + h^2)}{h} \\
&= \lim_{h \to 0} (12 - 6h + h^2),
\end{aligned}
$$

which goes to 12 as $h \to 0$. So $f'(-2) = 12$.

61.

$$
\begin{aligned}
g'(2) &= \lim_{h \to 0} \frac{g(2 + h) - g(2)}{h} = \lim_{h \to 0} \frac{\frac{1}{(2+h)^2} - \frac{1}{2^2}}{h} \\
&= \lim_{h \to 0} \frac{2^2 - (2 + h)^2}{2^2 (2 + h)^2 h} = \lim_{h \to 0} \frac{4 - 4 - 4h - h^2}{4h(2 + h)^2} \\
&= \lim_{h \to 0} \frac{-4h - h^2}{4h(2 + h)^2} = \lim_{h \to 0} \frac{-4 - h}{4(2 + h)^2} \\
&= \frac{-4}{4(2)^2} = -\frac{1}{4}.
\end{aligned}
$$

65. First find the derivative of $f(x) = 1/x^2$ at $x = 1$.

$$
\begin{aligned}
f'(1) &= \lim_{h \to 0} \frac{f(1 + h) - f(1)}{h} = \lim_{h \to 0} \frac{\frac{1}{(1+h)^2} - \frac{1}{1^2}}{h} \\
&= \lim_{h \to 0} \frac{1^2 - (1 + h)^2}{h(1 + h)^2} = \lim_{h \to 0} \frac{1 - (1 + 2h + h^2)}{h(1 + h)^2} \\
&= \lim_{h \to 0} \frac{-2h - h^2}{h(1 + h)^2} = \lim_{h \to 0} \frac{-2 - h}{(1 + h)^2} = -2
\end{aligned}
$$

Thus the tangent line has a slope of -2 and goes through the point $(1, 1)$, and so its equation is

$$
y - 1 = -2(x - 1) \quad \text{or} \quad y = -2x + 3.
$$

Strengthen Your Understanding

69. A linear function is of the form $f(x) = ax + b$. The derivative of this function is the slope of the line $y = ax + b$, so $f'(x) = a$, so $a = 2$. One such function is $f(x) = 2x + 1$.

73. (a). This is best observed graphically.

Solutions for Section 2.3

Exercises

1. (a) We use the interval to the right of $x = 2$ to estimate the derivative. (Alternately, we could use the interval to the left of 2, or we could use both and average the results.) We have

$$
f'(2) \approx \frac{f(4) - f(2)}{4 - 2} = \frac{24 - 18}{4 - 2} = \frac{6}{2} = 3.
$$

We estimate $f'(2) \approx 3$.

(b) We know that $f'(x)$ is positive when $f(x)$ is increasing and negative when $f(x)$ is decreasing, so it appears that $f'(x)$ is positive for $0 < x < 4$ and is negative for $4 < x < 12$.

5. See Figure 2.6.

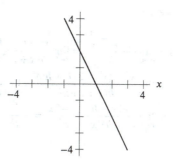

Figure 2.6

9. See Figure 2.7.

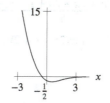

Figure 2.7

13. See Figure 2.8.

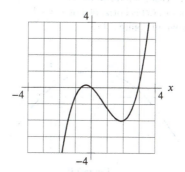

Figure 2.8

17. The graph of $f(x)$ and its derivative look the same, as in Figures 2.9 and 2.10.

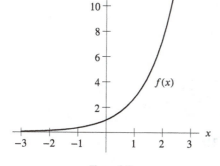

Figure 2.9

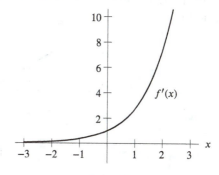

Figure 2.10

21. Since $1/x^2 = x^{-2}$, using the power rule gives

$$l'(x) = -2x^{-3} = -\frac{2}{x^3}.$$

Using the definition of the derivative, we have

$$l'(x) = \lim_{h \to 0} \frac{\frac{1}{(x+h)^2} - \frac{1}{x^2}}{h} = \lim_{h \to 0} \frac{x^2 - (x+h)^2}{h(x+h)^2 x^2}$$

$$= \lim_{h \to 0} \frac{x^2 - (x^2 + 2xh + h^2)}{h(x+h)^2 x^2} = \lim_{h \to 0} \frac{-2xh - h^2}{h(x+h)^2 x^2}$$

$$= \lim_{h \to 0} \frac{-2x - h}{(x+h)^2 x^2} = \frac{-2x}{x^2 x^2} = -\frac{2}{x^3}.$$

25. Using the definition of the derivative,

$$g'(x) = \lim_{h \to 0} \frac{g(x+h) - g(x)}{h} = \lim_{h \to 0} \frac{(x+h)^2 + 2(x+h) + 1 - (x^2 + 2x + 1)}{h}$$

$$= \lim_{h \to 0} \frac{x^2 + 2xh + h^2 + 2x + 2h + 1 - x^2 - 2x - 1}{h} = \lim_{h \to 0} \frac{2xh + h^2 + 2h}{h}$$

$$= \lim_{h \to 0} \frac{h(2x + h + 2)}{h} = \lim_{h \to 0}(2x + h + 2)$$

$$= 2x + 2.$$

Problems

29. Since $f'(x) > 0$ for $x < -1$, $f(x)$ is increasing on this interval.
Since $f'(x) < 0$ for $x > -1$, $f(x)$ is decreasing on this interval.
Since $f'(x) = 0$ at $x = -1$, the tangent to $f(x)$ is horizontal at $x = -1$.
One possible shape for $y = f(x)$ is shown in Figure 2.11.

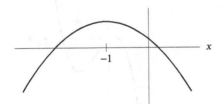

Figure 2.11

33. (a) x_3 (b) x_4 (c) x_5 (d) x_3

37. See Figure 2.12.

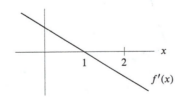

Figure 2.12

41. One possible graph is shown in Figure 2.13. Notice that as x gets large, the graph of $f(x)$ gets more and more horizontal. Thus, as x gets large, $f'(x)$ gets closer and closer to 0.

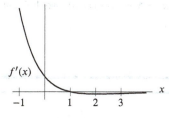

Figure 2.13

45. Since $f'(x) = 1$ for all x, the slope of f is constant; that is, f is linear. Therefore, our answer is (V).

49. (a) Graph II
(b) Graph I
(c) Graph III

53. On intervals where $f' = 0$, f is not changing at all, and is therefore constant. On the small interval where $f' > 0$, f is increasing; at the point where f' hits the top of its spike, f is increasing quite sharply. So f should be constant for a while, have a sudden increase, and then be constant again. A possible graph for f is shown in Figure 2.14.

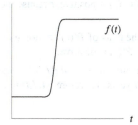

Figure 2.14: Step function

57. Since $f'(x) > 0$ for $1 < x < 3$, we see that $f(x)$ is increasing on this interval.
Since $f'(x) < 0$ for $x < 1$ and for $x > 3$, we see that $f(x)$ is decreasing on these intervals.
Since $f'(x) = 0$ for $x = 1$ and $x = 3$, the tangent to $f(x)$ will be horizontal at these x's.
One of many possible shapes of $y = f(x)$ is shown in Figure 2.15.

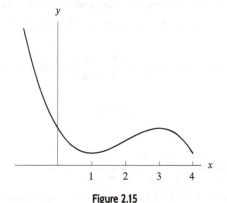

Figure 2.15

Strengthen Your Understanding

61. Since $f(x) = \cos x$ is decreasing on some intervals, its derivative $f'(x)$ is negative on those intervals, and the graph of $f'(x)$ is below the x-axis where $\cos x$ is decreasing.

65. Every linear function is of the form $f(x) = b + mx$ and has derivative $f'(x) = m$. One family of functions with the same derivative is $f(x) = b + 2x$.

69. False. A counterexample is given by $f(x) = 5$ and $g(x) = 10$, two different functions with the same derivatives: $f'(x) = g'(x) = 0$.

Solutions for Section 2.4

Exercises

1. (a) The statement $f(200) = 1300$ means that it costs \$1300 to produce 200 gallons of the chemical.
   (b) The statement $f'(200) = 6$ means that when the number of gallons produced is 200, costs are increasing at a rate of \$6 per gallon. In other words, it costs about \$6 to produce the next (the 201st) gallon of the chemical.

5. (a) The function f takes quarts of ice cream to cost in dollars, so 200 is the amount of ice cream, in quarts, and \$600 is the corresponding cost, in dollars. It costs \$600 to produce 200 quarts of ice cream.
   (b) Here, 200 is in quarts, but the 2 is in dollars/quart. After producing 200 quarts of ice cream, the cost to produce one additional quart is about \$2.

9. Units of $C'(r)$ are dollars/percent. Approximately, $C'(r)$ means the additional amount needed to pay off the loan when the interest rate is increased by 1%. The sign of $C'(r)$ is positive, because increasing the interest rate will increase the amount it costs to pay off a loan.

13. Since r is a percent and $P(r)$ is in dollars, the units of $P'(r)$ are dollars per percent. Because the monthly payment on the loan increases as the interest rate increases, $P'(r)$ is positive.

17. Since B is measured in dollars and t is measured in years, dB/dt is measured in dollars per year. We can interpret dB as the extra money added to your balance in dt years. Therefore dB/dt represents how fast your balance is growing, in units of dollars/year.

21. (a) We have $\Delta R \approx 3\Delta S$.
   (b) We have $\Delta S = 10.2 - 10 = 0.2$. Therefore $\Delta R \approx 3(0.2) = 0.6$.
   (c) Since $R = f(10) = 13$ and the change in $\Delta R = 0.6$, we have $R = f(10.2) = 13 + 0.6 = 13.6$.

Problems

25. (a) Since the derivative acts as a multiplier, the only function for which $\Delta y > \Delta x$ is f because its derivative is positive and greater than 1.
   (b) Here we want the functions whose derivatives have magnitude less than 1, so g and h.

29. Let p be the rating points earned by the CBS Evening News, let R be the revenue earned in millions of dollars, and let $R = f(p)$. When $p = 4.3$,

$$\text{Rate of change of revenue} \approx \frac{\$5.5 \text{ million}}{0.1 \text{ point}} = 55 \text{ million dollars/point.}$$

Thus

$$f'(4.3) \approx 55.$$

33. (a) The depth of the water is 3 feet at time $t = 5$ hours.
   (b) The depth of the water is increasing at 0.7 feet/hour at time $t = 5$ hours.
   (c) When the depth of the water is 5 feet, the time is $t = 7$ hours.
   (d) Since 5 is the depth in feet and $h^{-1}(5)$ is time in hours, the units of $(h^{-1})'$ are hours/feet. Thus, $(h^{-1})'(5) = 1.2$ tells us that when the water depth is 5 feet, the rate of change of time with depth is 1.2 hours per foot. In other words, when the depth is 5 feet, water is entering at a rate such that it takes 1.2 hours to add an extra foot of water.

37. (a) The pressure in dynes/cm^2 at a depth of 100 meters.
   (b) The depth of water in meters giving a pressure of $1.2 \cdot 10^6$ dynes/cm^2.
   (c) The pressure at a depth of h meters plus a pressure of 20 dynes/cm^2.
   (d) The pressure at a depth of 20 meters below the diver.

(e) The rate of increase of pressure with respect to depth, at 100 meters, in units of dynes/cm^2 per meter. Approximately, $p'(100)$ represents the increase in pressure in going from 100 meters to 101 meters.

(f) The depth, in meters, at which the rate of change of pressure with respect to depth is 100,000 dynes/cm^2 per meter.

41. Units of dP/dt are barrels/year. dP/dt is the change in quantity of petroleum per change in time (a year). This is negative. We could estimate it by finding the amount of petroleum used worldwide over a short period of time.

45. (a) The units of 0.073 are million square kilometers per day. This number tell us that on January 1, 2015, the area of Arctic Sea covered by ice was growing by about 73,000 km^2 per day.

(b) We know that, as long as Δt is small,

$$\frac{\Delta F}{\Delta t} \approx 0.073 \quad \text{so} \quad \Delta F \approx 0.073 \Delta t.$$

In our case, $\Delta t = 5$ because there are five days between January 1 and January 6, so we have:

$$\Delta F = 0.073(5) = 0.365 \text{ million km}^2.$$

Thus, between January 1 and January 6 2015, the area of Arctic Sea covered by ice grew by about 365,000 km^2.

49. (a) We have

$$\text{Units of } \frac{dR}{dC} = \frac{\text{Units of } R}{\text{Units of } C} = \frac{\text{cell divisions per hour}}{10^{-4} \text{ M}}.$$

The units are number of cell divisions per hour per 10^{-4} M of concentration.

(b) The units of the 1.5 are the units of the concentration, 10^{-4} M.

(c) We have

$$\Delta R \approx \frac{dR}{dC}\Big|_{C=1.5} \Delta C$$
$$= (0.1)(0.2) = 0.02 \text{ cell divisions per hour.}$$

The population in the more concentrated solution has 0.02 more cell divisions per hour.

53. Since

$$\frac{P(67) - P(66)}{67 - 66} \approx P'(66),$$

we may think of $P'(66)$ as an estimate of $P(67) - P(66)$, and the latter is the number of people between 66 and 67 inches tall. Alternatively, since

$$\frac{P(66.5) - P(65.5)}{66.5 - 65.5} \text{ is a better estimate of } P'(66),$$

we may regard $P'(66)$ as an estimate of the number of people of height between 65.5 and 66.5 inches. The units for $P'(x)$ are people per inch. Since there are about 300 million people in the US, we guess that there are about 250 million full-grown persons in the US whose heights are distributed between 60 inches (5 ft) and 75 inches (6 ft 3 in). There are probably quite a few people of height 66 inches—between one and two times what we would expect from an even, or uniform, distribution— because 66 inches is nearly average. An even distribution would yield

$$P'(66) = \frac{250 \text{ million}}{15 \text{ ins}} \approx 17 \text{ million people per inch,}$$

so we expect $P'(66)$ to be between 17 and 34 million people per inch.

The value of $P'(x)$ is never negative because $P(x)$ is never decreasing. To see this, let's look at an example involving a particular value of x, say $x = 70$. The value $P(70)$ represents the number of people whose height is less than or equal to 70 inches, and $P(71)$ represents the number of people whose height is less than or equal to 71 inches. Since everyone shorter than 70 inches is also shorter than 71 inches, $P(70) \le P(71)$. In general, $P(x)$ is 0 for small x, and increases as x increases, and is eventually constant (for large enough x).

Strengthen Your Understanding

57. Since T has units of minutes, its derivative with respect to P will have units of minutes/page.

61. True. The two sides of the equation are different frequently used notations for the very same quantity, the derivative of f at the point a.

65. (b) and (d) are equivalent, with (d) containing the most information. Notice that (a) and (c) are wrong.

Solutions for Section 2.5

Exercises

1. (a) Increasing, concave up
   (b) Decreasing, concave down

5.

(a)
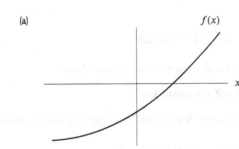

(b)
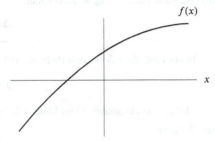

(c)
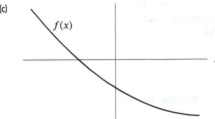

(d)
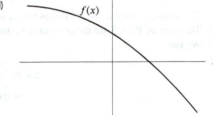

9. $f'(x) < 0$
   $f''(x) = 0$

13. $f'(x) < 0$
   $f''(x) < 0$

Problems

17. (a) Figure 2.16 shows the data:

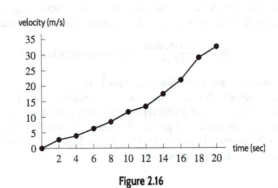

Figure 2.16

   (b) From the graph it appears that the greatest acceleration happens between 16 and 18 seconds. Since acceleration is the rate of change of velocity, we can estimate it by computing the average rate of change of the velocity between $t = 16$ and $t = 18$ seconds. We have:
$$\frac{29.1 - 21.9}{18 - 16} = 3.6 \text{ m/sec}^2$$

21. See Figure 2.17.

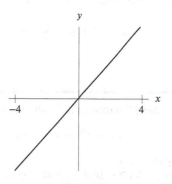

Figure 2.17

25. (a) $dP/dt > 0$ and $d^2P/dt^2 > 0$.

   (b) $dP/dt < 0$ and $d^2P/dt^2 > 0$ (but dP/dt is close to zero).

29. (a) The EPA will say that the rate of discharge is still rising. The industry will say that the rate of discharge is increasing less quickly, and may soon level off or even start to fall.

   (b) The EPA will say that the rate at which pollutants are being discharged is leveling off, but not to zero—so pollutants will continue to be dumped in the lake. The industry will say that the rate of discharge has decreased significantly.

33. No. Since f' measures the rate at which the plant stretches with respect to mass, an increasing f' means the amount the plant stretches for every unit increase in the plant's mass will get larger the larger the mass of the plant.

37. (a) At t_3, t_4, and t_5, because the graph is above the t-axis there.
   (b) At t_2 and t_3, because the graph is sloping up there.
   (c) At t_1, t_2, and t_5, because the graph is concave up there
   (d) At t_1, t_4, and t_5, because the graph is sloping down there.
   (e) At t_3 and t_4, because the graph is concave down there.

41. To the right of $x = 5$, the function starts by increasing, since $f'(5) = 2 > 0$ (though f may subsequently decrease) and is concave down, so its graph looks like the graph shown in Figure 2.18. Also, the tangent line to the curve at $x = 5$ has slope 2 and lies above the curve for $x > 5$. If we follow the tangent line until $x = 7$, we reach a height of 24. Therefore, $f(7)$ must be smaller than 24, meaning 22 is the only possible value for $f(7)$ from among the choices given.

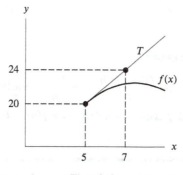

Figure 2.18

Strengthen Your Understanding

45. When the acceleration of a car is zero, the car is not speeding up or slowing down. This happens whenever the velocity is constant. The car does not have to be stationary for this to happen.

49. True. Instantaneous acceleration is a derivative, and all derivatives are limits of difference quotients. More precisely, instantaneous acceleration $a(t)$ is the derivative of the velocity $v(t)$, so

$$a(t) = \lim_{h \to 0} \frac{v(t+h) - v(t)}{h}.$$

Additional Problems (online only)

53. (a) Since $f'''(x) > 0$ for all x, we have that f'' is increasing for all x. Thus

$$f''(x) \geq f''(0) = 1 > 0 \text{ for } x \geq 0.$$

We conclude that f' is concave up for $x \geq 0$.
  (b) From part (a), we know that $f''(x) \geq 0$ for ≥ 0, so f' is increasing for $x \geq 0$. Thus

$$f'(x) \geq f(0) = 1 > 0 \text{ for } x \geq 0.$$

We conclude that f is increasing for $x \geq 0$.

Solutions for Section 2.6

Exercises

1. (a) Function f is not continuous at $x = 1$.
  (b) Function f appears not differentiable at $x = 1, 2, 3$.

5. Yes, f is differentiable at $x = 0$, since its graph does not have a "corner" at $x = 0$. See below.

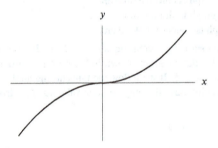

Another way to see this is by computing:

$$\lim_{h \to 0} \frac{f(h) - f(0)}{h} = \lim_{h \to 0} \frac{h|h|}{h} = \lim_{h \to 0} |h| = 0.$$

Since the limit exists and is equal to 0, f is differentiable at 0 and $f'(0) = 0$.

Problems

9. As we can see in Figure 2.19, f oscillates infinitely often between the x-axis and the line $y = 2x$ near the origin. This means a line from $(0, 0)$ to a point $(h, f(h))$ on the graph of f alternates between slope 0 (when $f(h) = 0$) and slope 2 (when $f(h) = 2h$) infinitely often as h tends to zero. Therefore, there is no limit of the slope of this line as h tends to zero, and thus there is no derivative at the origin. Another way to see this is by noting that

$$\lim_{h \to 0} \frac{f(h) - f(0)}{h} = \lim_{h \to 0} \frac{h \sin(\frac{1}{h}) + h}{h} = \lim_{h \to 0} \left(\sin\left(\frac{1}{h}\right) + 1 \right)$$

does not exist, since $\sin(\frac{1}{h})$ does not have a limit as h tends to zero. Thus, f is not differentiable at $x = 0$.

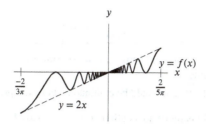

Figure 2.19

13. (a)

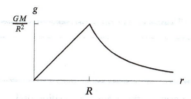

Figure 2.20

(b) The graph certainly looks continuous. The only point in question is $r = R$. Using the second formula with $r = R$ gives

$$g = \frac{GM}{R^2}.$$

Then, using the first formula with r approaching R from below, we see that as we get close to the surface of the earth

$$g \approx \frac{GMR}{R^3} = \frac{GM}{R^2}.$$

Since we get the same value for g from both formulas, g is continuous.

(c) For $r < R$, the graph of g is a line with a positive slope of $\dfrac{GM}{R^3}$. For $r > R$, the graph of g looks like $1/x^2$, and so has a negative slope. Therefore the graph has a "corner" at $r = R$ and so is not differentiable there.

17. (a) The graph of $g(r)$ does not have a break or jump at $r = 2$, and so $g(r)$ is continuous there. See Figure 2.21. This is confirmed by the fact that

$$g(2) = 1 + \cos(\pi 2/2) = 1 + (-1) = 0$$

so the value of $g(r)$ as you approach $r = 2$ from the left is the same as the value when you approach $r = 2$ from the right.

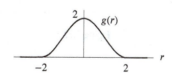

Figure 2.21

(b) The graph of $g(r)$ does not have a corner at $r = 2$, even after zooming in, so $g(r)$ appears to be differentiable at $r = 0$. This is confirmed by the fact that $\cos(\pi r/2)$ is at the bottom of a trough at $r = 2$, and so its slope is 0 there. Thus the slope to the left of $r = 2$ is the same as the slope to the right of $r = 2$.

Strengthen Your Understanding

21. There are several ways in which a function can fail to be differentiable at a point, one of which is because the graph has a sharp corner at the point. Other cases are when the function is not continuous at a point or if the graph has a vertical tangent line.

25. Let
$$f(x) = \frac{x^2 - 1}{x^2 - 4}.$$
Since $x^2 - 1 = (x - 1)(x + 1)$, this function has zeros at $x = \pm1$. However, at $x = \pm2$, the denominator $x^2 - 4 = 0$, so $f(x)$ is undefined and not differentiable.

29. True. If a function were differentiable, then it would be continuous. For example,
$$f(x) = \begin{cases} 1 & x \geq 0 \\ -1 & x < 0 \end{cases}$$
is neither differentiable nor continuous at $x = 0$. However, *one* example does not establish the truth of this statement; it merely illustrates the statement.

Solutions for Chapter 2 Review

Exercises

1. The average velocity over a time period is the change in position divided by the change in time. Since the function $s(t)$ gives the distance of the particle from a point, we find the values of $s(3) = 72$ and $s(10) = 144$. Using these values, we find
$$\text{Average velocity} = \frac{\Delta s(t)}{\Delta t} = \frac{s(10) - s(3)}{10 - 3} = \frac{144 - 72}{7} = \frac{72}{7} = 10.286 \text{ cm/sec.}$$

5. The average velocity over a time period is the change in position divided by the change in time. Since the function $s(t)$ gives the position of the particle, we find the values of $s(3) = 4$ and $s(1) = 4$. Using these values, we find
$$\text{Average velocity} = \frac{\Delta s(t)}{\Delta t} = \frac{s(3) - s(1)}{3 - 1} = \frac{4 - 4}{2} = 0 \text{ mm/sec.}$$

Though the particle moves, its average velocity over the interval is zero, since it is at the same position at $t = 1$ and $t = 3$.

9. See Figure 2.22.

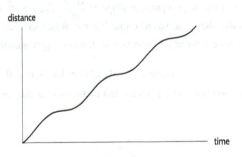

Figure 2.22

13. See Figure 2.23.

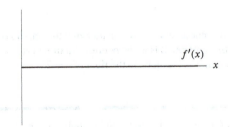

Figure 2.23

17. See Figure 2.24.

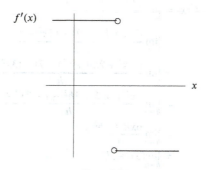

Figure 2.24

21. See Figure 2.25.

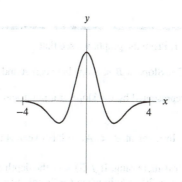

Figure 2.25

25. By joining consecutive points we get a line whose slope is the average rate of change. The steeper this line, the greater the average rate of change. See Figure 2.26.

   (a) (i) C and D. Steepest slope.

        (ii) B and C. Slope closest to 0.

   (b) A and B, and C and D. The two slopes are closest to each other.

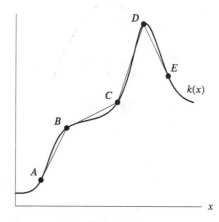

Figure 2.26

29. Using the definition of the derivative,

$$
\begin{aligned}
f'(x) &= \lim_{h \to 0} \frac{f(x+h) - f(x)}{h} \\
&= \lim_{h \to 0} \frac{(3(x+h)^2 - 7) - (3x^2 - 7)}{h} \\
&= \lim_{h \to 0} \frac{(3(x^2 + 2xh + h^2) - 7) - (3x^2 - 7)}{h} \\
&= \lim_{h \to 0} \frac{3x^2 + 6xh + 3h^2 - 7 - 3x^2 + 7}{h} \\
&= \lim_{h \to 0} \frac{6xh + 3h^2}{h} \\
&= \lim_{h \to 0} (6x + 3h) \\
&= 6x.
\end{aligned}
$$

33. $\lim\limits_{h \to 0} \dfrac{1}{h} \left(\dfrac{1}{(a+h)^2} - \dfrac{1}{a^2} \right) = \lim\limits_{h \to 0} \dfrac{a^2 - (a^2 + 2ah + h^2)}{(a+h)^2 a^2 h} = \lim\limits_{h \to 0} \dfrac{(-2a - h)}{(a+h)^2 a^2} = \dfrac{-2}{a^3}$

Problems

37. First note that the line $y = t$ has slope 1. From the graph, we see that

$$0 < \text{Slope at } C < \text{Slope at } B < \text{Slope between } A \text{ and } B < 1 < \text{Slope at } A.$$

Since instantaneous velocity is represented by the slope at a point and average velocity is represented by the slope between two points, we have

$$0 < \text{ Inst. vel. at } C \ < \text{ Inst. vel. at } B \ < \text{ Av. vel. between } A \text{ and } B \ < 1 < \text{ Inst. vel. at } A.$$

41. (a) If $f'(t) > 0$, the depth of the water is increasing. If $f'(t) < 0$, the depth of the water is decreasing.
  (b) The depth of the water is increasing at 20 cm/min when $t = 30$ minutes.
  (c) We use 1 meter = 100 cm, 1 hour = 60 min. At time $t = 30$ minutes

$$\text{Rate of change of depth} = 20\frac{\text{cm}}{\text{min}} = 20\frac{\text{cm}}{\text{min}} \cdot \frac{60 \text{ min}}{1 \text{ hr}} \cdot \frac{1 \text{ m}}{100 \text{ cm}} = 12 \text{ meters/hour.}$$

45. For $x < -2$, f is increasing and concave up. For $-2 < x < 1$, f is increasing and concave down. At $x = 1$, f has a maximum. For $x > 1$, f is decreasing and concave down. One such possible f is in Figure 2.27.

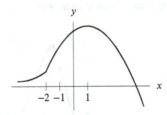

Figure 2.27

49. (a)

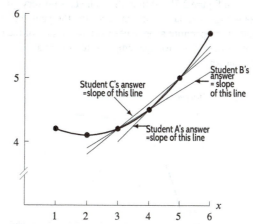

(b) The slope of f appears to be somewhere between student A's answer and student B's, so student C's answer, halfway in between, is probably the most accurate.

(c) Student A's estimate is $f'(x) \approx \frac{f(x+h)-f(x)}{h}$, while student B's estimate is $f'(x) \approx \frac{f(x)-f(x-h)}{h}$. Student C's estimate is the average of these two, or

$$f'(x) \approx \frac{1}{2}\left[\frac{f(x+h)-f(x)}{h} + \frac{f(x)-f(x-h)}{h}\right] = \frac{f(x+h)-f(x-h)}{2h}.$$

This estimate is the slope of the chord connecting $(x-h, f(x-h))$ to $(x+h, f(x+h))$. Thus, we estimate that the tangent to a curve is nearly parallel to a chord connecting points h units to the right and left, as shown below.

53. (a) Negative.

(b) $dw/dt = 0$ for t bigger than some t_0 (the time when the fire stops burning).

(c) $|dw/dt|$ increases, so dw/dt decreases since it is negative.

57. The rate of change of the US population is $P'(t)$, so

$$P'(t) = 0.8\% \cdot \text{Current population} = 0.008P(t).$$

61. (a) The slope of the tangent line at $(0, \sqrt{19})$ is zero: it is horizontal.

The slope of the tangent line at $(\sqrt{19}, 0)$ is undefined: it is vertical.

(b) The slope appears to be about $\frac{1}{2}$. (Note that when x is 2, y is about -4, but when x is 4, y is approximately -3.)

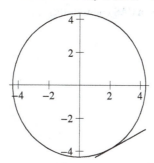

(c) Using symmetry we can determine: Slope at $(-2, \sqrt{15})$: about $\frac{1}{2}$. Slope at $(-2, -\sqrt{15})$: about $-\frac{1}{2}$. Slope at $(2, \sqrt{15})$: about $-\frac{1}{2}$.

65. (a) A possible graph is shown in Figure 2.28. At first, the yam heats up very quickly, since the difference in temperature between it and its surroundings is so large. As time goes by, the yam gets hotter and hotter, its rate of temperature increase slows down, and its temperature approaches the temperature of the oven as an asymptote. The graph is thus concave down. (We are considering the average temperature of the yam, since the temperature in its center and on its surface will vary in different ways.)

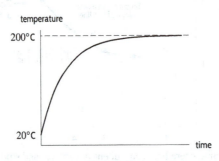

Figure 2.28

(b) If the rate of temperature increase were to remain 2°/min, in ten minutes the yam's temperature would increase 20°, from 120° to 140°. Since we know the graph is not linear, but concave down, the actual temperature is between 120° and 140°.

(c) In 30 minutes, we know the yam increases in temperature by 45° at an average rate of $45/30 = 1.5°$/min. Since the graph is concave down, the temperature at $t = 40$ is therefore between $120 + 1.5(10) = 135°$ and 140°.

(d) If the temperature increases at 2°/minute, it reaches 150° after 15 minutes, at $t = 45$. If the temperature increases at 1.5°/minute, it reaches 150° after 20 minutes, at $t = 50$. So t is between 45 and 50 minutes.

CAS Challenge Problems

69. The CAS says the derivative is zero. This can be explained by the fact that $f(x) = \sin^2 x + \cos^2 x = 1$, so $f'(x)$ is the derivative of the constant function 1. The derivative of a constant function is zero.

73. (a) The computer algebra system gives

$$\frac{d}{dx}(x^2 + 1)^2 = 4x(x^2 + 1)$$

$$\frac{d}{dx}(x^2 + 1)^3 = 6x(x^2 + 1)^2$$

$$\frac{d}{dx}(x^2 + 1)^4 = 8x(x^2 + 1)^3$$

(b) The pattern suggests that

$$\frac{d}{dx}(x^2 + 1)^n = 2nx(x^2 + 1)^{n-1}.$$

Taking the derivative of $(x^2 + 1)^n$ with a CAS confirms this.

CHAPTER THREE

Solutions for Section 3.1

Exercises

1. The derivative, $f'(x)$, is defined as

$$f'(x) = \lim_{h \to 0} \frac{f(x+h) - f(x)}{h}.$$

If $f(x) = 7$, then

$$f'(x) = \lim_{h \to 0} \frac{7 - 7}{h} = \lim_{h \to 0} \frac{0}{h} = 0.$$

5. $y' = \pi x^{\pi - 1}$.   (power rule)

9. $y' = -12x^{-13}$.

13. $y' = \frac{3}{4}x^{-1/4}$.

17. Since $g(t) = \frac{1}{t^5} = t^{-5}$, we have $g'(t) = -5t^{-6}$.

21. $f'(x) = \frac{1}{4}x^{-3/4}$.

25. $y' = 6x^{1/2} - \frac{5}{2}x^{-1/2}$.

29. The power rule gives $f'(x) = 20x^3 - \dfrac{2}{x^3}$.

33. $y' = 6t - \frac{6}{t^{3/2}} + \frac{2}{t^3}$.

37. Since $h(\theta) = \theta(\theta^{-1/2} - \theta^{-2}) = \theta\theta^{-1/2} - \theta\theta^{-2} = \theta^{1/2} - \theta^{-1}$, we have $h'(\theta) = \dfrac{1}{2}\theta^{-1/2} + \theta^{-2}$.

41. $y = \frac{\theta}{\sqrt{\theta}} - \frac{1}{\sqrt{\theta}} = \sqrt{\theta} - \frac{1}{\sqrt{\theta}}$

$y' = \frac{1}{2\sqrt{\theta}} + \frac{1}{2\theta^{3/2}}$.

45. Since $h(x) = \dfrac{ax+b}{c} = \dfrac{a}{c}x + \dfrac{b}{c}$, we have $h'(x) = \dfrac{a}{c}$.

49. Since a and b are constants, we have

$$\frac{dP}{dt} = 0 + b\frac{1}{2}t^{-1/2} = \frac{b}{2\sqrt{t}}.$$

53. We have $\Delta x = 1.95 - 2 = -0.05$. From the tangent line approximation we get:

$$\Delta y \approx f'(2)\Delta x = -3(-0.05) = 0.15.$$

Thus,

$$f(1.95) = f(2) + \Delta y \approx -4 + 0.15 = -3.85.$$

57. We have $\Delta x = 0.97 - 1 = -0.03$. Differentiating, we get $f'(x) = 3x^2 + 2x$, so $f'(1) = 5$.

$$\Delta y \approx f'(1)\Delta x = 5(-0.03) = -0.15.$$

Thus,

$$f(0.97) = f(1) + \Delta y \approx -4 + (-0.15) = -4.15.$$

Problems

61. First, we rewrite the given function as $w(x) = x^{1/2} + x^{-1/2}$. Differentiating, we have $dw/dx = (1/2)x^{-1/2} - (1/2)x^{-3/2}$, so

$$\frac{d^2w}{dx^2} = \frac{d}{dx}\left(\frac{dw}{dx}\right) = \frac{d}{dx}\left(\frac{1}{2}x^{-1/2} - \frac{1}{2}x^{-3/2}\right) = -\frac{1}{4}x^{-3/2} + \frac{3}{4}x^{-5/2},$$

and

$$\frac{d^3w}{dx^3} = \frac{d}{dx}\left(\frac{d^2w}{dx^2}\right) = \frac{d}{dx}\left(-\frac{1}{4}x^{-3/2} + \frac{3}{4}x^{-5/2}\right) = \frac{3}{8}x^{-5/2} - \frac{15}{8}x^{-7/2}.$$

65. $y' = 6x$, differentiating term-by-term and using the power rule.

69. Differentiating gives

$$f'(x) = 6x^2 - 4x \quad \text{so} \quad f'(1) = 6 - 4 = 2.$$

Thus the equation of the tangent line is $(y - 1) = 2(x - 1)$ or $y = 2x - 1$.

73. (a) The slope of f at $x = 1$ is negative, so $f'(1) < 0$. On the other hand, the slopes of f at $x = -1, x = 0$, and $x = 4$ are all positive, with the least steep slope occurring at $x = 0$ and the steepest slope occurring at $x = 4$. Therefore, we must have

$$f'(1) < f'(0) < f'(-1) < f'(4).$$

(b) The derivative function of f is given by

$$f'(x) = 3x^2 - 6x + 2.$$

Evaluating at $x = -1, 0, 1, 4$, we have $f'(1) = -1$, $f'(0) = 2$, $f'(-1) = 11$ and $f'(4) = 26$ which confirms the ordering of the four quantities.

77. (a) Since the power of x will go down by one every time you take a derivative (until the exponent is zero after which the derivative will be zero), we can see immediately that $f^{(8)}(x) = 0$.

(b) $f^{(7)}(x) = 7 \cdot 6 \cdot 5 \cdot 4 \cdot 3 \cdot 2 \cdot 1 \cdot x^0 = 5040$.

81. (a) Differentiating, we have $h'(x) = 8x^3 + 24x^2 + 30x + 14$, so

$$h'(-1) = 8(-1)^3 + 24(-1)^2 + 30(-1) + 14 = 0.$$

(b) Taking the second derivative, we have $h''(x) = 24x^2 + 48x + 30$, so

$$h''(-1) = 24(-1)^2 + 48(-1) + 30 = 6.$$

(c) Since $h'(-1) = 0$, we see that $h(x)$ has a horizontal tangent line at $x = -1$. This means it could match (I) or (IV). Since $h''(-1) = 6$, we see that $h(x)$ is concave up at $x = -1$, so $h(x)$ matches (I).

85. (a) The average velocity between $t = 0$ and $t = 2$ is given by

$$\text{Average velocity} = \frac{f(2) - f(0)}{2 - 0} = \frac{-4.9(2^2) + 25(2) + 3 - 3}{2 - 0} = \frac{33.4 - 3}{2} = 15.2 \text{ m/sec}.$$

(b) Since $f'(t) = -9.8t + 25$, we have

$$\text{Instantaneous velocity} = f'(2) = -9.8(2) + 25 = 5.4 \text{ m/sec}.$$

(c) Acceleration is given $f''(t) = -9.8$. The acceleration at $t = 2$ (and all other times) is the acceleration due to gravity, which is -9.8 m/sec^2.

(d) We can use a graph of height against time to estimate the maximum height of the tomato. See Figure 3.1. Alternately, we can find the answer analytically. The maximum height occurs when the velocity is zero and $v(t) = -9.8t + 25 = 0$ when $t = 2.6$ sec. At this time the tomato is at a height of $f(2.6) = 34.9$. The maximum height is 34.9 meters.

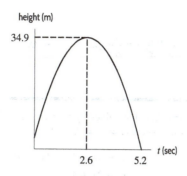

Figure 3.1

(e) We see in Figure 3.1 that the tomato hits ground at about $t = 5.2$ seconds. Alternately, we can find the answer analytically. The tomato hits the ground when

$$f(t) = -4.9t^2 + 25t + 3 = 0.$$

We solve for t using the quadratic formula:

$$t = \frac{-25 \pm \sqrt{(25)^2 - 4(-4.9)(3)}}{2(-4.9)}$$

$$t = \frac{-25 \pm \sqrt{683.8}}{-9.8}$$

$$t = -0.12 \quad \text{and} \quad t = 5.2.$$

We use the positive values, so the tomato hits the ground at $t = 5.2$ seconds.

89. We have $A = \pi r^2$. Thus, the multiplier for percent error is the power 2, so

$$\frac{\Delta A}{A} \approx 2\frac{\Delta r}{r} = 2(5\%) = 10\%.$$

The percent error in A is approximately $2(5\%) = 10\%$.

93. (a) Using the power rule we have

$$f'(x) = \frac{2}{3}(0.07)x^{-1/3} = 0.0467x^{-1/3}.$$

(b) We have

$$f'(30) = (0.0467)30^{-1/3} = 0.015 \text{ mm per meter.}$$

(c) We can approximate this difference using the derivative $f'(30)$ as follows

$$\Delta h \approx f'(30)\Delta x = 0.015 \cdot 6 = 0.09 \text{ mm.}$$

The runoff is about 0.09 mm deeper at a distance of 6 meters farther down the slope.
To check, we can calculate this difference exactly; it is

$$\Delta h = f(36) - f(30) = 0.07(36)^{2/3} - 0.07(30)^{2/3} = 0.0873 \text{ mm.}$$

97. If $f(x) = x^n$, then $f'(x) = nx^{n-1}$. This means $f'(1) = n \cdot 1^{n-1} = n \cdot 1 = n$, because any power of 1 equals 1.

101. (a) Differentiating $f(x)$ and $g(x)$, we have

$$f'(x) = 3x^2 + 6x - 2 \quad \text{and} \quad g'(x) = 3x^2 + 6x - 2.$$

(b) The graph of $g(x)$ is a vertical shift of the graph of $f(x)$ down by 5 units, so it has the same slope as $f(x)$ for every x. Since derivatives measure slope, it follows that $f'(x) = g'(x)$.

(c) Any function whose graph is a vertical shift of $f(x)$ or $g(x)$ has the same derivative as $f(x)$ or $g(x)$.

Strengthen Your Understanding

105. One possible example is $f(x) = x^2$ and $g(x) = 3x$. More generally, $f(x) = x^2 + c$ and $g(x) = 3x + k$ work for any c and k.

109. False, since

$$\frac{d}{dx}\left(\frac{\pi}{x^2}\right) = \frac{d}{dx}\left(\pi x^{-2}\right) = -2\pi x^{-3} = \frac{-2\pi}{x^3}.$$

Additional Problems (online only)

113. The slopes of the tangent lines to $y = x^2 - 2x + 4$ are given by $y' = 2x - 2$. A line through the origin has equation $y = mx$. So, at the tangent point, $x^2 - 2x + 4 = mx$ where $m = y' = 2x - 2$.

$$x^2 - 2x + 4 = (2x - 2)x$$
$$x^2 - 2x + 4 = 2x^2 - 2x$$
$$-x^2 + 4 = 0$$
$$-(x + 2)(x - 2) = 0$$
$$x = 2, -2.$$

Thus, the points of tangency are $(2, 4)$ and $(-2, 12)$. The lines through these points and the origin are $y = 2x$ and $y = -6x$, respectively. Graphically, this can be seen in Figure 3.2.

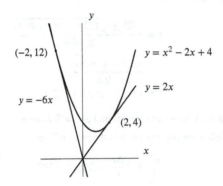

Figure 3.2

117. Yes. To see why, we substitute $y = x^n$ into the equation $13x\dfrac{dy}{dx} = y$. We first calculate $\dfrac{dy}{dx} = \dfrac{d}{dx}(x^n) = nx^{n-1}$. The differential equation becomes

$$13x(nx^{n-1}) = x^n$$

But $13x(nx^{n-1}) = 13n(x \cdot x^{n-1}) = 13nx^n$, so we have

$$13n(x^n) = x^n$$

This equality must hold for all x, so we get $13n = 1$, so $n = 1/13$. Thus, $y = x^{1/13}$ is a solution.

121. The only way for a polynomial $f(x)$ to be an even function is for all of its terms to be even powers of x. Using the power rule, we see that this means all the terms of $f'(x)$ are odd powers of x. Since $(-x)^n = -(x^n)$ when n is an odd power, the only way for $f'(-x) = f'(x)$ is if $f'(x)$ contains no powers of x, or $f'(x) = 0$. Any polynomial with $f'(x) = 0$ is of the form $f(x) = C$ for a constant C, and any such function is an even function. Therefore, there are infinitely many polynomial functions which are even functions whose derivatives are also even functions.

Solutions for Section 3.2

Exercises

1. $f'(x) = 2e^x + 2x$.

5. $y' = 10x + (\ln 2)2^x$.

9. $\dfrac{dy}{dx} = \dfrac{1}{3}(\ln 3)3^x - \dfrac{33}{2}(x^{-\frac{3}{2}}).$

13. $\dfrac{dy}{dx} = 5 \cdot 5^t \ln 5 + 6 \cdot 6^t \ln 6$

17. $f'(x) = (\ln \pi)\pi^x.$

21. $f(t) = e^t \cdot e^2$. Then, since e^2 is just a constant, $f'(t) = \frac{d}{dt}(e^t e^2) = e^2 \frac{d}{dt}e^t = e^2 e^t = e^{t+2}.$

25. $g'(x) = \dfrac{d}{dx}(2x - x^{-1/3} + 3^x - e) = 2 + \dfrac{1}{3x^{\frac{4}{3}}} + 3^x \ln 3.$

Problems

29. $y' = 2x + (\ln 2)2^x.$

33. Since $y = e^5 e^x$, $y' = e^5 e^x = e^{x+5}.$

37. We can't use our rules if the exponent is $\sqrt{\theta}$.

41. The derivative is
$$P'(t) = 300(\ln 1.044)(1.044)^t$$
so
$$P'(5) = 300(\ln 1.044)(1.044)^5 = 16.021.$$
The value
$$P'(5) = 16.021$$
means that when $t = 5$, the population is increasing by approximately 16 animals per year.

45. (a) Let $R = 17.9(1.025)^t$. Then the annual extraction is changing at a rate of
$$\frac{dR}{dt} = 17.9\frac{d}{dt}\left(1.025^t\right) = 17.9 \cdot \ln 1.025 \left(1.025^t\right) \text{ million tonnes per year.}$$

(b) At the start of 2025, we have $t = 11$, so
$$\frac{dR}{dt}\bigg|_{t=11} = 17.9 \cdot \ln 1.025 \left(1.025^{11}\right) = 0.580 \text{ million tonnes per year.}$$

(c) If the extraction rate continues to change at 0.580 million tonnes per year, in five years the extraction rate will have increased by $5(0.580) = 2.9$ million tonnes.

(d) Since the graph of $R = 17.9(1.025)^t$ is concave up, 2.9 million tonnes is smaller than the predicted increase from the model. In fact, the model predicts the increase to be
$$R(16) - R(11) = 17.9(1.025)^{16} - 17.9(1.025)^{11} = 3.09 \text{ million tonnes.}$$

49.

$g(x) = ax^2 + bx + c$ $f(x) = e^x$

$g'(x) = 2ax + b$ $f'(x) = e^x$

$g''(x) = 2a$ $f''(x) = e^x$

So, using $g''(0) = f''(0)$, etc., we have $2a = 1$, $b = 1$, and $c = 1$, and thus $g(x) = \frac{1}{2}x^2 + x + 1$, as shown in Figure 3.3.

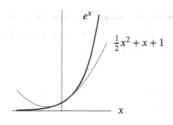

Figure 3.3

The two functions do look very much alike near $x = 0$. They both increase for large values of x, but e^x increases much more quickly. For very negative values of x, the quadratic goes to ∞ whereas the exponential goes to 0. By choosing a function whose first few derivatives agreed with the exponential when $x = 0$, we got a function which looks like the exponential for x-values near 0.

Strengthen Your Understanding

53. A possibility is $f(x) = e^x$. Then $f'(x) = e^x$, $f''(x) = e^x$, and $f'''(x) = e^x$, so $f'''(x) = f(x)$.

Additional Problems (online only)

57. The first and second derivatives of e^x are e^x. Thus, the graph of $y = e^x$ is concave up. The tangent line at $x = 0$ has slope $e^0 = 1$ and equation $y = x + 1$. A graph that is always concave up is always above any of its tangent lines. Thus $e^x \geq x + 1$ for all x, as shown in Figure 3.4.

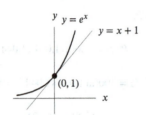

Figure 3.4

61. For $x = 0$, we have $y = a^0 = 1$ and $y = 1 + 0 = 1$, so both curves go through the point $(0, 1)$ for all values of a.
Differentiating gives

$$\frac{d(a^x)}{dx}\bigg|_{x=0} = a^x \ln a|_{x=0} = a^0 \ln a = \ln a$$

$$\frac{d(1+x)}{dx}\bigg|_{x=0} = 1.$$

The graphs are tangent at $x = 0$ if

$$\ln a = 1 \qquad \text{so} \qquad a = e.$$

65. In order to be decreasing everywhere, we must have $f'(x) < 0$ for all x and in order to be concave down everywhere we must have $f''(x) < 0$ for all x. Differentiating, we have

$$f'(x) = C(\ln a)a^x, \qquad \text{and} \qquad f''(x) = C(\ln a)^2 a^x.$$

Since a^x and $(\ln a)^2$ are both always positive, $f''(x)$ is negative only if $C < 0$. Since C is negative and a^x is positive, $f'(x)$ is negative only if $\ln a > 0$, which gives $a > 1$. This means we can choose any $a > 1$ and $C < 0$.

Solutions for Section 3.3

Exercises

1. By the product rule, $f'(x) = 2x(x^3 + 5) + x^2(3x^2) = 2x^4 + 3x^4 + 10x = 5x^4 + 10x$. Alternatively, $f'(x) = (x^5 + 5x^2)' = 5x^4 + 10x$. The two answers should, and do, match.

5. $y' = \frac{1}{2\sqrt{x}}2^x + \sqrt{x}(\ln 2)2^x$.

9. $f'(x) = \dfrac{e^x \cdot 1 - x \cdot e^x}{(e^x)^2} = \dfrac{e^x(1 - x)}{(e^x)^2} = \dfrac{1 - x}{e^x}$.

13. $q'(r) = \dfrac{3(5r + 2) - 3r(5)}{(5r + 2)^2} = \dfrac{15r + 6 - 15r}{(5r + 2)^2} = \dfrac{6}{(5r + 2)^2}$

17. $f'(t) = \dfrac{d}{dt}\left(2te^t - \dfrac{1}{\sqrt{t}}\right) = 2e^t + 2te^t + \dfrac{1}{2t^{3/2}}.$

21. $\dfrac{d}{dz}\left(\dfrac{z^2+1}{\sqrt{z}}\right) = \dfrac{d}{dz}(z^{\frac{3}{2}} + z^{-\frac{1}{2}}) = \dfrac{3}{2}z^{\frac{1}{2}} - \dfrac{1}{2}z^{-\frac{3}{2}} = \dfrac{\sqrt{z}}{2}(3 - z^{-2}).$

25. $w'(x) = \dfrac{17e^x(2^x) - (\ln 2)(17e^x)2^x}{2^{2x}} = \dfrac{17e^x(2^x)(1 - \ln 2)}{2^{2x}} = \dfrac{17e^x(1 - \ln 2)}{2^x}.$

29. $f'(x) = \dfrac{d}{dx}(2 - 4x - 3x^2)(6x^e - 3\pi) = (-4 - 6x)(6x^e - 3\pi) + (2 - 4x - 3x^2)(6ex^{e-1}).$

Problems

33. Using the quotient rule, we know that $j'(x) = (g'(x)\cdot f(x) - g(x)\cdot f'(x))/(f(x))^2$. We use slope to compute the derivatives. Since $f(x)$ is linear on the interval $0 < x < 2$, we compute the slope of the line to see that $f'(x) = 2$ on this interval. Similarly, we compute the slope on the interval $2 < x < 4$ to see that $f'(x) = -2$ on the interval $2 < x < 4$. Since $f(x)$ has a corner at $x = 2$, we know that $f'(2)$ does not exist.

Similarly, $g(x)$ is linear on the interval shown, and we see that the slope of $g(x)$ on this interval is -1 so we have $g'(x) = -1$ on this interval.

(a) We have
$$j'(1) = \frac{g'(1) \cdot f(1) - g(1) \cdot f'(1)}{(f(1))^2} = \frac{(-1)2 - 3 \cdot 2}{2^2} = \frac{-2 - 6}{4} = \frac{-8}{4} = -2.$$

(b) Since $f(x)$ has a corner at $x = 2$, so the quotient rule does not apply. We know that $f'(2)$ does not exist, so $j'(2)$ does not exist.

(c) We have
$$j'(3) = \frac{g'(3) \cdot f(3) - g(3) \cdot f'(3)}{(f(3))^2} = \frac{(-1)2 - 1(-2)}{2^2} = \frac{-2 + 2}{4} = 0.$$

37. Estimates may vary. From the graphs, we estimate $f(2) \approx 0.3$, $f'(2) \approx 1.1$, $g(2) \approx 1.6$, and $g'(2) \approx -0.5$. By the quotient rule, to one decimal place
$$k'(2) = \frac{f'(2) \cdot g(2) - f(2) \cdot g'(2)}{(g(2))^2} \approx \frac{1.1(1.6) - 0.3(-0.5)}{(1.6)^2} = 0.7.$$

41. $f(x) = e^x \cdot e^x$
$f'(x) = e^x \cdot e^x + e^x \cdot e^x = 2e^{2x}.$

45. Since $f(0) = 5(0)e^0 = 0$, the tangent line passes through the point $(0, 0)$, so its vertical intercept is 0. To find the slope of the tangent line, we find the derivative of $f(x)$ using the product rule:
$$f'(x) = (5x) \cdot e^x + 5 \cdot e^x.$$

At $x = 0$, the slope of the tangent line is $m = f'(0) = 5(0)e^0 + 5e^0 = 5$. The equation of the tangent line is $y = 5x$.

49. By the product rule, we have
$$\frac{d}{dx}(4^x(f(x) + g(x))) = \frac{d}{dx}(4^x) \cdot (f(x) + g(x)) + 4^x \cdot \frac{d}{dx}(f(x) + g(x))$$
$$= (\ln 4 \cdot 4^x)(f(x) + g(x)) + 4^x(f'(x) + g'(x))$$
$$= 4^x(\ln 4 \cdot f(x) + \ln 4 \cdot g(x) + f'(x) + g'(x)).$$

53. (a) $G'(z) = F'(z)H(z) + H'(z)F(z)$, so
$G'(3) = F'(3)H(3) + H'(3)F(3) = 4 \cdot 1 + 3 \cdot 5 = 19.$
(b) $G'(w) = \dfrac{F'(w)H(w) - H'(w)F(w)}{[H(w)]^2}$, so $G'(3) = \dfrac{4(1) - 3(5)}{1^2} = -11.$

57. (a) $f(140) = 15{,}000$ says that $15{,}000$ skateboards are sold when the cost is \$140 per board.
$f'(140) = -100$ means that if the price is increased from \$140, every dollar of increase will decrease the total sales by about 100 boards.

(b) $\dfrac{dR}{dp} = \dfrac{d}{dp}(p \cdot q) = \dfrac{d}{dp}(p \cdot f(p)) = f(p) + pf'(p)$.

So,

$$\left.\dfrac{dR}{dp}\right|_{p=140} = f(140) + 140f'(140)$$

$$= 15{,}000 + 140(-100) = 1000.$$

(c) From (b) we see that $\left.\dfrac{dR}{dp}\right|_{p=140} = 1000 > 0$. This means that the revenue will increase by about \$1000 if the price is raised by \$1.

61. (a) Since $A(t)$ gives the total value in millions of dollars, the total value in dollars is $1{,}000{,}000A(t)$ and since $N(t)$ is the number of shares in millions, the total number of shares is $1{,}000{,}000N(t)$. Therefore, the price per share on day t is

$$P(t) = \dfrac{1{,}000{,}000A(t)}{1{,}000{,}000N(t)} = \dfrac{A(t)}{N(t)}.$$

(b) Even though the price per share decreases, more shares can be purchased to increase $A(t)$.

(c) On January 1, 2016, we know $N(0) = 2$ and $A(0) = 32$. Using the quotient rule, the rate of change in dollars per day of $P(t) = A(t)/N(t)$ when $t = 0$ is

$$P'(0) = \dfrac{A'(0)N(0) - A(0)N'(0)}{(N(0))^2} = \dfrac{2A'(0) - 32N'(0)}{4}.$$

Since the value of each share is dropping at a rate of \$0.23 per day, we know $P'(0) = -0.23$ and since we want the total value $A(t)$ not to change, we want $A'(0) = 0$. Therefore, we get

$$-0.23 = \dfrac{0 - 32N'(0)}{4}$$

giving $N'(0) = 0.02875$ million shares per day. This means 28,750 additional shares must be purchased per day in order for $A(t)$ not to change.

65. Since $f(x) = g(x)h(x)$, the product rule gives

$$f'(0) = g'(0)h(0) + g(0)h'(0).$$

From the graph, we see that $g(0) = g'(0) = 0$, so

$$f'(0) = 0 \cdot h(0) + 0 \cdot h'(0) = 0.$$

Notice that we do not need $h(0)$ or $h'(0)$ in order to find this value.

Strengthen Your Understanding

69. Rewrite $f(x)$ as $f(x) = e^x(x+1)$ then the product rule gives $f'(x) = e^x(x+1) + e^x \cdot 1 = (x+2)e^x$.

73. True; looking at the statement from the other direction, if both $f(x)$ and $g(x)$ are differentiable at $x = 1$, then so is their quotient, $f(x)/g(x)$, as long as it is defined there, which requires that $g(1) \neq 0$. So the only way in which $f(x)/g(x)$ can be defined but not differentiable at $x = 1$ is if either $f(x)$ or $g(x)$, or both, is not differentiable there.

77. This is false as we may choose $g(2)$ and $h(2)$ to have opposite signs with appropriate values so they cancel when applying the product rule. For example, if we choose $g(2) = -1/0.7$ and $h(2) = 2$, the product rule gives

$$f'(2) = g'(2) \cdot h(2) + g(2) \cdot h'(2) = 1 - 1 = 0.$$

Additional Problems (online only)

81. (a) Although the answer you would get by using the quotient rule is equivalent, the answer looks simpler in this case if you just use the product rule:

$$\frac{d}{dx}\left(\frac{e^x}{x}\right) = \frac{d}{dx}\left(e^x \cdot \frac{1}{x}\right) = \frac{e^x}{x} - \frac{e^x}{x^2}$$

$$\frac{d}{dx}\left(\frac{e^x}{x^2}\right) = \frac{d}{dx}\left(e^x \cdot \frac{1}{x^2}\right) = \frac{e^x}{x^2} - \frac{2e^x}{x^3}$$

$$\frac{d}{dx}\left(\frac{e^x}{x^3}\right) = \frac{d}{dx}\left(e^x \cdot \frac{1}{x^3}\right) = \frac{e^x}{x^3} - \frac{3e^x}{x^4}.$$

(b) $\dfrac{d}{dx}\dfrac{e^x}{x^n} = \dfrac{e^x}{x^n} - \dfrac{ne^x}{x^{n+1}}.$

85. Assume for $g(x) \neq f(x)$, $g'(x) = g(x)$ and $g(0) = 1$. Then for

$$h(x) = \frac{g(x)}{e^x}$$

$$h'(x) = \frac{g'(x)e^x - g(x)e^x}{(e^x)^2} = \frac{e^x(g'(x) - g(x))}{(e^x)^2} = \frac{g'(x) - g(x)}{e^x}.$$

But, since $g(x) = g'(x)$, $h'(x) = 0$, so $h(x)$ is constant. Thus, the ratio of $g(x)$ to e^x is constant. Since $\dfrac{g(0)}{e^0} = \dfrac{1}{1} = 1$, $\dfrac{g(x)}{e^x}$ must equal 1 for all x. Thus $g(x) = e^x = f(x)$ for all x, so f and g are the same function.

89. (a) Since $x = a$ is a double zero of a polynomial $P(x)$, we can write $P(x) = (x-a)^2 Q(x)$, so $P(a) = 0$. Using the product rule, we have

$$P'(x) = 2(x-a)Q(x) + (x-a)^2 Q'(x).$$

Substituting in $x = a$, we see $P'(a) = 0$ also.

(b) Since $P(a) = 0$, we know $x = a$ is a zero of P, so that $x - a$ is a factor of P and we can write

$$P(x) = (x-a)Q(x),$$

where Q is some polynomial. Differentiating this expression for P using the product rule, we get

$$P'(x) = Q(x) + (x-a)Q'(x).$$

Since we are told that $P'(a) = 0$, we have

$$P'(a) = Q(a) + (a-a)Q'(a) = 0$$

and so $Q(a) = 0$. Therefore $x = a$ is a zero of Q, so again we can write

$$Q(x) = (x-a)R(x),$$

where R is some other polynomial. As a result,

$$P(x) = (x-a)Q(x) = (x-a)^2 R(x),$$

so that $x = a$ is a double zero of P.

Solutions for Section 3.4

Exercises

1. $f'(x) = 99(x+1)^{98} \cdot 1 = 99(x+1)^{98}.$

5. $\dfrac{dy}{dx} = \dfrac{d}{dx}(\sqrt{e^x + 1}) = \dfrac{d}{dx}(e^x + 1)^{1/2} = \dfrac{1}{2}(e^x + 1)^{-1/2}\dfrac{d}{dx}(e^x + 1) = \dfrac{e^x}{2\sqrt{e^x + 1}}.$

9. We can write $w(r) = (r^4 + 1)^{1/2}$, so
$$w'(r) = \frac{1}{2}(r^4 + 1)^{-1/2}(4r^3) = \frac{2r^3}{\sqrt{r^4 + 1}}.$$

13. $g(x) = \pi e^{\pi x}$.

17. $y' = (\ln \pi)\pi^{(x+2)}$.

21. Using the product rule gives $v'(t) = 2te^{-ct} - ce^{-ct}t^2 = (2t - ct^2)e^{-ct}$.

25. $y' = \dfrac{3s^2}{2\sqrt{s^3 + 1}}.$

29. $z' = 5 \cdot \ln 2 \cdot 2^{5t-3}$.

33. $y' = \dfrac{\frac{2^z}{2\sqrt{z}} - (\sqrt{z})(\ln 2)(2^z)}{2^{2z}} = \dfrac{1 - 2z\ln 2}{2^{z+1}\sqrt{z}}.$

37. $y' = \dfrac{-(3e^{3x} + 2x)}{(e^{3x} + x^2)^2}.$

41. $w' = (2t + 3)(1 - e^{-2t}) + (t^2 + 3t)(2e^{-2t})$.

45.
$$f'(w) = (e^{w^2})(10w) + (5w^2 + 3)(e^{w^2})(2w)$$
$$= 2we^{w^2}(5 + 5w^2 + 3)$$
$$= 2we^{w^2}(5w^2 + 8).$$

49. $f'(y) = e^{e^{(y^2)}}\left[(e^{y^2})(2y)\right] = 2ye^{[e^{(y^2)}+y^2]}$.

53. We use the product rule. We have
$$f'(x) = (ax)(e^{-bx}(-b)) + (a)(e^{-bx}) = -abxe^{-bx} + ae^{-bx}.$$

57. Using the product and chain rules, we have
$$\frac{dy}{dx} = 3(x^2 + 5)^2(2x)(3x^3 - 2)^2 + (x^2 + 5)^3[2(3x^3 - 2)(9x^2)]$$
$$= 3(2x)(x^2 + 5)^2(3x^3 - 2)[(3x^3 - 2) + (x^2 + 5)(3x)]$$
$$= 6x(x^2 + 5)^2(3x^3 - 2)[6x^3 + 15x - 2].$$

Problems

61. When f and g are differentiable, the chain rule gives $w'(x) = g'(g(x)) \cdot g'(x)$. We use slope to compute the derivatives. Since $g(x)$ is linear on the interval shown, with slope equal to -1, we have $g'(x) = -1$ on this interval.

(a) We have $w'(1) = g'(g(1)) \cdot g'(1) = (g'(3))(-1) = (-1)(-1) = 1$.
(b) We have $w'(2) = g'(g(2)) \cdot g'(2) = (g'(2))(-1) = (-1)(-1) = 1$.
(c) We have $w'(3) = g'(g(3)) \cdot g'(3) = (g'(1))(-1) = (-1)(-1) = 1$.

65. The chain rule gives
$$\left.\frac{d}{dx}g(f(x))\right|_{x=70} = g'(f(70))f'(70) = g'(30)f'(70) = (1)(\tfrac{1}{2}) = \frac{1}{2}.$$

69. The graph is concave down when $f''(x) < 0$.
$$f'(x) = e^{-x^2}(-2x)$$
$$f''(x) = \left[e^{-x^2}(-2x)\right](-2x) + e^{-x^2}(-2)$$
$$= \frac{4x^2}{e^{x^2}} - \frac{2}{e^{x^2}}$$
$$= \frac{4x^2 - 2}{e^{x^2}} < 0$$

The graph is concave down when $4x^2 < 2$. This occurs when $x^2 < \frac{1}{2}$, or $-\frac{1}{\sqrt{2}} < x < \frac{1}{\sqrt{2}}$.

73. (a) $H(x) = F(G(x))$
$H(4) = F(G(4)) = F(2) = 1$

(b) $H(x) = F(G(x))$
$H'(x) = F'(G(x)) \cdot G'(x)$
$H'(4) = F'(G(4)) \cdot G'(4) = F'(2) \cdot 6 = 5 \cdot 6 = 30$

(c) $H(x) = G(F(x))$
$H(4) = G(F(4)) = G(3) = 4$

(d) $H(x) = G(F(x))$
$H'(x) = G'(F(x)) \cdot F'(x)$
$H'(4) = G'(F(4)) \cdot F'(4) = G'(3) \cdot 7 = 8 \cdot 7 = 56$

(e) $H(x) = \frac{F(x)}{G(x)}$
$H'(x) = \frac{G(x) \cdot F'(x) - F(x) \cdot G'(x)}{[G(x)]^2}$
$H'(4) = \frac{G(4) \cdot F'(4) - F(4) \cdot G'(4)}{[G(4)]^2} = \frac{2 \cdot 7 - 3 \cdot 6}{2^2} = \frac{14-18}{4} = \frac{-4}{4} = -1$

77. (a) $P(12) = 10e^{0.6(12)} = 10e^{7.2} \approx 13{,}394$ fish. There are 13,394 fish in the area after 12 months.

(b) We differentiate to find $P'(t)$, and then substitute in to find $P'(12)$:

$$P'(t) = 10(e^{0.6t})(0.6) = 6e^{0.6t}$$
$$P'(12) = 6e^{0.6(12)} \approx 8037 \text{ fish/month.}$$

The population is growing at a rate of approximately 8037 fish per month.

81. (a) We see from the formula that \$5000 was deposited initially, and that the money is earning interest at 2% compounded continuously.

(b) We have $f(10) = 5000e^{0.02(10)} = 6107.01$ dollars. Since $f'(t) = 5000e^{0.02t} \cdot 0.02$, we have

$$f'(10) = 5000e^{0.02(10)}(0.02) = 122.14 \text{ dollars per year.}$$

Ten years after the money was deposited, the balance is \$6107.01 and is growing at a rate of \$122.14 per year.

85. (a) $\frac{dB}{dt} = P\left(1 + \frac{r}{100}\right)^t \ln\left(1 + \frac{r}{100}\right)$. The expression $\frac{dB}{dt}$ tells us how fast the amount of money in the bank is changing with respect to time for fixed initial investment P and interest rate r.

(b) $\frac{dB}{dr} = Pt\left(1 + \frac{r}{100}\right)^{t-1} \frac{1}{100}$. The expression $\frac{dB}{dr}$ indicates how fast the amount of money changes with respect to the interest rate r, assuming fixed initial investment P and time t.

89. Since $f(x)$ is increasing, $f'(x) > 0$ for all x, and since $g(x)$ is decreasing, $g'(x) < 0$ for all x. By the chain rule, we have

$$h'(x) = \underbrace{f'(g(x))}_{+} \cdot \underbrace{g'(x)}_{-} .$$

This means that $h'(x) < 0$, so $h(x)$ is decreasing for all x.

Strengthen Your Understanding

93. The derivative of the inside function, e^x, is missing.

Let $z = h(x) = e^x + 2$, so $g(z) = z^5$ and $g'(z) = 5z^4 \cdot h'(x)$. Taking the derivative of h, we have $h'(x) = e^x$, so

$$g'(x) = 5(e^x + 2)^4 e^x = 5e^x(e^x + 2)^4.$$

97. One possibility is $f(x) = (x^2 + 1)^2$, which can be differentiated using the chain rule with $x^2 + 1$ as the inside function and x^2 as the outside function:

$$f'(x) = 2(x^2 + 1)^1 \cdot 2x.$$

In addition, we can expand the function into a polynomial: $f(x) = x^4 + 2x^2 + 1$, and now differentiate term-by-term:

$$f'(x) = 4x^3 + 4x.$$

101. False. Let $f(x) = e^{-x}$ and $g(x) = x^2$. Let $h(x) = f(g(x)) = e^{-x^2}$. Then $h'(x) = -2xe^{-x^2}$ and $h''(x) = (-2 + 4x^2)e^{-x^2}$. Since $h''(0) < 0$, clearly h is not concave up for all x.

Additional Problems (online only)

105. We have $h(d) = f(g(d)) = f(-d) = d$ so $h(d)$ is positive. From the chain rule,

$$h'(d) = f'(g(d))g'(d).$$

We have

$$f'(g(d)) = f'(-d).$$

From the graph of f, we see that $f'(-d) < 0$, and from the graph of g, we see that $g'(d) < 0$. This means the sign of $h'(d)$ is the product of two negative numbers, so $h'(d) > 0$.

109. (a) If

$$p(x) = k(2x),$$

then

$$p'(x) = k'(2x) \cdot 2.$$

When $x = \frac{1}{2}$,

$$p'\left(\frac{1}{2}\right) = k'\left(2 \cdot \frac{1}{2}\right)(2) = 2 \cdot 2 = 4.$$

(b) If

$$q(x) = k(x + 1),$$

then

$$q'(x) = k'(x + 1) \cdot 1.$$

When $x = 0$,

$$q'(0) = k'(0 + 1)(1) = 2 \cdot 1 = 2.$$

(c) If

$$r(x) = k\left(\frac{1}{4}x\right),$$

then

$$r'(x) = k'\left(\frac{1}{4}x\right) \cdot \frac{1}{4}.$$

When $x = 4$,

$$r'(4) = k'\left(\frac{1}{4}4\right)\frac{1}{4} = 2 \cdot \frac{1}{4} = \frac{1}{2}.$$

113. Let f have a zero of multiplicity m at $x = a$ so that

$$f(x) = (x - a)^m h(x), \quad h(a) \neq 0.$$

Differentiating this expression gives

$$f'(x) = (x - a)^m h'(x) + m(x - a)^{(m-1)} h(x)$$

and both terms in the sum are zero when $x = a$ so $f'(a) = 0$. Taking another derivative gives

$$f''(x) = (x - a)^m h''(x) + 2m(x - a)^{(m-1)} h'(x) + m(m - 1)(x - a)^{(m-2)} h(x).$$

Again, each term in the sum contains a factor of $(x - a)$ to some positive power, so at $x = a$ this will evaluate to 0. Differentiating repeatedly, all derivatives will have positive integer powers of $(x - a)$ until the mth and will therefore vanish. However,

$$f^{(m)}(a) = m!h(a) \neq 0.$$

Solutions for Section 3.5

Exercises

1.

Table 3.1

x	$\cos x$	Difference Quotient	$-\sin x$
0	1.0	−0.0005	0.0
0.1	0.995	−0.10033	−0.099833
0.2	0.98007	−0.19916	−0.19867
0.3	0.95534	−0.296	−0.29552
0.4	0.92106	−0.38988	−0.38942
0.5	0.87758	−0.47986	−0.47943
0.6	0.82534	−0.56506	−0.56464

5. $f'(x) = \cos(3x) \cdot 3 = 3\cos(3x)$.

9. Using the chain rule gives $R'(x) = 3\pi \sin(\pi x)$.

13. $w' = e^t \cos(e^t)$.

17. Using the chain rule gives $R'(\theta) = 3\cos(3\theta)e^{\sin(3\theta)}$.

21. $f'(x) = \dfrac{1}{2}(3 + \sin(8x))^{-\frac{1}{2}} \cdot (\cos(8x) \cdot 8) = 4\cos(8x)(3 + \sin(8x))^{-0.5}$.

25. $f'(x) = 2 \cdot [\sin(3x)] + 2x[\cos(3x)] \cdot 3 = 2\sin(3x) + 6x\cos(3x)$

29. $y' = 5\sin^4 \theta \cos \theta$.

33. $\dfrac{dQ}{dx} = -\sin(e^{2x}) \cdot (e^{2x} \cdot 2) = -2e^{2x}\sin(e^{2x})$.

37. $f'(\theta) = 3\theta^2 \cos \theta - \theta^3 \sin \theta$.

41. We use the quotient rule. We have

$$\frac{dP}{dt} = \frac{(-\sin t) \cdot t^3 - (3t^2) \cdot \cos t}{(t^3)^2} = \frac{-t^3 \sin t - 3t^2 \cos t}{t^6} = \frac{-t \sin t - 3\cos t}{t^4}.$$

45. The quotient rule gives $G'(x) = \dfrac{2\sin x \cos x(\cos^2 x + 1) + 2\sin x \cos x(\sin^2 x + 1)}{(\cos^2 x + 1)^2}$

or, using $\sin^2 x + \cos^2 x = 1$,
$$G'(x) = \frac{6\sin x \cos x}{(\cos^2 x + 1)^2}.$$

49. Although there are three factors in this expression, we can use parentheses to consider it as a product of two factors and then use the product rule twice. Considering the expression as $(x^3 e^{5x}) \cdot \sin(2x)$ and using the product rule on these two factors first, we have:

$$y' = (x^3 e^{5x}) \cdot 2\cos(2x) + (\text{Derivative of } x^3 e^{5x}) \cdot \sin(2x)$$
$$= x^3 e^{5x} \cdot 2\cos(2x) + (x^3 \cdot 5e^{5x} + 3x^2 \cdot e^{5x}) \cdot \sin(2x)$$
$$= 2x^3 e^{5x} \cos(2x) + 5x^3 e^{5x} \sin(2x) + 3x^2 e^{5x} \sin(2x).$$

We could also have started by writing the expression as, for example, $x^3 \cdot (e^{5x} \sin(2x))$.

Problems

53. We begin by taking the derivative of $y = \sin(x^4)$ and evaluating at $x = 10$:

$$\frac{dy}{dx} = \cos(x^4) \cdot 4x^3.$$

Evaluating cos(10,000) on a calculator (in radians) we see cos(10,000) < 0, so we know that $dy/dx < 0$, and therefore the function is decreasing.

Next, we take the second derivative and evaluate it at $x = 10$, giving sin(10,000) < 0:

$$\frac{d^2y}{dx^2} = \underbrace{\cos(x^4) \cdot (12x^2)}_{\text{negative}} + \underbrace{4x^3 \cdot (-\sin(x^4))(4x^3)}_{\substack{\text{positive, but much} \\ \text{larger in magnitude}}}.$$

From this we can see that $d^2y/dx^2 > 0$, thus the graph is concave up.

57. The pattern in the table below allows us to generalize and say that the $(4n)^{\text{th}}$ derivative of $\cos x$ is $\cos x$, i.e.,

$$\frac{d^4y}{dx^4} = \frac{d^8y}{dx^8} = \cdots = \frac{d^{4n}y}{dx^{4n}} = \cos x.$$

Thus we can say that $d^{48}y/dx^{48} = \cos x$. From there we differentiate twice more to obtain $d^{50}y/dx^{50} = -\cos x$.

n	1	2	3	4	...	48	49	50
n^{th} derivative	$-\sin x$	$-\cos x$	$\sin x$	$\cos x$		$\cos x$	$-\sin x$	$-\cos x$

61. (a) $v(t) = \dfrac{dy}{dt} = \dfrac{d}{dt}(15 + \sin(2\pi t)) = 2\pi \cos(2\pi t).$

(b)

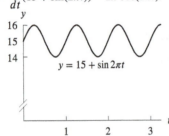

65. (a) The function $d(t)$ is increasing at a constant rate for the period $0 \le t \le 2$, when the derivative of $d(t)$ is k.

(b) The functions $d(t)$ must be continuous, since the depth of water cannot shift suddenly and instantly (even the fastest change takes some amount of time), so we know that

$$2k = 50 + \sin(0.2), \quad \text{so} \quad k = 25.099.$$

This means that the derivative for $0 < t < 2$ is 25.099, whereas the derivative for $t > 2$ is $0.1 \cos(0.1t)$. In other words

$$d(t) = \begin{cases} 25.099t & 0 \le t \le 2 \\ 50 + \sin(0.1t) & t > 2, \end{cases}$$

so

$$d'(t) = \begin{cases} 25.099 & 0 \le t < 2 \\ 0.1 \cos(0.1t) & t > 2, \end{cases}$$

At $t = 2$, the derivative is undefined, since $0.1 \cos(0.1 \cdot 2) \ne 25.099$.

69. We have:

$$C'(t) = 3.5\frac{d}{dt}\left(\sin\left(\frac{\pi}{6}t\right)\right) + 0 + \frac{1}{6} = \frac{3.5\pi}{6}\cos\left(\frac{\pi}{6}t\right) + \frac{1}{6} = \frac{7\pi}{12}\cos\left(\frac{\pi}{6}t\right) + \frac{1}{6}.$$

Thus,

$$C'(30) = \frac{7\pi}{12}\cos(5\pi) + \frac{1}{6} = -\frac{7\pi}{12} + \frac{1}{6} = -1.666 \text{ ppm/month}.$$

This means that on June 1, 2008, the concentration of CO_2 in the air was decreasing by about 1.666 parts per million per month.

73. (a) Using the product rule we have
$$f'(x) = -e^{-x} \sin x + e^{-x} \cos x.$$
(b) From part (a), we know that $f'(x) = -e^{-x} \sin x + e^{-x} \cos x$. Setting to zero and solving, we get
$$-e^{-x} \sin x + e^{-x} \cos x = 0$$
$$e^{-x} \cos x = e^{-x} \sin x$$
$$\cos x = \sin x \qquad \text{(Since } e^{-x} \neq 0\text{)}$$
$$1 = \tan x.$$

So $f'(x) = 0$ when $\tan x = 1$.

Since e^{-x} is never 0, the derivative $-e^{-x} \sin x + e^{-x} \cos x = 0$ only when $\tan x = 1$. Thus, $f'(x) = 0$ *precisely* when $\tan x = 1$.

Strengthen Your Understanding

77. The function $f(x) = \sin x$ satisfies this condition because
$$f'(x) = \frac{d}{dx} \sin x = \cos x$$
$$f''(x) = \frac{d}{dx} \cos x = -\sin x = -f(x).$$

There are many other possibilities, including $f(x) = \cos x$.

Additional Problems (online only)

81. Differentiating with respect to t using the chain rule and substituting for dx/dt gives
$$\frac{d^2x}{dt^2} = \frac{d}{dt}\left(\frac{dx}{dt}\right) = \frac{d}{dx}(x \sin x) \cdot \frac{dx}{dt} = (\sin x + x \cos x)x \sin x.$$

85. If the graphs of $y = \sin x$ and $y = ke^{-x}$ are tangent, then the y-values and the derivatives, $\dfrac{dy}{dx} = \cos x$ and $\dfrac{dy}{dx} = -ke^{-x}$, are equal at that point, so
$$\sin x = ke^{-x} \qquad \text{and} \qquad \cos x = -ke^{-x}.$$

Thus $\sin x = -\cos x$ so $\tan x = -1$. The smallest x-value is $x = 3\pi/4$, which leads to the smallest k value
$$k = \frac{\sin(3\pi/4)}{e^{-3\pi/4}} = 7.46.$$

When $x = \dfrac{3\pi}{4}$, we have $y = \sin\left(\dfrac{3\pi}{4}\right) = \dfrac{1}{\sqrt{2}}$ so the point is $\left(\dfrac{3\pi}{4}, \dfrac{1}{\sqrt{2}}\right)$.

89. (a) Sector OAQ is a sector of a circle with radius $\dfrac{1}{\cos\theta}$ and angle $\Delta\theta$. Thus its area is the left side of the inequality. Similarly, the area of Sector OBR is the right side of the equality. The area of the triangle OQR is $\frac{1}{2}\Delta \tan\theta$ since it is a triangle with base $\Delta \tan\theta$ (the segment QR) and height 1 (if you turn it sideways, it is easier to see this). Thus, using the given fact about areas (which is also clear from looking at the picture), we have
$$\frac{\Delta\theta}{2\pi} \cdot \pi \left(\frac{1}{\cos\theta}\right)^2 \leq \frac{1}{2} \cdot \Delta(\tan\theta) \leq \frac{\Delta\theta}{2\pi} \cdot \pi \left(\frac{1}{\cos(\theta + \Delta\theta)}\right)^2.$$

(b) Dividing the inequality through by $\frac{\Delta\theta}{2}$ and canceling the π's gives:
$$\left(\frac{1}{\cos\theta}\right)^2 \leq \frac{\Delta \tan\theta}{\Delta\theta} \leq \left(\frac{1}{\cos(\theta + \Delta\theta)}\right)^2$$

Then as $\Delta\theta \to 0$, the right and left sides both tend toward $\left(\frac{1}{\cos\theta}\right)^2$ while the middle (which is the difference quotient for tangent) tends to $(\tan\theta)'$. Thus, the derivative of tangent is "squeezed" between two values heading toward the same thing and must, itself, also tend to that value. Therefore, $(\tan\theta)' = \left(\frac{1}{\cos\theta}\right)^2$.

(c) Take the identity $\sin^2\theta + \cos^2\theta = 1$ and divide through by $\cos^2\theta$ to get $(\tan\theta)^2 + 1 = \left(\frac{1}{\cos\theta}\right)^2$. Differentiating with respect to θ yields:

$$2(\tan\theta)\cdot(\tan\theta)' = 2\left(\frac{1}{\cos\theta}\right)\cdot\left(\frac{1}{\cos\theta}\right)'$$

$$2\left(\frac{\sin\theta}{\cos\theta}\right)\cdot\left(\frac{1}{\cos\theta}\right)^2 = 2\left(\frac{1}{\cos\theta}\right)\cdot(-1)\left(\frac{1}{\cos\theta}\right)^2(\cos\theta)'$$

$$2\frac{\sin\theta}{\cos^3\theta} = (-1)2\frac{1}{\cos^3\theta}(\cos\theta)'$$

$$-\sin\theta = (\cos\theta)'.$$

(d)

$$\frac{d}{d\theta}\left(\sin^2\theta + \cos^2\theta\right) = \frac{d}{d\theta}(1)$$

$$2\sin\theta\cdot(\sin\theta)' + 2\cos\theta\cdot(\cos\theta)' = 0$$

$$2\sin\theta\cdot(\sin\theta)' + 2\cos\theta\cdot(-\sin\theta) = 0$$

$$(\sin\theta)' - \cos\theta = 0$$

$$(\sin\theta)' = \cos\theta.$$

Solutions for Section 3.6

Exercises

1. $f'(t) = \dfrac{2t}{t^2 + 1}$.

5. $\dfrac{dy}{dx} = \dfrac{1}{\sqrt{1 - (x + 1)^2}}$.

9. Since $\ln(e^{2x}) = 2x$, the derivative $f'(x) = 2$.

13. $f'(x) = \dfrac{1}{e^x + 1}\cdot e^x$.

17. Using the product and chain rules gives $h'(w) = 3w^2\ln(10w) + w^3\dfrac{10}{10w} = 3w^2\ln(10w) + w^2$.

21. $f(t) = \ln t$ (because $\ln e^x = x$ or because $e^{\ln t} = t$), so $f'(t) = \frac{1}{t}$.

25. $g'(t) = e^{\arctan(3t^2)}\left(\dfrac{1}{1 + (3t^2)^2}\right)(6t) = e^{\arctan(3t^2)}\left(\dfrac{6t}{1 + 9t^4}\right)$.

29. Note that $f(x) = kx$ so, $f'(x) = k$.

33. $f'(z) = -1(\ln z)^{-2}\cdot\dfrac{1}{z} = \dfrac{-1}{z(\ln z)^2}$.

37. $f'(w) = 3w^{-1/2} - 2w^{-3} + 5\dfrac{1}{w} = \dfrac{3}{\sqrt{w}} - \dfrac{2}{w^3} + \dfrac{5}{w}$.

41. Using the chain rule gives

$$T'(u) = \left[\frac{1}{1 + \left(\frac{u}{1+u}\right)^2}\right]\left[\frac{(1+u) - u}{(1+u)^2}\right]$$

$$= \frac{(1+u)^2}{(1+u)^2 + u^2}\left[\frac{1}{(1+u)^2}\right]$$

$$= \frac{1}{1 + 2u + 2u^2}.$$

Problems

45. Let
$$g(x) = \log x.$$
Then
$$10^{g(x)} = x.$$
Differentiating,
$$(\ln 10)[10^{g(x)}]g'(x) = 1$$
$$g'(x) = \frac{1}{(\ln 10)[10^{g(x)}]}$$
$$g'(x) = \frac{1}{(\ln 10)x}.$$

49. (a) We have
$$f(x) = 32.7 \ln\left(\frac{x}{244.5}\right) = 32.7 \ln(x) - 32.7 \ln(244.5)$$
so
$$f'(x) = \frac{32.7}{x}.$$

(b) We have
$$f'(2000) = \frac{32.7}{2000} = 0.01635 \text{ mm of leaf width per mm of rain.}$$

(c) We can approximate this difference using the derivative $f'(2000)$ as follows:
$$\Delta w \approx f'(2000)\Delta x = (0.01635)(150) = 2.4525 \text{ mm.}$$

The average leaf width in the rainier forest is about 2.5 mm greater. To check, we can calculate this difference exactly; it is
$$\Delta w = f(2150) - f(2000) = 32.7 \ln(2150/244.5) - 32.7 \ln(2000/244.5) = 2.365 \text{ mm.}$$

53. (a) Since $f(x) = 2x^5 + 3x^3 + x$, we differentiate to get $f'(x) = 10x^4 + 9x^2 + 1$.

(b) Because $f'(x)$ is always positive, we know that $f(x)$ is increasing everywhere. Thus, $f(x)$ is a one-to-one function and is invertible.

(c) To find $f(1)$, substitute 1 for x into $f(x)$. We get $f(1) = 2(1)^5 + 3(1)^3 + 1 = 2 + 3 + 1 = 6$.

(d) To find $f'(1)$, substitute 1 for x into $f'(x)$. We get $f'(1) = 10(1)^4 + 9(1)^2 + 1 = 20$.

(e) Since $f(1) = 6$, we have $f^{-1}(6) = 1$, so
$$(f^{-1})'(6) = \frac{1}{f'(f^{-1}(6))} = \frac{1}{f'(1)} = \frac{1}{20}.$$

57. Since the chain rule gives $h'(x) = n'(m(x))m'(x) = 1$ we must find values a and x such that $a = m(x)$ and $n'(a)m'(x) = 1$. Calculating slopes from the graph of n gives
$$n'(a) = \begin{cases} 1 & \text{if } 0 < a < 50 \\ 1/2 & \text{if } 50 < a < 100. \end{cases}$$

Calculating slopes from the graph of m gives
$$m'(x) = \begin{cases} -2 & \text{if } 0 < x < 50 \\ 2 & \text{if } 50 < x < 100. \end{cases}$$

The only values of the derivative n' are 1 and 1/2 and the only values of the derivative m' are 2 and -2. In order to have $n'(a)m'(x) = 1$ we must therefore have $n'(a) = 1/2$ and $m'(x) = 2$. Thus $50 < a < 100$ and $50 < x < 100$.

Now $a = m(x)$ and from the graph of m we see that $50 < m(x) < 100$ for $0 < x < 25$ or $75 < x < 100$.

The two conditions on x we have found are both satisfied when $75 < x < 100$. Thus $h'(x) = 1$ for all x in the interval $75 < x < 100$. The question asks for just one of these x values, for example $x = 80$.

61. We have

$$(f^{-1})'(15) = \frac{1}{f'(f^{-1}(15))}.$$

From the graph of $f(x)$ we see that $f^{-1}(15) = 30$. From the graph of $f'(x)$ we see that $f'(30) = 0.73$. Thus $(f^{-1})'(15) = 1/0.73 = 1.4$.

65. To find $(f^{-1})'(3)$, we first look in the table to find that $3 = f(9)$, so $f^{-1}(3) = 9$. Thus,

$$(f^{-1})'(3) = \frac{1}{f'(f^{-1}(3))} = \frac{1}{f'(9)} = \frac{1}{5}.$$

69. We have $(f^{-1})'(8) = 1/f'(f^{-1}(8))$. From the graph we see $f^{-1}(8) = 4$. Thus $(f^{-1})'(8) = \frac{1}{f'(4)} = \frac{1}{3.0}$.

73. A continuous invertible function $f(x)$ cannot be increasing on one interval and decreasing on another because it would fail the horizontal line test. The same is true of the inverse function $f^{-1}(x)$. Either $f^{-1}(x)$ is increasing and $(f^{-1})'(x) \geq 0$ for all x, or $f^{-1}(x)$ is decreasing and $(f^{-1})'(x) \leq 0$ for all x. We can not have both $(f^{-1})'(10) = 8$ and $(f^{-1})'(20) = -6$.

Strengthen Your Understanding

77. The formula for $(f^{-1})'(2)$ is wrong. We need $f'(f^{-1}(2))$, and we are not given $f^{-1}(2)$.

81. For the statement to be true, we need $f'(x) = 1$, so $f(x) = x$ is a function to try. Then $f^{-1}(x) = x$, so

$$(f^{-1})'(x) = \frac{1}{f'(f^{-1}(x))} = \frac{1}{f'(x)} = \frac{1}{1} = 1.$$

Additional Problems (online only)

85. Since $f(20) = 10$, we have $f^{-1}(10) = 20$, so $(f^{-1})'(10) = \frac{1}{f'(f^{-1}(10))} = \frac{1}{f'(20)}$. Therefore $(f^{-1})'(10)f'(20) = 1$. Option (b) is wrong.

Solutions for Section 3.7

Exercises

1. We differentiate implicitly both sides of the equation with respect to x.

$$2x + 2y\frac{dy}{dx} = 0,$$

$$\frac{dy}{dx} = -\frac{2x}{2y} = -\frac{x}{y}.$$

5. Implicit differentiation gives

$$1 \cdot y + x \cdot \frac{dy}{dx} + 1 + \frac{dy}{dx} = 0.$$

Solving for dy/dx, we have

$$\frac{dy}{dx} = -\frac{1+y}{1+x}.$$

9. We differentiate implicitly both sides of the equation with respect to x.

$$x^{\frac{1}{2}} + y^{\frac{1}{2}} = 25,$$

$$\frac{1}{2}x^{-\frac{1}{2}} + \frac{1}{2}y^{-\frac{1}{2}}\frac{dy}{dx} = 0,$$

$$\frac{dy}{dx} = -\frac{\frac{1}{2}x^{-\frac{1}{2}}}{\frac{1}{2}y^{-\frac{1}{2}}} = -\frac{x^{-\frac{1}{2}}}{y^{-\frac{1}{2}}} = -\frac{\sqrt{y}}{\sqrt{x}} = -\sqrt{\frac{y}{x}}.$$

13. We differentiate implicitly both sides of the equation with respect to x.

$$\ln x + \ln(y^2) = 3$$
$$\frac{1}{x} + \frac{1}{y^2}(2y)\frac{dy}{dx} = 0$$
$$\frac{dy}{dx} = \frac{-1/x}{2y/y^2} = -\frac{y}{2x}.$$

17. We differentiate implicitly both sides of the equation with respect to x.

$$\arctan(x^2 y) = xy^2$$
$$\frac{1}{1+x^4 y^2}(2xy + x^2\frac{dy}{dx}) = y^2 + 2xy\frac{dy}{dx}$$
$$2xy + x^2\frac{dy}{dx} = [1+x^4 y^2][y^2 + 2xy\frac{dy}{dx}]$$
$$\frac{dy}{dx}[x^2 - (1+x^4 y^2)(2xy)] = (1+x^4 y^2)y^2 - 2xy$$
$$\frac{dy}{dx} = \frac{y^2 + x^4 y^4 - 2xy}{x^2 - 2xy - 2x^5 y^3}.$$

21. Differentiating implicitly on both sides with respect to x,

$$a\cos(ay)\frac{dy}{dx} - b\sin(bx) = y + x\frac{dy}{dx}$$
$$(a\cos(ay) - x)\frac{dy}{dx} = y + b\sin(bx)$$
$$\frac{dy}{dx} = \frac{y + b\sin(bx)}{a\cos(ay) - x}.$$

25. The slope is given by dy/dx, which we find using implicit differentiation. Notice that the product rule is needed for the second term. We differentiate to obtain:

$$3x^2 + 5x^2\frac{dy}{dx} + 10xy + 4y\frac{dy}{dx} = 4\frac{dy}{dx}$$
$$(5x^2 + 4y - 4)\frac{dy}{dx} = -3x^2 - 10xy$$
$$\frac{dy}{dx} = \frac{-3x^2 - 10xy}{5x^2 + 4y - 4}.$$

At the point $(1, 2)$, we have $dy/dx = (-3 - 20)/(5 + 8 - 4) = -23/9$. The slope of this curve at the point $(1, 2)$ is $-23/9$.

29. First, we must find the slope of the tangent at the origin, that is $\left.\dfrac{dy}{dx}\right|_{(0,0)}$. Rewriting $y = \dfrac{x}{y+a}$ as $y(y+a) = x$ so that we have

$$y^2 + ay = x$$

and differentiating implicitly gives

$$2y\frac{dy}{dx} + a\frac{dy}{dx} = 1$$
$$\frac{dy}{dx}(2y + a) = 1$$
$$\frac{dy}{dx} = \frac{1}{2y + a}.$$

Substituting $x = 0$, $y = 0$ yields $\left.\dfrac{dy}{dx}\right|_{(0,0)} = \dfrac{1}{a}$. Using the point-slope formula for a line, we have that the equation for the tangent line is

$$y - 0 = \frac{1}{a}(x - 0) \quad \text{or} \quad y = \frac{x}{a}.$$

Problems

33. (a) At $x = 1$, the corresponding y-values satisfy

$$y^2 + y + 1 = 1$$
$$y^2 + y = 0$$
$$y(y + 1) = 0$$
so $y = -1$ or $y = 0$.

Thus, there are two points on the graph of $y^2 + xy + x^2 = 1$ with $x = 1$, these being $(1, -1)$ and $(1, 0)$.

(b) Solving for dy/dx by implicit differentiation yields

$$2y\frac{dy}{dx} + y + x\frac{dy}{dx} + 2x = 0$$
$$\frac{dy}{dx} = -\frac{y + 2x}{x + 2y}.$$

(c) From part (a) we know $(1, -1)$ and $(1, 0)$ are the only two points on the graph of $y^2 + xy + x^2 = 1$ with $x = 1$. Evaluating dy/dx at $(1, -1)$ yields $-(-1 + 2)/(1 + (-2)) = 1$ and at $(1, 0)$ yields $-(0 + 2)/(1 + 0) = -2$.

37. Using implicit differentiation we have

$$1 = (\cos y)\frac{dy}{dx}.$$

Therefore,

$$\frac{dy}{dx} = \frac{1}{\cos y}.$$

Solving the trig identity $\sin^2 y + \cos^2 y = 1$ for $\cos y$ and substituting $x = \sin y$ gives $\cos y = \pm\sqrt{1 - x^2}$. However, because $y = \arcsin x$, we have $-\pi/2 \le y \le \pi/2$, so $\cos y \ge 0$ and thus we take the positive root:

$$\frac{dy}{dx} = \frac{1}{\cos y} = \frac{1}{\sqrt{1 - \sin^2 y}} = \frac{1}{\sqrt{1 - x^2}}.$$

Strengthen Your Understanding

41. Since we cannot solve for y in terms of x, we need to differentiate implicitly. This gives

$$\frac{d}{dx}(y) = \frac{d}{dx}(\sin(xy)),$$

so

$$\frac{dy}{dx} = \cos(xy)\left(1 \cdot y + x \cdot \frac{dy}{dx}\right).$$

Solving for dy/dx gives

$$\frac{dy}{dx} = \frac{y\cos(xy)}{1 - x\cos(xy)}.$$

45. True; differentiating the equation with respect to x, we get

$$2y\frac{dy}{dx} + y + x\frac{dy}{dx} = 0.$$

Solving for dy/dx, we get that

$$\frac{dy}{dx} = \frac{-y}{2y + x}.$$

Thus dy/dx exists where $2y + x \neq 0$. Now if $2y + x = 0$, then $x = -2y$. Substituting for x in the original equation, $y^2 + xy - 1 = 0$, we get

$$y^2 - 2y^2 - 1 = 0.$$

This simplifies to $y^2 + 1 = 0$, which has no solutions. Thus dy/dx exists everywhere.

Solutions for Section 3.8

Exercises

1. Using the chain rule, $\dfrac{d}{dz}(\sinh(3z+5)) = \cosh(3z+5) \cdot 3 = 3\cosh(3z+5)$.

5. Using the product rule,

$$\frac{d}{dt}\left(t^3 \sinh t\right) = 3t^2 \sinh t + t^3 \cosh t.$$

9. Using the chain rule,

$$\frac{d}{d\theta}(\ln(\cosh(1+\theta))) = \frac{1}{\cosh(1+\theta)} \cdot \sinh(1+\theta) = \frac{\sinh(1+\theta)}{\cosh(1+\theta)} = \tanh(1+\theta).$$

13. Substitute $x = 0$ into the formula for $\sinh x$. This yields

$$\sinh 0 = \frac{e^0 - e^{-0}}{2} = \frac{1-1}{2} = 0.$$

Problems

17. The graph of $\sinh x$ in the text suggests that

$$\text{As } x \to \infty, \quad \sinh x \to \frac{1}{2}e^x.$$
$$\text{As } x \to -\infty, \quad \sinh x \to -\frac{1}{2}e^{-x}.$$

Using the facts that

$$\text{As } x \to \infty, \quad e^{-x} \to 0,$$
$$\text{As } x \to -\infty, \quad e^{x} \to 0,$$

we can obtain the same results analytically:

$$\text{As } x \to \infty, \quad \sinh x = \frac{e^x - e^{-x}}{2} \to \frac{1}{2}e^x.$$
$$\text{As } x \to -\infty, \quad \sinh x = \frac{e^x - e^{-x}}{2} \to -\frac{1}{2}e^{-x}.$$

21. Recall that

$$\sinh A = \frac{1}{2}(e^A - e^{-A}) \quad \text{and} \quad \cosh A = \frac{1}{2}(e^A + e^{-A}).$$

Now substitute, expand and collect terms:

$$\cosh A \cosh B + \sinh A \sinh B = \frac{1}{2}(e^A + e^{-A}) \cdot \frac{1}{2}(e^B + e^{-B}) + \frac{1}{2}(e^A - e^{-A}) \cdot \frac{1}{2}(e^B - e^{-B})$$

$$= \frac{1}{4}\left(e^{A+B} + e^{A-B} + e^{-A+B} + e^{-(A+B)}\right.$$

$$\left. + e^{B+A} - e^{B-A} - e^{-B+A} + e^{-A-B}\right)$$

$$= \frac{1}{2}\left(e^{A+B} + e^{-(A+B)}\right)$$

$$= \cosh(A + B).$$

25. Using the definition of $\cosh x$ and $\sinh x$, we have $\cosh x^2 = \dfrac{e^{x^2} + e^{-x^2}}{2}$ and $\sinh x^2 = \dfrac{e^{x^2} - e^{-x^2}}{2}$. Therefore

$$\lim_{x\to\infty} \frac{\sinh(x^2)}{\cosh(x^2)} = \lim_{x\to\infty} \frac{e^{x^2} - e^{-x^2}}{e^{x^2} + e^{-x^2}}$$

$$= \lim_{x\to\infty} \frac{e^{x^2}(1 - e^{-2x^2})}{e^{x^2}(1 + e^{-2x^2})}$$

$$= \lim_{x\to\infty} \frac{1 - e^{-2x^2}}{1 + e^{-2x^2}}$$

$$= 1.$$

29. (a) Since the cosh function is even, the height, y, is the same at $x = -T/w$ and $x = T/w$. The height at these endpoints is

$$y = \frac{T}{w}\cosh\left(\frac{w}{T} \cdot \frac{T}{w}\right) = \frac{T}{w}\cosh 1 = \frac{T}{w}\left(\frac{e^1 + e^{-1}}{2}\right).$$

At the lowest point, $x = 0$, and the height is

$$y = \frac{T}{w}\cosh 0 = \frac{T}{w}.$$

Thus the "sag" in the cable is given by

$$\text{Sag} = \frac{T}{w}\left(\frac{e + e^{-1}}{2}\right) - \frac{T}{w} = \frac{T}{w}\left(\frac{e + e^{-1}}{2} - 1\right) \approx 0.54\frac{T}{w}.$$

(b) To show that the differential equation is satisfied, take derivatives

$$\frac{dy}{dx} = \frac{T}{w} \cdot \frac{w}{T}\sinh\left(\frac{wx}{T}\right) = \sinh\left(\frac{wx}{T}\right)$$

$$\frac{d^2y}{dx^2} = \frac{w}{T}\cosh\left(\frac{wx}{T}\right).$$

Therefore, using the fact that $1 + \sinh^2 a = \cosh^2 a$ and that cosh is always positive, we have:

$$\frac{w}{T}\sqrt{1 + \left(\frac{dy}{dx}\right)^2} = \frac{w}{T}\sqrt{1 + \sinh^2\left(\frac{wx}{T}\right)} = \frac{w}{T}\sqrt{\cosh^2\left(\frac{wx}{T}\right)}$$

$$= \frac{w}{T}\cosh\left(\frac{wx}{T}\right).$$

So

$$\frac{w}{T}\sqrt{1 + \left(\frac{dy}{dx}\right)^2} = \frac{d^2y}{dx^2}.$$

33. (a) Substituting $x = 0$ gives

$$\tanh 0 = \frac{e^0 - e^{-0}}{e^0 + e^{-0}} = \frac{1-1}{2} = 0.$$

(b) Since $\tanh x = \dfrac{e^x - e^{-x}}{e^x + e^{-x}}$ and $e^x + e^{-x}$ is always positive, $\tanh x$ has the same sign as $e^x - e^{-x}$. For $x > 0$, we have $e^x > 1$ and $e^{-x} < 1$, so $e^x - e^{-x} > 0$. For $x < 0$, we have $e^x < 1$ and $e^{-x} > 1$, so $e^x - e^{-x} < 0$. For $x = 0$, we have $e^x = 1$ and $e^{-x} = 1$, so $e^x - e^{-x} = 0$. Thus, $\tanh x$ is positive for $x > 0$, negative for $x < 0$, and zero for $x = 0$.

(c) Taking the derivative, we have

$$\frac{d}{dx}(\tanh x) = \frac{1}{\cosh^2 x}.$$

Thus, for all x,

$$\frac{d}{dx}(\tanh x) > 0.$$

Thus, $\tanh x$ is increasing everywhere.

(d) As $x \to \infty$ we have $e^{-x} \to 0$; as $x \to -\infty$, we have $e^x \to 0$. Thus

$$\lim_{x \to \infty} \tanh x = \lim_{x \to \infty} \left(\frac{e^x - e^{-x}}{e^x + e^{-x}} \right) = 1,$$

$$\lim_{x \to -\infty} \tanh x = \lim_{x \to -\infty} \left(\frac{e^x - e^{-x}}{e^x + e^{-x}} \right) = -1.$$

Thus, $y = 1$ and $y = -1$ are horizontal asymptotes to the graph of $\tanh x$. See Figure 3.5.

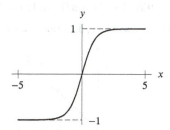

Figure 3.5: Graph of $y = \tanh x$

(e) The graph of $\tanh x$ suggests that $\tanh x$ is increasing everywhere; the fact that the derivative of $\tanh x$ is positive for all x confirms this. Since $\tanh x$ is increasing for all x, different values of x lead to different values of y, and therefore $\tanh x$ does have an inverse.

Strengthen Your Understanding

37. Since $\cosh x$ and $\sinh x$ behave like $e^x/2$ as $x \to \infty$ then $\tanh x = \sinh x / \cosh x \to 1$ as $x \to \infty$.

41. True. We have $\tanh x = (\sinh x) / \cosh x = (e^x - e^{-x})/(e^x + e^{-x})$. Replacing x by $-x$ in this expression gives $(e^{-x} - e^x)/(e^{-x} + e^x) = -\tanh x$.

45. False. Since $(\sinh^2 x)' = 2 \sinh x \cosh x$ and $(2 \sinh x \cosh x)' = 2 \sinh^2 x + 2 \cosh^2 x > 0$, the function $\sinh^2 x$ is concave up everywhere.

Solutions for Section 3.9

Exercises

1. With $f(x) = \sqrt{1 + x}$, the chain rule gives $f'(x) = 1/(2\sqrt{1 + x})$, so $f(0) = 1$ and $f'(0) = 1/2$. Therefore the tangent line approximation of f near $x = 0$,

$$f(x) \approx f(0) + f'(0)(x - 0),$$

becomes

$$\sqrt{1 + x} \approx 1 + \frac{x}{2}.$$

This means that, near $x = 0$, the function $\sqrt{1+x}$ can be approximated by its tangent line $y = 1 + x/2$. (See Figure 3.6.)

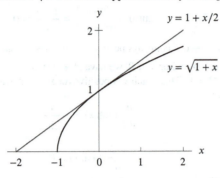

Figure 3.6

5. With $f(x) = e^{x^2}$, we get a tangent line approximation of $f(x) \approx f(1) + f'(1)(x-1)$ which becomes $e^{x^2} \approx e + \left(2xe^{x^2}\right)\Big|_{x=1} (x-1) = e + 2e(x-1) = 2ex - e$. Thus, our local linearization of e^{x^2} near $x = 1$ is $e^{x^2} \approx 2ex - e$.

9. From Figure 3.7, we see that the error has its maximum magnitude at the end points of the interval, $x = \pm 1$. The magnitude of the error can be read off the graph as less than 0.2 or estimated as

$$|\text{Error}| \leq |1 - \sin 1| = 0.159 < 0.2.$$

The approximation is an overestimate for $x > 0$ and an underestimate for $x < 0$.

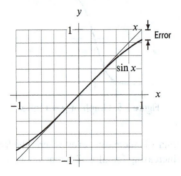

Figure 3.7

Problems

13. (a) The tangent line approximation near $x = 50$ is

$$f(x) \approx f(50) + f'(50)(x - 50)$$
$$f(x) \approx 99.5 + 0.2(x - 50).$$

(b) Since we do not have enough information to solve the equation $f(x) = 100$ exactly, we replace $f(x)$ by its tangent line approximation:

$$\underbrace{99.5 + 0.2(x - 50)}_{f(x)} = 100.$$

Then we solve the linear equation:

$$0.2(x - 50) = 0.5$$
$$x - 50 = 2.5$$
$$x = 52.5.$$

We estimate that $x = 52.5$ is an approximation of a solution to the equation $f(x) = 100$.

(c) The procedure is valid if the tangent line approximation is a good approximation to f over the interval between $x = 50$ and 52.5, the solution for the equation $f(x) = 100$. The graph of f needs to be approximately straight on this interval.

17. Since the line meets the curve at $x = 1$, we have $a = 1$. Since the point with $x = 1$ lies on both the line and the curve, we have

$$f(a) = f(1) = 2 \cdot 1 - 1 = 1.$$

The approximation is an underestimate because the line lies under the curve. Since the linear function approximates $f(x)$, we have

$$f(1.2) \approx 2(1.2) - 1 = 1.4.$$

21. (a) Using a difference quotient to approximate $g'(3)$ gives

$$g'(3) \approx \frac{g(3.001) - g(3)}{0.001} = \frac{0.0025}{0.001} = 2.5.$$

(b) With this estimate for $g'(3)$, the tangent line approximation is given by

$$g(x) \approx g(3) + g'(3)(x - 3) = 7 + 2.5(x - 3).$$

Using this approximation, we get $g(3.1) \approx 7.25$.

(c) Since $g''(x) > 0$, we know the graph of g is concave up on the interval $3 \le x \le 3.1$. This means the tangent line at $x = 3$ lies below the graph of g and so the value given by tangent line approximation is an underestimate.

25. (a) Using the power rule, we have

$$\frac{dG}{dL} = (40.9)(0.579)L^{0.579-1} = 23.681L^{-0.421}.$$

(b) We have

$$\left.\frac{dG}{dL}\right|_{L=1} = (23.681)1^{-0.421} = 23.681 \text{ days per cm.}$$

For marine organisms with lengths near 1 cm, an increase of 0.1 cm corresponds to a generation time about 2.3681 days longer. This is about 2 days per millimeter. The bigger organisms live longer before reproducing.

(c) We have

$$\Delta G \approx \left.\frac{dG}{dL}\right|_{L=1} \Delta L = 23.681(1.1 - 0.9) = 4.736 \text{ days.}$$

The organism that is 2 mm longer takes about 5 days longer to reproduce.

(To check, we calculate the exact value of the difference $= (40.9)1.1^{0.579} - (40.9)0.9^{0.579} = 4.741$.)

(d) At $L = 1$ cm we have $G = 40.9$ days. Since $dG/dL = 23.681$ at $L = 1$, we have the tangent line approximation

$$G \approx 40.9 + 23.681(L - 1) \text{ days}$$

This approximation is valid for marine organisms with lengths near 1 cm.

29. (a) Let $a = f(s)$ so that $f(s) = 120s^{-1.118}$. We have

$$f(10) = 9.14$$
$$f(20) = 4.21$$
$$f(50) = 1.51.$$

An individual bug is expected to suffer about 9 attacks per hour if swimming in a group of 10 bugs, about 4 attacks per hour in a group of 20 bugs, and about 1.5 attacks per hour in a group of 50 bugs. Notice that the larger the group size the safer the bug is from attack.

(b) Using the power rule we have

$$\frac{da}{ds} = 120(-1.118)s^{-2.118} = -134.16s^{-2.118}.$$

(c) We have

$$\left.\frac{da}{ds}\right|_{s=50} = (-134.16)50^{2.118} = -0.0338 \text{ attacks per bug per hour per number of bugs.}$$

(d) We have

$$a \approx f(50) + f'(50)(s - 50)$$
$$= 1.51 - 0.0338(s - 50).$$

(e) Using the linear approximation, we have

$$f(53) - f(48) \approx (1.51 - 0.0338(53 - 50)) - (1.51 - 0.0338(48 - 50))$$
$$= -0.0338(53 - 48) = -0.169.$$

Using the power function model, we have

$$f(53) - f(48) = 120 \cdot 53^{-1.118} - 120 \cdot 48^{-1.118} = -0.166.$$

The linear approximation model predicts that individual bugs in the larger group suffer 0.169 fewer attacks per hour than those in the smaller group. The power function models predicts 0.166 fewer attacks per hour for those in the larger group. The two predictions are extremely close.

33. Since f has a positive second derivative, its graph is concave up, as in Figure 3.8 or 3.9. This means that the graph of $f(x)$ is above its tangent line. We see that in both cases

$$f(1 + \Delta x) \geq f(1) + f'(1)\Delta x.$$

(The diagrams show Δx positive, but the result is also true if Δx is negative.)

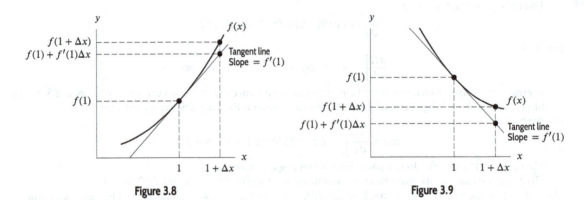

Figure 3.8   **Figure 3.9**

37. (a) The peak altitude is $h(20) = 1492$ meters
    (b) We have

$$h'(\theta) = 2 \cdot \frac{\pi}{180} 12755 \sin \frac{\pi\theta}{180} \cos \frac{\pi\theta}{180}$$
$$h'(20) = 143.$$

Thus, for angles, θ, near $20°$, we have

$$\text{Peak altitude} = h(\theta) \approx h(20) + h'(20)(\theta - 20) = 1490 + 143(\theta - 20) \text{ meters.}$$

(c) The true peak altitude for $21°$ is $h(21) = 1638$ meters. The linear approximation gives

$$\text{Approximate peak altitude} = 1492 + 143(21 - 20) = 1635 \text{ meters}$$

which is a little too low.

41. We have $f(1) = 1$ and $f'(1) = 4$. Thus
$$E(x) = x^4 - (1 + 4(x - 1)).$$

Values of $E(x)/(x - 1)$ near $x = 1$ are in Table 3.2.

Table 3.2

x	1.1	1.01	1.001
$E(x)/(x - 1)$	0.641	0.060401	0.006004

From the table, we can see that

$$\frac{E(x)}{(x-1)} \approx 6(x-1),$$

so $k = 6$ and

$$E(x) \approx 6(x-1)^2.$$

In addition, $f''(1) = 12$, so

$$E(x) \approx 6(x-1)^2 = \frac{f''(1)}{2}(x-1)^2.$$

The same result can be obtained by rewriting the function x^4 using $x = 1 + (x-1)$ and expanding:

$$x^4 = (1 + (x-1))^4 = 1 + 4(x-1) + 6(x-1)^2 + 4(x-1)^3 + (x-1)^4.$$

Thus,

$$E(x) = x^4 - (1 + 4(x-1)) = 6(x-1)^2 + 4(x-1)^3 + (x-1)^4.$$

For x near 1, the value of $x - 1$ is small, so we ignore powers of $x - 1$ higher than the first, giving

$$E(x) \approx 6(x-1)^2.$$

45. We have $f(1) = 0$ and $f'(1) = 1$. Thus

$$E(x) = \ln x - (x-1).$$

Values of $E(x)/(x-1)$ near $x = 1$ are in Table 3.3.

Table 3.3

x	1.1	1.01	1.001
$E(x)/(x-1)$	-0.047	-0.0050	-0.00050

From the table, we see that

$$\frac{E(x)}{(x-1)} \approx -0.5(x-1),$$

so $k = -1/2$ and

$$E(x) \approx -\frac{1}{2}(x-1)^2.$$

In addition, $f''(1) = -1$, so

$$E(x) \approx -\frac{1}{2}(x-1)^2 = \frac{f''(1)}{2}(x-1)^2.$$

49. Note that

$$[f(x)g(x)]' = \lim_{h \to 0} \frac{f(x+h)g(x+h) - f(x)g(x)}{h}.$$

We use the hint: For small h, $f(x+h) \approx f(x) + f'(x)h$, and $g(x+h) \approx g(x) + g'(x)h$. Therefore

$$f(x+h)g(x+h) - f(x)g(x) \approx [f(x) + hf'(x)][g(x) + hg'(x)] - f(x)g(x)$$
$$= f(x)g(x) + hf'(x)g(x) + hf(x)g'(x)$$
$$+ h^2 f'(x)g'(x) - f(x)g(x)$$
$$= hf'(x)g(x) + hf(x)g'(x) + h^2 f'(x)g'(x).$$

Therefore

$$\lim_{h \to 0} \frac{f(x+h)g(x+h) - f(x)g(x)}{h} = \lim_{h \to 0} \frac{hf'(x)g(x) + hf(x)g'(x) + h^2 f'(x)g'(x)}{h}$$
$$= \lim_{h \to 0} \frac{h\left(f'(x)g(x) + f(x)g'(x) + hf'(x)g'(x)\right)}{h}$$
$$= \lim_{h \to 0} \left(f'(x)g(x) + f(x)g'(x) + hf'(x)g'(x)\right)$$
$$= f'(x)g(x) + f(x)g'(x).$$

A more complete derivation can be given using the error term discussed in the section on Differentiability and Linear Approximation in Chapter 2. Adapting the notation of that section to this problem, we write

$$f(x + h) = f(x) + f'(x)h + E_f(h) \quad \text{and} \quad g(x + h) = g(x) + g'(x)h + E_g(h),$$

where $\lim_{h \to 0} \dfrac{E_f(h)}{h} = \lim_{h \to 0} \dfrac{E_g(h)}{h} = 0$. (This implies that $\lim_{h \to 0} E_f(h) = \lim_{h \to 0} E_g(h) = 0$.)

We have

$$\frac{f(x + h)g(x + h) - f(x)g(x)}{h} = \frac{f(x)g(x)}{h} + f(x)g'(x) + f'(x)g(x) + f(x)\frac{E_g(h)}{h} + g(x)\frac{E_f(h)}{h}$$

$$+ f'(x)g'(x)h + f'(x)E_g(h) + g'(x)E_f(h) + \frac{E_f(h)E_g(h)}{h} - \frac{f(x)g(x)}{h}$$

The terms $f(x)g(x)/h$ and $-f(x)g(x)/h$ cancel out. All the remaining terms on the right, with the exception of the second and third terms, go to zero as $h \to 0$. Thus, we have

$$[f(x)g(x)]' = \lim_{h \to 0} \frac{f(x + h)g(x + h) - f(x)g(x)}{h} = f(x)g'(x) + f'(x)g(x).$$

Strengthen Your Understanding

53. The line $y = x + 1$ is the linear approximation for $f(x) = e^x$ near $x = 0$. If we move far from $x = 0$, the approximation is useless.

For example, for $x = 1$, the approximation gives $e^1 \approx 2$ (instead of 2.718). For $x = 2$, our estimate of 3 is not a good approximation for $e^2 = 7.389$.

This linear approximation is only useful near $x = 0$.

57. The linear approximation of a function f, for values of x near a, is given by $f(x) \approx f(a) + f'(a)(x - a)$. Since $f(x) = |x + 1|$ does not have a derivative at $x = -1$ this function does not have a linear approximation for x near -1. Other answers are possible.

Solutions for Section 3.10

Exercises

1. False. The derivative, $f'(x)$, is not equal to zero everywhere, because the function is not continuous at integral values of x, so $f'(x)$ does not exist there. Thus, the Constant Function Theorem does not apply.

5. True. If $g(x)$ is the position of the slower horse at time x and $h(x)$ is the position of the faster, then $g'(x) \le h'(x)$ for $a < x < b$. Since the horses start at the same time, $g(a) = h(a)$, so, by the Racetrack Principle, $g(x) \le h(x)$ for $a \le x \le b$. Therefore, $g(b) \le h(b)$, so the slower horse loses the race.

9. No, it does not satisfy the hypotheses. This function does not appear to be continuous.

No, it does not satisfy the conclusion as there is no horizontal tangent.

Problems

13. A polynomial $p(x)$ satisfies the conditions of Rolle's Theorem for all intervals $a \le x \le b$.

Suppose $a_1, a_2, a_3, a_4, a_5, a_6, a_7$ are the seven distinct zeros of $p(x)$ in increasing order. Thus $p(a_1) = p(a_2) = 0$, so by Rolle's Theorem, $p'(x)$ has a zero, c_1, between a_1 and a_2.

Similarly, $p'(x)$ has 6 distinct zeros, $c_1, c_2, c_3, c_4, c_5, c_6$, where

$$a_1 < c_1 < a_2$$
$$a_2 < c_2 < a_3$$
$$a_3 < c_3 < a_4$$
$$a_4 < c_4 < a_5$$
$$a_5 < c_5 < a_6$$
$$a_6 < c_6 < a_7.$$

The polynomial $p'(x)$ is of degree 6, so $p'(x)$ cannot have more than 6 zeros.

17. The Decreasing Function Theorem is: Suppose that f is continuous on $[a, b]$ and differentiable on (a, b). If $f'(x) < 0$ on (a, b), then f is decreasing on $[a, b]$. If $f'(x) \leq 0$ on (a, b), then f is nonincreasing on $[a, b]$.

To prove the theorem, we note that if f is decreasing then $-f$ is increasing and vice-versa. Similarly, if f is nonincreasing, then $-f$ is nondecreasing. Thus if $f'(x) < 0$, then $-f'(x) > 0$, so $-f$ is increasing, which means f is decreasing. And if $f'(x) \leq 0$, then $-f'(x) \geq 0$, so $-f$ is nondecreasing, which means f is nonincreasing.

21. Apply the Constant Function Theorem, Theorem 3.9, to $h(x) = f(x) - g(x)$. Then $h'(x) = 0$ for all x, so $h(x)$ is constant for all x. Since $h(5) = f(5) - g(5) = 0$, we have $h(x) = 0$ for all x. Therefore $f(x) - g(x) = 0$ for all x, so $f(x) = g(x)$ for all x.

25. If $f'(x) = 0$, then both $f'(x) \geq 0$ and $f'(x) \leq 0$. By the Increasing and Decreasing Function Theorems, f is both nondecreasing and nonincreasing, so f is constant.

29. (a) Since $f''(x) \geq 0$, $f'(x)$ is nondecreasing on (a, b). Thus $f'(c) \leq f'(x)$ for $c \leq x < b$ and $f'(x) \leq f'(c)$ for $a < x \leq c$.
   (b) Let $g(x) = f(c) + f'(c)(x - c)$ and $h(x) = f(x)$. Then $g(c) = f(c) = h(c)$, and $g'(x) = f'(c)$ and $h'(x) = f'(x)$. If $c \leq x < b$, then $g'(x) \leq h'(x)$, and if $a < x \leq c$, then $g'(x) \geq h'(x)$, by (a). By the Racetrack Principle, $g(x) \leq h(x)$ for $c \leq x < b$ and for $a < x \leq c$, as we wanted.

Strengthen Your Understanding

33. The function f must be continuous on $a \leq x \leq b$ and differentiable on $a < x < b$. Any interval avoiding $x \leq 0$ will suffice, so, for example $1 \leq x \leq 2$.

37. Let f be defined by

$$f(x) = \begin{cases} x^2 & \text{if } 0 \leq x < 1 \\ 1/2 & \text{if } x = 1. \end{cases}$$

Then f is not continuous at $x = 1$, but f is differentiable on $(0, 1)$ and $f'(x) = 2x$ for $0 < x < 1$. Thus, $c = 1/4$ satisfies

$$f'(c) = \frac{f(1) - f(0)}{1 - 0} = \frac{1}{2}, \quad \text{since} \quad f'\left(\frac{1}{4}\right) = 2 \cdot \frac{1}{4} = \frac{1}{2}.$$

41. False. For example, if $f(x) = -x$, then $f'(x) \leq 1$ for all x, but $f(-2) = 2$, so $f(-2) > -2$.

Solutions for Chapter 3 Review

Exercises

1. $w' = 100(t^2 + 1)^{99}(2t) = 200t(t^2 + 1)^{99}$.

5. Using the quotient rule,

$$h'(t) = \frac{(-1)(4 + t) - (4 - t)}{(4 + t)^2} = -\frac{8}{(4 + t)^2}.$$

9. Using the chain rule, $g'(\theta) = (\cos \theta)e^{\sin \theta}$.

13. $g'(x) = \dfrac{d}{dx}\left(x^k + k^x\right) = kx^{k-1} + k^x \ln k$.

17.

$$M'(\alpha) = 2\tan(2 + 3\alpha) \cdot \frac{1}{\cos^2(2 + 3\alpha)} \cdot 3$$

$$= 6 \cdot \frac{\tan(2 + 3\alpha)}{\cos^2(2 + 3\alpha)}$$

21. $w'(\theta) = \dfrac{1}{\sin^2 \theta} - \dfrac{2\theta \cos \theta}{\sin^3 \theta}$

25. Using the quotient rule and the chain rule,

$$h'(z) = \frac{1}{2}\left(\frac{\sin(2z)}{\cos(2z)}\right)^{-1/2}\left[\frac{2\cos(2z)\cos(2z) - \sin(2z)(-2\sin(2z))}{\cos^2(2z)}\right]$$

$$= \left(\frac{\cos(2z)}{\sin(2z)}\right)^{1/2}\left[\frac{\cos^2(2z) + \sin^2(2z)}{\cos^2(2z)}\right]$$

$$= \frac{(\cos(2z))^{1/2}}{(\sin(2z))^{1/2}\cos^2(2z)} = \frac{1}{\sqrt{\sin(2z)}\sqrt{\cos^3(2z)}}.$$

29. $r'(\theta) = \frac{d}{d\theta}\left(e^{(e^\theta + e^{-\theta})}\right) = e^{(e^\theta + e^{-\theta})}\left(e^\theta - e^{-\theta}\right).$

33. $f'(r) = e(\tan 2 + \tan r)^{e-1}(\tan 2 + \tan r)' = e(\tan 2 + \tan r)^{e-1}\left(\frac{1}{\cos^2 r}\right)$

37. $\frac{dy}{dx} = (\ln 2)2^{\sin x}\cos x \cdot \cos x + 2^{\sin x}(-\sin x) = 2^{\sin x}\left((\ln 2)\cos^2 x - \sin x\right)$

41. Using the product rule gives

$$H'(t) = 2ate^{-ct} - c(at^2 + b)e^{-ct}$$

$$= (-cat^2 + 2at - bc)e^{-ct}.$$

45. Using the quotient rule gives

$$w'(r) = \frac{2ar(b + r^3) - 3r^2(ar^2)}{(b + r^3)^2}$$

$$= \frac{2abr - ar^4}{(b + r^3)^2}.$$

49. Since $g(w) = 5(a^2 - w^2)^{-2}$, $g'(w) = -10(a^2 - w^2)^{-3}(-2w) = \frac{20w}{(a^2 - w^2)^3}$

53. Using the quotient rule gives

$$f'(x) = \frac{1 + \ln x - x(\frac{1}{x})}{(1 + \ln x)^2}$$

$$= \frac{\ln x}{(1 + \ln x)^2}.$$

57. $y' = 18x^2 + 8x - 2.$

61. $f'(x) = 3x^2 + 3^x \ln 3$

65. It is easier to do this by multiplying it out first, rather than using the product rule first: $z = s^4 - s$, $z' = 4s^3 - 1$.

69. $f'(t) = 4(\sin(2t) - \cos(3t))^3[2\cos(2t) + 3\sin(3t)]$

73. Note: $f(z) = (5z)^{1/2} + 5z^{1/2} + 5z^{-1/2} - \sqrt{5}z^{-1/2} + \sqrt{5}$, so $f'(z) = \frac{5}{2}(5z)^{-1/2} + \frac{5}{2}z^{-1/2} - \frac{5}{2}z^{-3/2} + \frac{\sqrt{5}}{2}z^{-3/2}$.

77. Taking derivatives implicitly, we find

$$\frac{dy}{dx} + \cos y\frac{dy}{dx} + 2x = 0$$

$$\frac{dy}{dx} = \frac{-2x}{1 + \cos y}$$

So, at the point $x = 3, y = 0$,

$$\frac{dy}{dx} = \frac{(-2)(3)}{1 + \cos 0} = \frac{-6}{2} = -3.$$

Problems

81. Since $f(x) = x^3 - 6x^2 - 15x + 20$, we have $f'(x) = 3x^2 - 12x - 15$. To find the points at which $f'(x) = 0$, we solve

$$3x^2 - 12x - 15 = 0$$
$$3(x^2 - 4x - 5) = 0$$
$$3(x + 1)(x - 5) = 0.$$

We see that $f'(x) = 0$ at $x = -1$ and at $x = 5$. The graph of $f(x)$ in Figure 3.10 appears to be horizontal at $x = -1$ and at $x = 5$, confirming what we found analytically.

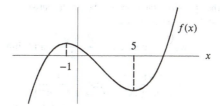

Figure 3.10

85. We need to find all values for x such that

$$\frac{dy}{dx} = s'(s(x)) \cdot s'(x) = 0.$$

This is the case when either $s'(s(x)) = 0$ or $s'(x) = 0$. From the graph we see that $s'(x) = 0$ when $x \approx 1$. Also, $s'(s(x)) = 0$ when $s(x) \approx 1$, which happens when $x \approx -0.4$ or $x \approx 2.4$.

To find $s'(a)$, for any a, draw a line tangent to the curve at the point $(a, s(a))$, and estimate the slope.

89. Estimates may vary. From the graphs, we estimate $f(1) \approx -0.4$, $f'(1) \approx 0.5$, and $g'(-0.4) \approx 2$. Thus, by the chain rule,

$$k'(1) = g'(f(1)) \cdot f'(1) \approx g'(-0.4) \cdot 0.5 \approx 2 \cdot 0.5 = 1.$$

93. (a) $H'(2) = r'(2)s(2) + r(2)s'(2) = -1 \cdot 1 + 4 \cdot 3 = 11.$

(b) $H'(2) = \dfrac{r'(2)}{2\sqrt{r(2)}} = \dfrac{-1}{2\sqrt{4}} = -\dfrac{1}{4}.$

(c) $H'(2) = r'(s(2))s'(2) = r'(1) \cdot 3$, but we don't know $r'(1)$.

(d) $H'(2) = s'(r(2))r'(2) = s'(4)r'(2) = -3.$

97. It makes sense to define the angle between two curves to be the angle between their tangent lines. (The tangent lines are the best linear approximations to the curves). See Figure 3.11. The functions $\sin x$ and $\cos x$ are equal at $x = \frac{\pi}{4}$.

$$\text{For } f_1(x) = \sin x, \quad f_1'(\tfrac{\pi}{4}) = \cos(\tfrac{\pi}{4}) = \frac{\sqrt{2}}{2}$$

$$\text{For } f_2(x) = \cos x, \quad f_2'(\tfrac{\pi}{4}) = -\sin(\tfrac{\pi}{4}) = -\frac{\sqrt{2}}{2}.$$

Using the point $(\frac{\pi}{4}, \frac{\sqrt{2}}{2})$ for each tangent line we get $y = \frac{\sqrt{2}}{2}x + \frac{\sqrt{2}}{2}(1 - \frac{\pi}{4})$ and $y = -\frac{\sqrt{2}}{2}x + \frac{\sqrt{2}}{2}(1 + \frac{\pi}{4})$, respectively.

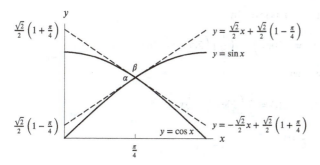

Figure 3.11

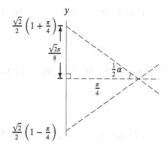

Figure 3.12

There are two possibilities of how to define the angle between the tangent lines, indicated by α and β above. The choice is arbitrary, so we will solve for both. To find the angle, α, we consider the triangle formed by these two lines and the y-axis. See Figure 3.12.

$$\tan\left(\frac{1}{2}\alpha\right) = \frac{\sqrt{2}\pi/8}{\pi/4} = \frac{\sqrt{2}}{2}$$

$$\frac{1}{2}\alpha = 0.61548 \text{ radians}$$

$$\alpha = 1.231 \text{ radians, or } 70.5°.$$

Now let us solve for β, the other possible measure of the angle between the two tangent lines. Since α and β are supplementary, $\beta = \pi - 1.231 = 1.909$ radians, or $109.4°$.

101. Using the definition of $\cosh x$ and $\sinh x$, we have $\cosh 2x = \dfrac{e^{2x} + e^{-2x}}{2}$ and $\sinh 3x = \dfrac{e^{3x} - e^{-3x}}{2}$. Therefore

$$\lim_{x\to-\infty} \frac{\cosh(2x)}{\sinh(2x)} = \lim_{x\to-\infty} \frac{e^{2x} + e^{-2x}}{e^{3x} - e^{-3x}}$$

$$= \lim_{x\to-\infty} \frac{e^{-2x}(e^{4x} + 1)}{e^{-2x}(e^{5x} - e^{-x})}$$

$$= \lim_{x\to-\infty} \frac{e^{4x} + 1}{e^{5x} - e^{-x}}$$

$$= 0.$$

105. (a) From the figure, we see $a = 2$. The point with $x = 2$ lies on both the line and the curve. Since

$$y = -3\cdot 2 + 7 = 1,$$

we have

$$f(a) = 1.$$

Since the slope of the line is -3, we have

$$f'(a) = -3.$$

(b) We use the line to approximate the function, so

$$f(2.1) \approx -3(2.1) + 7 = 0.7.$$

This is an underestimate, because the line is beneath the curve for $x > 2$. Similarly,

$$f(1.98) \approx -3(1.98) + 7 = 1.06.$$

This is an overestimate because the line is above the curve for $x < 2$.

The approximation $f(1.98) \approx 1.06$ is likely to be more accurate because 1.98 is closer to 2 than 2.1 is. Since the graph of $f(x)$ appears to bend away from the line at approximately the same rate on either side of $x = 2$, in this example, the error is larger for points farther from $x = 2$.

109. (a) If the distance $s(t) = 20e^{\frac{t}{2}}$, then the velocity, $v(t)$, is given by

$$v(t) = s'(t) = \left(20e^{\frac{t}{2}}\right)' = \left(\frac{1}{2}\right)\left(20e^{\frac{t}{2}}\right) = 10e^{\frac{t}{2}}.$$

(b) Observing the differentiation in (a), we note that

$$s'(t) = v(t) = \frac{1}{2}\left(20e^{\frac{t}{2}}\right) = \frac{1}{2}s(t).$$

Substituting $s(t)$ for $20e^{\frac{t}{2}}$, we obtain $s'(t) = \frac{1}{2}s(t)$.

113. (a) We solve for t to find the time it takes for the population to reach 10 billion.

$$P(t) = 10$$
$$7.17e^{kt} = 10$$
$$e^{kt} = \frac{10}{7.17}$$
$$t = \frac{\ln(10/7.17)}{k} = \frac{0.33268}{k} \text{ years.}$$

Thus the time is

$$f(k) = \frac{0.33268}{k} \text{ years.}$$

(b) The time to reach 10 billion with a growth rate of 1.1% is

$$f(0.011) = \frac{0.33268}{0.011} = 30.24 \text{ years.}$$

(c) We have

$$f'(k) = \frac{-0.33268}{k^2}$$
$$f'(0.011) = -2749.42$$

Thus, for growth rates, k, near 1.1% we have

Time for world population to reach 10 billion $= f(k) \approx f(0.011) + f'(0.011)(k - 0.011)$
$$f(k) = 30.24 - 2749.42(k - 0.011) \text{ years}$$

(d) Time to reach 10 billion for the exponential growth model with growth rate is 1.0% is

$$f(0.01) = 0.33268/0.01 = 33.3 \text{ years.}$$

The linear approximation gives

Approximate time $= 30.24 - 2749.42(0.01 - 0.011) = 32.99$ years.

117. We know that the velocity is given by

$$\frac{dx}{dt} = v(x).$$

By the chain rule,

$$\text{Acceleration} = \frac{dv}{dt} = \frac{dv}{dx} \cdot \frac{dx}{dt} = v'(x)v(x).$$

121. This problem can be solved by using either the quotient rule or the fact that

$$\frac{f'}{f} = \frac{d}{dx}(\ln f) \quad \text{and} \quad \frac{g'}{g} = \frac{d}{dx}(\ln g).$$

We use the second method. The relative rate of change of f/g is $(f/g)'/(f/g)$, so

$$\frac{(f/g)'}{f/g} = \frac{d}{dx}\ln\left(\frac{f}{g}\right) = \frac{d}{dx}(\ln f - \ln g) = \frac{d}{dx}(\ln f) - \frac{d}{dx}(\ln g) = \frac{f'}{f} - \frac{g'}{g}.$$

Thus, the relative rate of change of f/g is the difference between the relative rates of change of f and of g.

CAS Challenge Problems

125. (a) A CAS gives $h'(t) = 0$
(b) By the chain rule

$$h'(t) = \frac{\frac{d}{dt}\left(1 - \frac{1}{t}\right)}{1 - \frac{1}{t}} + \frac{\frac{d}{dt}\left(\frac{t}{t-1}\right)}{\frac{t}{t-1}} = \frac{\frac{1}{t^2}}{\frac{t-1}{t}} + \frac{\frac{1}{t-1} - \frac{t}{(t-1)^2}}{\frac{t}{t-1}}$$

$$= \frac{1}{t^2 - t} + \frac{(t-1) - t}{t^2 - t} = \frac{1}{t^2 - t} + \frac{-1}{t^2 - t} = 0.$$

(c) The expression inside the first logarithm is $1 - (1/t) = (t-1)/t$. Using the property $\log A + \log B = \log(AB)$, we get

$$\ln\left(1 - \frac{1}{t}\right) + \ln\left(\frac{t}{t-1}\right) = \ln\left(\frac{t-1}{t}\right) + \ln\left(\frac{t}{t-1}\right)$$
$$= \ln\left(\frac{t-1}{t} \cdot \frac{t}{1-t}\right) = \ln 1 = 0.$$

Thus $h(t) = 0$, so $h'(t) = 0$ also.

CHAPTER FOUR

Solutions for Section 4.1

Exercises

1. See Figure 4.1.

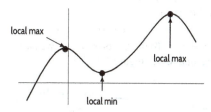

Figure 4.1

5. To find the critical points, we set $f'(x) = 0$. Since $f'(x) = 5x^4 - 30x^2$, we have

$$5x^4 - 30x^2 = 0$$
$$5x^2(x^2 - 6) = 0$$
$$x = 0, -\sqrt{6}, \sqrt{6}.$$

There are three critical points: $x = 0$, $x = -\sqrt{6}$, and $x = \sqrt{6}$.

To find the inflection points, we look for points where f'' is undefined or zero. Since $f''(x) = 20x^3 - 60x$, it is defined everywhere. Setting it equal to zero, we get

$$20x^3 - 60x = 0$$
$$20x(x^2 - 3) = 0$$
$$x = 0, -\sqrt{3}, \sqrt{3}.$$

Furthermore, $20x^3 - 60x$ is positive when $-\sqrt{3} < x < 0$ or $x > \sqrt{3}$ and negative when $x < -\sqrt{3}$ or $0 < x < \sqrt{3}$, so f'' changes sign at each of the solutions. Thus, there are three inflection points: $x = 0$, $x = -\sqrt{3}$, and $x = \sqrt{3}$.

9. $f'(x) = 12x^3 - 12x^2$. To find critical points, we set $f'(x) = 0$. This implies $12x^2(x-1) = 0$. So the critical points of f are $x = 0$ and $x = 1$. To the left of $x = 0$, $f'(x) < 0$. Between $x = 0$ and $x = 1$, $f'(x) < 0$. To the right of $x = 1$, $f'(x) > 0$. Therefore, $f(1)$ is a local minimum, but $f(0)$ is not a local extremum. See Figure 4.2.

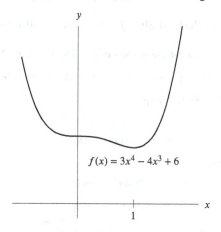

$$f(x) = 3x^4 - 4x^3 + 6$$

Figure 4.2

13. To find the critical points, we set $f'(x) = 0$. Since $f'(x) = 12x - 3x^2$, we have

$$12x - 3x^2 = 0$$
$$3x(4 - x) = 0$$
$$x = 0, \quad x = 4.$$

There are two critical points: at $x = 0$ and $x = 4$.

The second derivative is $f''(x) = 12 - 6x$. Using the second-derivative test, we see that $f''(0) = 12 > 0$, so f has a local minimum at $x = 0$, and $f''(4) = 12 - 6 \cdot 4 = -12 < 0$, so f has a local maximum at $x = 4$.

17. We have

$$g'(x) = e^{-3x} - 3xe^{-3x} = (1 - 3x)e^{-3x}.$$

To find critical points, we set $g'(x) = 0$. Then

$$(1 - 3x)e^{-3x} = 0.$$

Therefore, the critical point of g is $x = 1/3$. To the left of $x = 1/3$, we have $g'(x) > 0$. To the right of $x = 1/3$, we have $g'(x) < 0$. Thus $g(1/3)$ is a local maximum. See Figure 4.3.

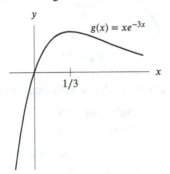

Figure 4.3

21. (a) A graph of $f(x) = e^{-x^2}$ is shown in Figure 4.4. It appears to have one critical point, at $x = 0$, and two inflection points, one between 0 and 1 and the other between 0 and -1.

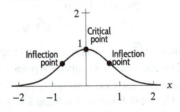

Figure 4.4

(b) To find the critical points, we set $f'(x) = 0$. Since $f'(x) = -2xe^{-x^2} = 0$, there is one solution, $x = 0$. The only critical point is at $x = 0$.

To find the inflection points, we first use the product rule to find $f''(x)$. We have

$$f''(x) = (-2x)(e^{-x^2}(-2x)) + (-2)(e^{-x^2}) = 4x^2e^{-x^2} - 2e^{-x^2}.$$

We set $f''(x) = 0$ and solve for x by factoring:

$$4x^2e^{-x^2} - 2e^{-x^2} = 0$$
$$(4x^2 - 2)e^{-x^2} = 0.$$

Since e^{-x^2} is never zero, we have

$$4x^2 - 2 = 0$$
$$x^2 = \frac{1}{2}$$
$$x = \pm 1/\sqrt{2}.$$

There are exactly two inflection points, at $x = 1/\sqrt{2} \approx 0.707$ and $x = -1/\sqrt{2} \approx -0.707$.

25. (a) Increasing for $x > 0$, decreasing for $x < 0$.
(b) $f(0)$ is a local and global minimum, and f has no global maximum.

Problems

29. (a) To find the critical points, we set the derivative equal to zero and solve for x:

$$f'(x) = 3x^2 - a = 0$$
$$x^2 = \frac{a}{3}$$
$$x = \pm\sqrt{\frac{a}{3}}.$$

(b) We want the local extrema to occur at ±2, so we need the critical points $\pm\sqrt{a/3}$ to occur at ±2. Thus, we need to have $\sqrt{a/3} = 2$. We solve for a:

$$\sqrt{\frac{a}{3}} = 2$$
$$\frac{a}{3} = 4$$
$$a = 12.$$

33. (a) The function $f(x)$ is defined for $x \geq 0$. We set the derivative equal to zero to find critical points:

$$f'(x) = 1 + \frac{1}{2}ax^{-1/2} = 0$$
$$1 + \frac{a}{2\sqrt{x}} = 0.$$

Since $a > 0$ and $\sqrt{x} > 0$ for $x > 0$, we have $f'(x) > 0$ for $x > 0$, so there is no critical point with $x > 0$. The only critical point is at $x = 0$ where $f'(x)$ is undefined.
(b) We see in part (a) that the derivative is positive for $x > 0$ so the function is increasing for all $x > 0$. The second derivative is

$$f''(x) = -\frac{1}{4}ax^{-3/2}.$$

Since a and $x^{-3/2}$ are positive for $x > 0$, the second derivative is negative for all $x > 0$. Thus the graph of f is concave down for all $x > 0$.

37. The inflection points of f are the points where f'' changes sign. See Figure 4.5.

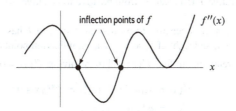

Figure 4.5

41. See Figure 4.6.

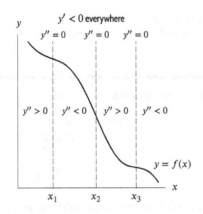

Figure 4.6

45. (a) It appears that this function has a local maximum at about $x = 1$, a local minimum at about $x = 4$, and a local maximum at about $x = 8$.

(b) The table now gives values of the derivative, so critical points occur where $f'(x) = 0$. Since f' is continuous, this occurs between 2 and 3, so there is a critical point somewhere around 2.5. Since f' is positive for values less than 2.5 and negative for values greater than 2.5, it appears that f has a local maximum at about $x = 2.5$. Similarly, it appears that f has a local minimum at about $x = 6.5$ and another local maximum at about $x = 9.5$.

49. By the product rule

$$f'(x) = \frac{d}{dx}(x^m(1-x)^n) = mx^{m-1}(1-x)^n - nx^m(1-x)^{n-1}$$
$$= x^{m-1}(1-x)^{n-1}(m(1-x) - nx)$$
$$= x^{m-1}(1-x)^{n-1}(m - (m+n)x).$$

We have $f'(x) = 0$ at $x = 0$, $x = 1$, and $x = m/(m+n)$, so these are the three critical points of f.

We can classify the critical points by determining the sign of $f'(x)$.

If $x < 0$, then $f'(x)$ has the same sign as $(-1)^{m-1}$: negative if m is even, positive if m is odd.

If $0 < x < m/(m+n)$, then $f'(x)$ is positive.

If $m/(m+n) < x < 1$, then $f'(x)$ is negative.

If $1 < x$, then $f'(x)$ has the same sign as $(-1)^n$: positive if n is even, negative if n is odd.

If m is even, then $f'(x)$ changes from negative to positive at $x = 0$, so f has a local minimum at $x = 0$.

If m is odd, then $f'(x)$ is positive to both the left and right of 0, so $x = 0$ is an inflection point of f.

At $x = m/(m+n)$, the derivative $f'(x)$ changes from positive to negative, so $x = m/(m+n)$ is a local maximum of f.

If n is even, then $f'(x)$ changes from negative to positive at $x = 1$, so f has a local minimum at $x = 1$. If n is odd, then $f'(x)$ is negative to both the left and right of 1, so $x = 1$ is an inflection point of f.

53. We wish to have $f'(3) = 0$. Differentiating to find $f'(x)$ and then solving $f'(3) = 0$ for a gives:

$$f'(x) = x(ae^{ax}) + 1(e^{ax}) = e^{ax}(ax + 1)$$
$$f'(3) = e^{3a}(3a + 1) = 0$$
$$3a + 1 = 0$$
$$a = -\frac{1}{3}.$$

Thus, $f(x) = xe^{-x/3}$.

57. Figure 4.7 contains the graph of $f(x) = x^2 + \cos x$.

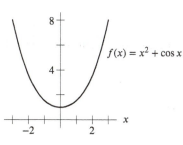

Figure 4.7

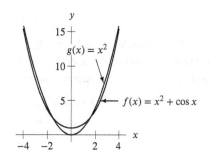

Figure 4.8

The graph looks like a parabola with no waves because $f''(x) = 2 - \cos x$, which is always positive. Thus, the graph of f is concave up everywhere; there are no waves. If you plot the graph of $f(x)$ together with the graph of $g(x) = x^2$, you see that the graph of f does wave back and forth across the graph of g, but never enough to change the concavity of f. See Figure 4.8.

61. Since the derivative of an even function is odd and the derivative of an odd function is even, f and f'' are either both odd or both even, and f' is the opposite. Graphs I and II represent odd functions; III represents an even function, so III is f'. Since the maxima and minima of III occur where I crosses the x-axis, I must be the derivative of f', that is, f''. In addition, the maxima and minima of II occur where III crosses the x-axis, so II is f.

Strengthen Your Understanding

65. Consider $f(x) = x^4$. We have $f'(x) = 4x^3$, so there is a critical point at $x = 0$, and $f'(x)$ is negative for $x < 0$ and positive for $x > 0$. We have $f''(x) = 12x^2$ so $f''(0)$ exists, but $f''(0) = 0$.

69. One possible answer is $f(x) = \cos x$ which has local maxima at $0, \pm 2\pi, \pm 4\pi, \ldots$ (that is, at $x = 2n\pi$ for all integers n). The value at all the local maxima is 1. The local minima are at $\pm\pi, \pm 3\pi, \ldots$ (that is, at $x = (2n + 1)\pi$), all with value -1.
 The function $f(x) = \sin x$ is another possibility, and there are many more.

73. False. A local maximum of f might occur at a point where f' does not exist. For example, $f(x) = -|x|$ has a local maximum at $x = 0$, but the derivative is not 0 (or defined) there.

77. True. Since f'' changes sign at the inflection point $x = p$, by the Intermediate Value Theorem, $f''(p) = 0$.

81. This is impossible. If $f(a) > 0$, then the downward concavity forces the graph of f to cross the x-axis to the right or left of $x = a$, which means $f(x)$ cannot be positive for all values of x. More precisely, suppose that $f(x)$ is positive for all x and f is concave down. Thus there must be some value $x = a$ where $f(a) > 0$ and $f'(a)$ is not zero, since a constant function is not concave down. The tangent line at $x = a$ has nonzero slope and hence must cross the x-axis somewhere to the right or left of $x = a$. Since the graph of f must lie below this tangent line, it must also cross the x-axis, contradicting the assumption that $f(x)$ is positive for all x.

Additional Problems (online only)

85. We see that:

$$g(3) = \sin(f(3))$$
$$= \sin(0) \qquad \text{since } f(3) = 0$$
$$= 0.$$

Thus, the statement is true.

89. We know that $f'(\pi) = 0$ because at a critical point $f' = 0$ or is undefined, but we are given that f is differentiable everywhere. By the product rule, we have

$$y' = f'(x)\cos x - f(x)\sin x.$$

This means that at $x = \pi$, we have:

$$y' = \underbrace{f'(\pi)}_{0} \underbrace{\cos \pi}_{-1} - \underbrace{f(\pi)}_{} \underbrace{\sin \pi}_{0}$$

$$= 0.$$

Thus, $y' = 0$ at $x = \pi$, so this is a critical point.

93. We know that $x = 1$ is the only solution to $f(x) = 0$, and that $x = 2$ is the only solution to $f'(x) = 0$.

(a) To find the zeros of this function, we solve the equation $y = 0$:

$$f\left(x^2 - 3\right) = 0$$
$$x^2 - 3 = 1 \qquad \text{because } f(1) = 0$$
$$x = \pm 2.$$

Thus, the zeros are $x = \pm 2$,

(b) To find the critical points of this function, we solve the equation $y' = 0$:

$$\left(f\left(x^2 - 3\right)\right)' = 0$$
$$f'\left(x^2 - 3\right)\left(x^2 - 3\right)' = 0 \qquad \text{Chain rule}$$
$$f'\left(x^2 - 3\right) \cdot 2x = 0$$
$$\text{so either} \quad 2x = 0$$
$$x = 0$$
$$\text{or} \quad f'\left(x^2 - 3\right) = 0$$
$$x^2 - 3 = 2 \qquad \text{because } f'(2) = 0$$
$$x = \pm\sqrt{5}.$$

Thus, the critical points are $x = 0, \pm\sqrt{5}$.

97. This is impossible. Since f'' exists, so must f', which means that f is differentiable and hence continuous. If $f(x)$ were positive for some values of x and negative for other values, then by the Intermediate Value Theorem, $f(x)$ would have to be zero somewhere, but this is impossible since $f(x)f''(x) < 0$ for all x. Thus either $f(x) > 0$ for all values of x, in which case $f''(x) < 0$ for all values of x, that is f is concave down. But this is impossible by Problem 81. Or else $f(x) < 0$ for all x, in which case $f''(x) > 0$ for all x, that is f is concave up. But this is impossible by Problem 83.

Solutions for Section 4.2

Exercises

1. See Figure 4.9.

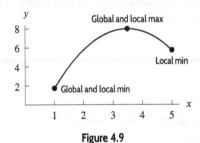

Figure 4.9

5. Since $f(x) = x^4 - 8x^2$ is continuous and the interval $-3 \le x \le 1$ is closed, there must be a global maximum and minimum. The possible candidates are critical points in the interval and endpoints. Since there are no points where $f'(x)$ is undefined, we solve $f'(x) = 0$ to find all the critical points:

$$f'(x) = 4x^3 - 16x = 4x(x^2 - 4) = 0,$$

so $x = -2, 0, 2$ are the critical points; only $x = -2, 0$ are in the interval. We then compare the values of f at the critical points and the endpoints:

$$f(-3) = 9, \quad f(-2) = -16, \quad f(0) = 0, \quad f(1) = -7.$$

Thus the global maximum is 9 at $x = -3$ and the global minimum is -16 at $x = -2$.

9. Since $f(x) = x^2 - 2|x|$ is continuous and the interval $-3 \leq x \leq 4$ is closed, there must be a global maximum and minimum. The possible candidates are critical points in the interval and endpoints. The derivative f' is not defined at $x = 0$. To find the other critical points we solve $f'(x) = 0$. For $x > 0$ we have

$$f(x) = x^2 - 2x, \text{ so } f'(x) = 2x - 2 = 0.$$

Thus, $x = 1$ is the only critical point for $0 < x < 4$. For $x < 0$, we have:

$$f(x) = x^2 + 2x, \text{ so } f'(x) = 2x + 2 = 0.$$

Thus $x = -1$ is the only critical point for $-3 < x < 0$. We then compare the values of f at the critical points and the endpoints:

$$f(-3) = 3, \quad f(-1) = -1, \quad f(0) = 0, \quad f(1) = -1, \quad f(4) = 8.$$

Thus the global maximum is 8 at $x = 4$ and the global minimum is -1 at $x = -1$ and $x = 1$.

13. (a) Differentiating gives

$$f(x) = \sin^2 x - \cos x \quad \text{for } 0 \leq x \leq \pi$$
$$f'(x) = 2 \sin x \cos x + \sin x = (\sin x)(2 \cos x + 1),$$

so $f'(x) = 0$ when $\sin x = 0$ or when $2 \cos x + 1 = 0$. Now, $\sin x = 0$ when $x = 0$ or when $x = \pi$. On the other hand, $2 \cos x + 1 = 0$ when $\cos x = -1/2$, which happens when $x = 2\pi/3$. So the critical points are $x = 0$, $x = 2\pi/3$, and $x = \pi$.

Note that $\sin x > 0$ for $0 < x < \pi$. Also, $2 \cos x + 1 < 0$ if $2\pi/3 < x \leq \pi$ and $2 \cos x + 1 > 0$ if $0 < x < 2\pi/3$. Therefore,

$$f'(x) < 0 \quad \text{for} \quad \frac{2\pi}{3} < x < \pi$$
$$f'(x) > 0 \quad \text{for} \quad 0 < x < \frac{2\pi}{3}.$$

Thus f has a local maximum at $x = 2\pi/3$ and local minima at $x = 0$ and $x = \pi$.

(b) We have

$$f(0) = [\sin(0)]^2 - \cos(0) = -1$$
$$f\left(\frac{2\pi}{3}\right) = \left[\sin\left(\frac{2\pi}{3}\right)\right]^2 - \cos\frac{2\pi}{3} = 1.25$$
$$f(\pi) = [\sin(\pi)]^2 - \cos(\pi) = 1.$$

Thus, the global maximum is at $x = 2\pi/3$, and the global minimum is at $x = 0$.

17. Differentiating gives

$$f'(x) = 1 - \frac{1}{x},$$

so the critical points satisfy

$$1 - \frac{1}{x} = 0$$
$$\frac{1}{x} = 1$$
$$x = 1.$$

Since f' is negative for $0 < x < 1$ and f' is positive for $x > 1$, there is a local minimum at $x = 1$.

Since $f(x) \to \infty$ as $x \to 0^+$ and as $x \to \infty$, the local minimum at $x = 1$ is a global minimum; there is no global maximum. See Figure 4.10. Thus, the global minimum is $f(1) = 1$.

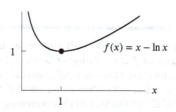

$f(x) = x - \ln x$

Figure 4.10

21. We have

$$f'(x) = 3e^{3x} - 2e^{2x}.$$

To find critical points, we set $f'(x) = 0$. Then

$$3e^{3x} - 2e^{2x} = 0$$
$$3e^{3x} = 2e^{2x}$$
$$e^x = 2/3$$
$$x = \ln(2/3).$$

There is just one critical point of f, at $x = \ln(2/3)$. We have

$$f''(x) = 9e^{3x} - 4e^{2x}.$$

At the critical point the second derivative

$$f''\left(\ln\frac{2}{3}\right) = 9\left(\frac{2}{3}\right)^3 - 4\left(\frac{2}{3}\right)^2 = \frac{8}{9} > 0$$

is positive. Hence the critical point is a local minimum. Since it is the only critical point, it is also the global minimum.
The minimum value is

$$f\left(\ln\frac{2}{3}\right) = \left(\frac{2}{3}\right)^3 - \left(\frac{2}{3}\right)^2 = -\frac{4}{27}.$$

25. The graph of $y = x + \sin x$ in Figure 4.11 suggests that the function is nondecreasing over the entire interval. You can confirm this by looking at the derivative:

$$y' = 1 + \cos x$$

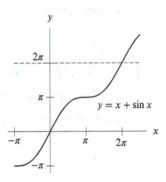

Figure 4.11: Graph of $y = x + \sin x$

Since $\cos x \geq -1$, we have $y' \geq 0$ everywhere, so y never decreases. This means that a lower bound for y is 0 (its value at the left endpoint of the interval) and an upper bound is 2π (its value at the right endpoint). That is, if $0 \leq x \leq 2\pi$:

$$0 \leq y \leq 2\pi.$$

These are the best bounds for y over the interval.

Problems

29. We see in the graph of $f'(x)$ that f' is positive or zero for all x-values to the left of G, and is negative for all x-values to the right of G. This tells us that the function f is increasing or horizontal everywhere to the left of G and is decreasing everywhere to the right of G. Thus, the function $f(x)$ has a global maximum at the point with x-value G.

To see that the function does not have a global minimum, we see that, to the right of G, the derivative gets more and more negative. Therefore, to the right of G, the function $f(x)$ is decreasing at a more and more rapid rate. The values of the function get more and more negative without bound, so there is no global minimum.

33. To find values of T that give critical points of Q with all other quantities staying constant, we set dQ/dT equal to zero and solve for T. We can rewrite Q by multiplying through to get

$$Q = AT(S - T) = AST - AT^2.$$

We then have:

$$\frac{dQ}{dT} = AS - 2AT = 0$$

$$T = \frac{-AS}{-2A} = \frac{S}{2}.$$

37. (a) Since a/q decreases with q, this term represents the ordering cost. Since bq increases with q, this term represents the storage cost.
  (b) At the minimum,
$$\frac{dC}{dq} = \frac{-a}{q^2} + b = 0$$

giving

$$q^2 = \frac{a}{b} \quad \text{so} \quad q = \sqrt{\frac{a}{b}}.$$

Since

$$\frac{d^2C}{dq^2} = \frac{2a}{q^3} > 0 \quad \text{for} \quad q > 0,$$

we know that $q = \sqrt{a/b}$ gives a local minimum. Since $q = \sqrt{a/b}$ is the only critical point, this must be the global minimum.

41. We set $f'(r) = 0$ to find the critical points:

$$\frac{2A}{r^3} - \frac{3B}{r^4} = 0$$

$$\frac{2Ar - 3B}{r^4} = 0$$

$$2Ar - 3B = 0$$

$$r = \frac{3B}{2A}.$$

The only critical point is at $r = 3B/(2A)$. If $r > 3B/(2A)$, we have $f' > 0$ and if $r < 3B/(2A)$, we have $f' < 0$. Thus, the force between the atoms is minimized at $r = 3B/(2A)$.

45. Suppose the points are given by x and $-x$, where $x \geq 0$. The function is odd, since

$$y = \frac{(-x)^3}{1 + (-x)^4} = -\frac{x^3}{1 + x^4},$$

so the corresponding y-coordinates are also opposite. See Figure 4.12. For $x > 0$, we have

$$m = \frac{\frac{x^3}{1+x^4} - \left(-\frac{x^3}{1+x^4}\right)}{x - (-x)} = \frac{1}{2x} \cdot \frac{2x^3}{1 + x^4} = \frac{x^2}{1 + x^4}.$$

For the maximum slope,

$$\frac{dm}{dx} = \frac{2x}{1 + x^4} - \frac{x^2(4x^3)}{(1 + x^4)^2} = 0$$

$$\frac{2x(1 + x^4) - 4x^5}{(1 + x^4)^2} = 0$$

$$\frac{2x(1 - x^4)}{(1 + x^4)^2} = 0$$

$$x\left(1 - x^4\right) = 0$$

$$x = 0, \pm 1.$$

For $x > 0$, there is one critical point, $x = 1$. Since m tends to 0 when $x \to 0$ and when $x \to \infty$, the critical point $x = 1$ gives the maximum slope. Thus, the maximum slope occurs when the line has endpoints

$$\left(-1, -\frac{1}{2}\right) \quad \text{and} \quad \left(1, \frac{1}{2}\right).$$

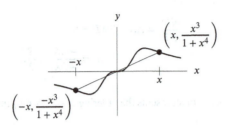

Figure 4.12

49. (a) The graph of f is always increasing and the graph of g is always decreasing, so the intersection point shown at $x = 0$ is the only one.
    (b) To find local maxima and minima, we find critical points of $h(x)$. We have

    $$\begin{aligned} h'(x) &= 3 \cdot 2e^{3x} + (-2) \cdot 3e^{-2x} \\ &= 6(e^{3x} - e^{-2x}) \\ &= 6(f(x) - g(x)). \end{aligned}$$

    So $h'(x) = 0$ when $f(x) = g(x)$, which by part (a) occurs only when $x = 0$. To the left of zero the graph of f is below the graph of g, so $f(x) - g(x) < 0$ for $x < 0$, thus the derivative of h is negative there. To the right of $x = 0$, $h'(x)$ is positive because the graph of f is above the graph of g. So h has a minimum at $x = 0$, and its value is $2f(0) + 3g(0) = 2 + 3 = 5$.

Strengthen Your Understanding

53. Consider the interval $1 \le x \le 2$. Since f is increasing on this interval, its global minimum occurs at the left end-point of the interval, $x = 1$. If $x = 0$ is inside the interval $a \le x \le b$, then the global minimum of f occurs at $x = 0$. Otherwise, the global minimum of f will occur at an endpoint.

57. Solving $x^2 = 5$ and $x^2 = 2$, we see that one such interval is $\sqrt{2} \le x \le \sqrt{5}$. A second possible interval is $-\sqrt{5} \le x \le -\sqrt{2}$.

61. True. The values of $f(x)$ get arbitrarily close to 0, but $f(x) > 0$ for all x in the interval $(0, 2)$.

65. False, since $f(x) = 1/x$ takes on arbitrarily large values as $x \to 0^+$. The Extreme Value Theorem requires the interval to be closed as well as bounded.

69. True. If $f'(0) > 0$, then f would be increasing at 0 and so $f(0) < f(x)$ for x just to the right of 0. Then $f(0)$ would not be a maximum for f on the interval $0 \le x \le 10$.

Additional Problems (online only)

73. (a) If both the global minimum and the global maximum are at the endpoints, then $f(x) = 0$ everywhere in $[a, b]$, since $f(a) = f(b) = 0$. In that case $f'(x) = 0$ everywhere as well, so any point in (a, b) will do for c.
    (b) Suppose that either the global maximum or the global minimum occurs at an interior point of the interval. Let c be that point. Then c must be a local extremum of f, so, by the theorem concerning local extrema on page 194, we have $f'(c) = 0$, as required.

Solutions for Section 4.3

Exercises

1. Let the numbers be x and y. Then $x + y = 100$, so $y = 100 - x$.
   Since both numbers are nonnegative, we restrict to $0 \le x \le 100$.
   The product is

   $$P = xy = x(100 - x) = 100x - x^2.$$

   Differentiating to find the maximum,

   $$\frac{dP}{dx} = 100 - 2x = 0$$

   $$x = \frac{100}{2} = 50.$$

   So there is a critical point at $x = 50$; the end points are at $x = 0, 100$.
   Evaluating gives
   At $x = 0$, we have $P = 0$.
   At $x = 50$, we have $P = 50(100 - 50) = 2500$.
   At $x = 100$, we have $P = 100(100 - 100) = 0$.
   Thus the maximum value is 2500.

5. Let the sides be x and y cm. Then $2x + 2y = 64$, so $y = 32 - x$.
   Since both sides are nonnegative, we restrict to $0 \le x \le 32$.
   The area in cm^2 is

   $$A = xy = x(32 - x) = 32x - x^2.$$

   Differentiating to find the maximum,

   $$\frac{dA}{dx} = 32 - 2x = 0$$

   $$x = \frac{32}{2} = 16.$$

   So there is a critical point at $x = 16$ cm; the end points are at $x = 0$ and $x = 32$ cm.
   Evaluating gives:
   At $x = 0$, we have $A = 0$ cm^2.
   At $x = 16$, we have $A = 16(32 - 16) = 256$ cm^2.
   At $x = 32$, we have $A = 32(32 - 32) = 0$ cm^2.
   Thus, the maximum area occurs in a square of side 16 cm.

9. We want to minimize the surface area S of the cylinder, shown in Figure 4.13. The cylinder has 3 pieces: the top and bottom disks, each of which has area πr^2 and the tube, which has area $2\pi rh$. Thus we want to minimize

   $$S = 2\pi r^2 + 2\pi rh.$$

   The volume of the cylinder is $8 = \pi r^2 h$, so $h = 8/\pi r^2$. Substituting this expression for h in the formula for S gives

   $$S = 2\pi r^2 + 2\pi r \cdot \frac{8}{\pi r^2} = 2\pi r^2 + \frac{16}{r}.$$

   Differentiating gives

   $$\frac{dS}{dr} = 4\pi r - \frac{16}{r^2}.$$

   To minimize S we look for critical points, so we solve $0 = 4\pi r - 16/r^2$. Multiplying by r^2 gives

   $$0 = 4\pi r^3 - 16,$$

   so $r = (4/\pi)^{1/3}$ cm. Then we can find

   $$h = \frac{8}{\pi r^2} = \frac{8}{\pi \left(\frac{4}{\pi}\right)^{2/3}} = \frac{2\left(\frac{4}{\pi}\right)}{\left(\frac{4}{\pi}\right)^{2/3}} = 2\left(\frac{4}{\pi}\right)^{1/3}.$$

We can check that this critical point is a minimum of S by checking the sign of

$$\frac{d^2S}{dr^2} = 4\pi + \frac{32}{r^3}$$

which is positive when $r > 0$. So S is concave up at the critical point and therefore $r = (4/\pi)^{1/3}$ cm is a minimum.

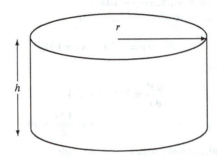

Figure 4.13

13. The rectangle in Figure 4.14 has area, A, given by

$$A = 2xy = \frac{2x}{1 + x^2} \qquad \text{for } x \geq 0.$$

At a critical point,

$$\frac{dA}{dx} = \frac{2}{1 + x^2} + 2x\left(\frac{-2x}{(1 + x^2)^2}\right) = 0$$

$$\frac{2(1 + x^2 - 2x^2)}{(1 + x^2)^2} = 0$$

$$1 - x^2 = 0$$

$$x = \pm 1.$$

Since $A = 0$ for $x = 0$ and $A \to 0$ as $x \to \infty$, the critical point $x = 1$ is a local and global maximum for the area. Then $y = 1/2$, so the vertices are

$$(-1, 0), \ (1, 0), \ \left(1, \frac{1}{2}\right), \ \left(-1, \frac{1}{2}\right).$$

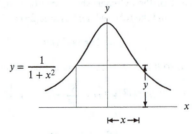

Figure 4.14

Problems

17. Let the sides of the rectangle have lengths a and b. We shall look for the minimum of the square s of the length of either diagonal, i.e. $s = a^2 + b^2$. The area is $A = ab$, so $b = A/a$. This gives

$$s(a) = a^2 + \frac{A^2}{a^2}.$$

To find the minimum squared length we need to find the critical points of s. Differentiating s with respect to a gives

$$\frac{ds}{da} = 2a + (-2)A^2 a^{-3} = 2a\left(1 - \frac{A^2}{a^4}\right)$$

The derivative $ds/da = 0$ when $a = \sqrt{A}$, that is when $a = b$ and so the rectangle is a square. Because $\dfrac{d^2 s}{da^2} = 2\left(1 + \dfrac{3A^2}{a^4}\right) > 0$, this is a minimum.

21. (a) The rectangle on the left has area xy, and the semicircle on the right with radius $y/2$ has area $(1/2)\pi(y/2)^2 = \pi y^2/8$. We have

$$\text{Area of entire region} = xy + \frac{\pi}{8}y^2.$$

(b) The perimeter of the figure is made of two horizontal line segments of length x each, one vertical segment of length y, and a semicircle of radius $y/2$ of length $\pi y/2$. We have

$$\text{Perimeter of entire region} = 2x + y + \frac{\pi}{2}y = 2x + (1 + \frac{\pi}{2})y.$$

(c) We want to maximize the area $xy + \pi y^2/8$, with the perimeter condition $2x + (1 + \pi/2)y = 100$. Substituting

$$x = 50 - \left(\frac{1}{2} + \frac{\pi}{4}\right)y = 50 - \frac{2 + \pi}{4}y$$

into the area formula, we must maximize

$$A(y) = \left(50 - \left(\frac{1}{2} + \frac{\pi}{4}\right)y\right)y + \frac{\pi}{8}y^2 = 50y - \left(\frac{1}{2} + \frac{\pi}{8}\right)y^2$$

on the interval

$$0 \le y \le 200/(2 + \pi) = 38.8985$$

where $y \ge 0$ because y is a length and $y \le 200/(2 + \pi)$ because $x \ge 0$ is a length.
The critical point of A occurs where

$$A'(y) = 50 - \left(1 + \frac{\pi}{4}\right)y = 0$$

at

$$y = \frac{200}{4 + \pi} = 28.005.$$

A maximum for A must occur at the critical point $y = 200/(4 + \pi)$ or at one of the endpoints $y = 0$ or $y = 200/(2 + \pi)$. Since

$$A(0) = 0 \qquad A\left(\frac{200}{4 + \pi}\right) = 700.124 \qquad A\left(\frac{200}{2 + \pi}\right) = 594.2$$

the maximum is at

$$y = \frac{200}{4 + \pi}.$$

Hence

$$x = 50 - \frac{2 + \pi}{4}y = 50 - \frac{2 + \pi}{4}\frac{200}{4 + \pi} = \frac{100}{4 + \pi}.$$

The dimensions giving maximum area with perimeter 100 are

$$x = \frac{100}{4 + \pi} = 14.0 \qquad y = \frac{200}{4 + \pi} = 28.0.$$

The length y is twice as great as x.

25. The minimum value of x in the interval $0 \le x \le 10$ is $x = 0$ and the maximum is $x = 10$.

29. The distance $d(x)$ from the point $(x, \sqrt{1 - x})$ on the curve to the origin is given by

$$d(x) = \sqrt{x^2 + (\sqrt{1 - x})^2} = \sqrt{x^2 + (1 - x)}.$$

Since x is in the domain of $y = \sqrt{1 - x}$, we have $-\infty < x \le 1$. Differentiating gives

$$d'(x) = \frac{2x - 1}{2\sqrt{x^2 - x + 1}},$$

so $x = 1/2$ is the only critical point. Since $d'(x) < 0$ for $x < 1/2$ and $d'(x) > 0$ for $x > 1/2$, the point $x = 1/2$ is a minimum for x. The point $(1/2, 1/\sqrt{2})$ is the closest point on the curve to the origin.

33. Let the radius of the can be r and let its height be h. The surface area, S, and volume, V, of the can are given by

$$S = \text{Area of the sides of can} + \text{Area of top and bottom}$$
$$S = 2\pi rh + 2\pi r^2$$
$$V = \pi r^2 h.$$

Since $S = 280$, we have $2\pi rh + 2\pi r^2 = 280$, we have

$$h = \frac{140 - \pi r^2}{\pi r}$$

and

$$V = \pi r^2 \frac{140 - \pi r^2}{\pi r} = 140r - \pi r^3.$$

Since $h \geq 0$, we have $\pi r^2 \leq 140$ and thus $0 < r \leq \sqrt{140/\pi} = 6.676$.

We have

$$V'(r) = 140 - 3\pi r^2,$$

so the only critical point of V with $r > 0$ is $r = \sqrt{140/3\pi} = 3.854$. The critical point is in the domain $0 < r \leq 6.676$.

Since $V''(r) = -6\pi r$ is negative for all $r > 0$, the critical point $r = 3.854$ cm gives the maximum value of the volume. For this value of r, we have $h = 7.708$ cm and $V = 359.721$ cm^3.

37. We only consider $\lambda > 0$. For such λ, the value of $v \to \infty$ as $\lambda \to \infty$ and as $\lambda \to 0^+$. Thus, v does not have a maximum velocity. It will have a minimum velocity. To find it, we set $dv/d\lambda = 0$:

$$\frac{dv}{d\lambda} = k\frac{1}{2}\left(\frac{\lambda}{c} + \frac{c}{\lambda}\right)^{-1/2}\left(\frac{1}{c} - \frac{c}{\lambda^2}\right) = 0.$$

Solving, and remembering that $\lambda > 0$, we obtain

$$\frac{1}{c} - \frac{c}{\lambda^2} = 0$$
$$\frac{1}{c} = \frac{c}{\lambda^2}$$
$$\lambda^2 = c^2,$$

so

$$\lambda = c.$$

Thus, we have one critical point. Since

$$\frac{dv}{d\lambda} < 0 \quad \text{for } \lambda < c$$

and

$$\frac{dv}{d\lambda} > 0 \quad \text{for } \lambda > c,$$

the first derivative test tells us that we have a local minimum of v at $x = c$. Since $\lambda = c$ is the only critical point, it gives the global minimum. Thus the minimum value of v is

$$v = k\sqrt{\frac{c}{c} + \frac{c}{c}} = \sqrt{2}k.$$

41. Let x equal the number of chairs ordered in excess of 300, so $0 \leq x \leq 100$.

$$\text{Revenue} = R = (90 - 0.25x)(300 + x)$$
$$= 27,000 - 75x + 90x - 0.25x^2 = 27,000 + 15x - 0.25x^2$$

At a critical point $dR/dx = 0$. Since $dR/dx = 15 - 0.5x$, we have $x = 30$, and the maximum revenue is $\$27,225$ since the graph of R is a parabola which opens downward. The minimum is $\$0$ (when no chairs are sold).

45. Let x be as indicated in the figure in the text. Then the distance from S to Town 1 is $\sqrt{1 + x^2}$ and the distance from S to Town 2 is $\sqrt{(4 - x)^2 + 4^2} = \sqrt{x^2 - 8x + 32}$.

$$\text{Total length of pipe} = f(x) = \sqrt{1 + x^2} + \sqrt{x^2 - 8x + 32}.$$

We want to look for critical points of f. The easiest way is to graph f and see that it has a local minimum at about $x = 0.8$ miles. Alternatively, we can use the formula:

$$
\begin{aligned}
f'(x) &= \frac{2x}{2\sqrt{1 + x^2}} + \frac{2x - 8}{2\sqrt{x^2 - 8x + 32}} \\
&= \frac{x}{\sqrt{1 + x^2}} + \frac{x - 4}{\sqrt{x^2 - 8x + 32}} \\
&= \frac{x\sqrt{x^2 - 8x + 32} + (x - 4)\sqrt{1 + x^2}}{\sqrt{1 + x^2}\sqrt{x^2 - 8x + 32}} = 0.
\end{aligned}
$$

$f'(x)$ is equal to zero when the numerator is equal to zero.

$$
\begin{aligned}
x\sqrt{x^2 - 8x + 32} + (x - 4)\sqrt{1 + x^2} &= 0 \\
x\sqrt{x^2 - 8x + 32} &= (4 - x)\sqrt{1 + x^2}.
\end{aligned}
$$

Squaring both sides and simplifying, we get

$$
\begin{aligned}
x^2(x^2 - 8x + 32) &= (x^2 - 8x + 16)(14x^2) \\
x^4 - 8x^3 + 32x^2 &= x^4 - 8x^3 + 17x^2 - 8x + 16 \\
15x^2 + 8x - 16 &= 0, \\
(3x + 4)(5x - 4) &= 0.
\end{aligned}
$$

So $x = 4/5$. (Discard $x = -4/3$ since we are only interested in x between 0 and 4, between the two towns.) Using the second derivative test, we can verify that $x = 4/5$ is a local minimum.

49. (a) We have

$$x^{1/x} = e^{\ln(x^{1/x})} = e^{(1/x)\ln x}.$$

Thus

$$
\begin{aligned}
\frac{d(x^{1/x})}{dx} &= \frac{d(e^{(1/x)\ln x})}{dx} = \frac{d(\frac{1}{x}\ln x)}{dx}e^{(1/x)\ln x} \\
&= \left(-\frac{\ln x}{x^2} + \frac{1}{x^2}\right)x^{1/x} \\
&= \frac{x^{1/x}}{x^2}(1 - \ln x)\begin{cases} = 0 & \text{when } x = e \\ < 0 & \text{when } x > e \\ > 0 & \text{when } x < e. \end{cases}
\end{aligned}
$$

Hence $e^{1/e}$ is the global maximum for $x^{1/x}$, by the first derivative test.

(b) Since $x^{1/x}$ is increasing for $0 < x < e$ and decreasing for $x > e$, and 2 and 3 are the closest integers to e, either $2^{1/2}$ or $3^{1/3}$ is the maximum for $n^{1/n}$. We have $2^{1/2} \approx 1.414$ and $3^{1/3} \approx 1.442$, so $3^{1/3}$ is the maximum.

(c) Since $e < 3 < \pi$, and $x^{1/x}$ is decreasing for $x > e$, $3^{1/3} > \pi^{1/\pi}$.

53. (a) To maximize benefit (surviving young), we pick 10, because that's the highest point of the benefit graph.

(b) To optimize the vertical distance between the curves, we can either do it by inspection or note that the slopes of the two curves will be the same where the difference is maximized. Either way, one gets approximately 9.

57. The independent variable is y, the extension length of the spring. We differentiate to find critical points:

$$\frac{dE}{dy} = ky - mg = 0$$

$$y = \frac{mg}{k}.$$

To decide if this critical point is a local minimum, we find

$$\frac{d^2E}{dy^2} = k.$$

Since $k > 0$, we know that $y = mg/k$ corresponds to a a local minimum of E. Thus, $y = mg/k$ gives a stable equilibrium. When the spring is extended by length mg/k, the downward pull of gravity and the upward pull of the spring on the mass are exactly equal, so the mass will hang there motionless, in stable equilibrium.

Strengthen Your Understanding

61. The solution of an optimization problem depends on both the modeling function and the interval over which that function is optimized. The solution can occur either at a critical point of the function or at an endpoint of the interval. For example, if the modeling function is $f(x) = -(x - 2)^2 + 6$ on the interval $5 \leq x \leq 10$, the absolute minimum occurs at $x = 10$ and the maximum occurs at $x = 5$; neither occurs at the vertex $x = 2$.

Solutions for Section 4.4

Exercises

1. (a) See Figure 4.15.

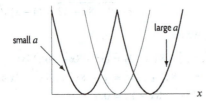

Figure 4.15

(b) We see in Figure 4.15 that in each case the graph of f is a parabola with one critical point, its vertex, on the positive x-axis. The critical point moves to the right along the x-axis as a increases.

(c) To find the critical points, we set the derivative equal to zero and solve for x.

$$f'(x) = 2(x - a) = 0$$
$$x = a.$$

The only critical point is at $x = a$. As we saw in the graph, and as a increases, the critical point moves to the right.

5. (a) See Figure 4.16.

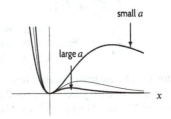

Figure 4.16

(b) We see in Figure 4.16 that in each case f appears to have two critical points. One critical point is a local minimum at the origin and the other is a local maximum in quadrant I. As the parameter a increases, the critical point in quadrant I appears to move down and to the left, closer to the origin.

(c) To find the critical points, we set the derivative equal to zero and solve for x. Using the product rule, we have:

$$f'(x) = x^2 \cdot e^{-ax}(-a) + 2x \cdot e^{-ax} = 0$$
$$xe^{-ax}(-ax + 2) = 0$$
$$x = 0 \quad \text{and} \quad x = \frac{2}{a}.$$

There are two critical points, at $x = 0$ and $x = 2/a$. As we saw in the graph, as a increases the nonzero critical point moves to the left.

9. (a) The larger the value of $|A|$, the steeper the graph (for the same x-value).
    (b) The graph is shifted horizontally by B. The shift is to the left for positive B, to the right for negative B. There is a vertical asymptote at $x = -B$. See Figure 4.17.
    (c)

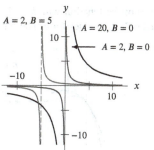

Figure 4.17

13. (a) Figure 4.18 shows the effect of varying a with $b = 1$.
    (b) Figure 4.19 shows the effect of varying b with $a = 1$.

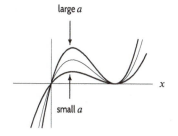

Figure 4.18

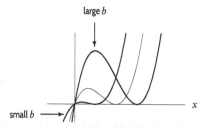

Figure 4.19

    (c) In each case f appears to have two critical points, a local maximum in quadrant I and a local minimum on the positive x-axis. From Figure 4.18 it appears that increasing a moves the local maximum up and does not move the local minimum. From Figure 4.19 it appears that increasing b moves the local maximum up and to the right and moves the local minimum to the right along the x-axis.
    (d) To find the critical points, we set the derivative equal to zero and solve for x. Using the product rule, we have:

$$f'(x) = ax \cdot 2(x - b) + a \cdot (x - b)^2 = 0$$
$$a(x - b)(2x + (x - b)) = 0$$
$$a(x - b)(3x - b) = 0$$
$$x = b \quad \text{or} \quad x = \frac{b}{3}.$$

    There are two critical points, at $x = b$ and at $x = b/3$. Increasing a does not move either critical point horizontally. Increasing b moves both critical points to the right. This confirms what we saw in the graphs.

Problems

17. We have $f'(x) = 1 - a^2/x^2$, so the critical points occur at $1 - a^2/x^2 = 0$, that is, at $x = \pm a$. We are shown only the first quadrant.

    The graph suggests that A has a minimum at $x = 3$, B at $x = 2$, and C at $x = 1$. Thus, C corresponds to $a = 1$, and B corresponds to $a = 2$, and A corresponding to $a = 3$. The third value of a is $a = 3$.

21. To find the critical points, we set the derivative equal to zero and solve for x. Using the product rule, we have

$$f'(x) = x \cdot n(x - b)^{n-1} + 1 \cdot (x - b)^n = 0$$
$$(x - b)^{n-1}(xn + (x - b)) = 0$$
$$(x - b)^{n-1}(x(n + 1) - b) = 0$$
$$x = b \quad \text{and} \quad x = \frac{b}{n + 1}.$$

There are two critical points, at $x = b$ and at $x = b/(n+1)$. The parameter b affects both critical points, and as b increases, both critical points increase. The parameter n affects only one of the critical points, and that critical point gets smaller as n increases.

Figure 4.20 shows several graphs of the function with $b = 1$ and different values for n. Figure 4.21 shows several graphs of the function with $n = 2$ and different values for b. These graphs confirm what we see analytically.

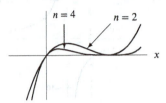

Figure 4.20: Various n with $b = 1$

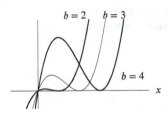

Figure 4.21: Various b with $n = 2$

25. (a) The critical point will occur where $f'(x) = 0$. But by the product rule,

$$f'(x) = be^{1+bx} + b^2 x e^{1+bx} = 0$$
$$be^{1+bx}(1 + bx) = 0$$
$$1 + bx = 0$$
$$x = -\frac{1}{b}.$$

Thus the critical point of f is located at $x = -\frac{1}{b}$.

(b) To determine whether this critical point is a local minimum or local maximum, we can use the first derivative test. Since both b and e^{1+bx} are positive for all x, the sign of $f'(x) = be^{1+bx}(1 + bx)$ depends on the sign of $1 + bx$.

$$\text{for } x < -\tfrac{1}{b}, \quad f'(x) \text{ is negative}$$
$$\text{for } x = \tfrac{1}{b}, \qquad f'(x) = 0$$
$$\text{for } x > -\tfrac{1}{b}, \quad f'(x) \text{ is positive}$$

Therefore, f goes from decreasing to increasing at $x = -\frac{1}{b}$, making this point a local minimum.

(c) We only need to substitute the critical point $x = -\frac{1}{b}$ into the original function f:

$$f\left(-\frac{1}{b}\right) = b\left(-\frac{1}{b}\right)e^{1+b\left(-\frac{1}{b}\right)}$$
$$= -1e^{1-1}$$
$$= -1.$$

This answer does not depend on the value of b. Though the x-coordinate of the critical point depends on b, the y-coordinate does not.

29. (a) The x-intercept occurs where $f(x) = 0$, so

$$ax - x \ln x = 0$$
$$x(a - \ln x) = 0.$$

Since $x > 0$, we must have

$$a - \ln x = 0$$
$$\ln x = a$$
$$x = e^a.$$

(b) See Figures 4.22 and 4.23.

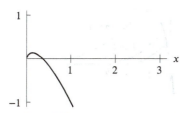

Figure 4.22: Graph of $f(x)$ with
$a = -1$

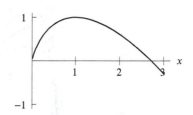

Figure 4.23: Graph of $f(x)$ with $a = 1$

(c) Differentiating gives $f'(x) = a - \ln x - 1$. Critical points are obtained by solving

$$a - \ln x - 1 = 0$$
$$\ln x = a - 1$$
$$x = e^{a-1}.$$

Since $e^{a-1} > 0$ for all a, there is no restriction on a. Now,

$$f(e^{a-1}) = ae^{a-1} - e^{a-1} \ln(e^{a-1}) = ae^{a-1} - (a-1)e^{a-1} = e^{a-1},$$

so the coordinates of the critical point are (e^{a-1}, e^{a-1}). From the graphs, we see that this critical point is a local maximum; this can be confirmed using the second derivative:

$$f''(x) = -\frac{1}{x} < 0 \qquad \text{for } x = e^{a-1}.$$

33. To determine concavity, we use the second derivative:

$$f'(x) = -ae^{-ax} + be^{bx}$$
$$f''(x) = (-a)^2 e^{-ax} + b^2 e^{bx} = a^2 e^{-ax} + b^2 e^{bx}.$$

Since a and b are non-zero, a^2 and b^2 are both positive. Also e^{-ax} and e^{bx} are both always positive, so the second derivative is always positive, which means that the graph is always concave up.

37. To determine concavity, we use the second derivative:

$$f''(x) = 12ax^2 - 2b.$$

Because a is positive, the graph of f'' is a parabola opening upward; because b is positive, the parabola has a negative vertical intercept. Thus, f'' has two zeros and is negative between those two zeros and positive outside them. So the graph of f is concave up on any interval outside the two zeros, and concave down on any interval between them.

41. (a) Let $f(x) = axe^{-bx}$. To find the local maxima and local minima of f, we solve

$$f'(x) = ae^{-bx} - abxe^{-bx} = ae^{-bx}(1 - bx) \begin{cases} = 0 & \text{if } x = 1/b \\ < 0 & \text{if } x > 1/b \\ > 0 & \text{if } x < 1/b. \end{cases}$$

Therefore, f is increasing ($f' > 0$) for $x < 1/b$ and decreasing ($f' > 0$) for $x > 1/b$. A local maximum occurs at $x = 1/b$. There are no local minima. To find the points of inflection, we write

$$f''(x) = -abe^{-bx} + ab^2 xe^{-bx} - abe^{-bx}$$
$$= -2abe^{-bx} + ab^2 xe^{-bx}$$
$$= ab(bx - 2)e^{-bx},$$

so $f'' = 0$ at $x = 2/b$. Therefore, f is concave up for $x < 2/b$ and concave down for $x > 2/b$, and the inflection point is $x = 2/b$.

(b) Varying a stretches or flattens the graph but does not affect the critical point $x = 1/b$ and the inflection point $x = 2/b$. Since the critical and inflection points are depend on b, varying b will change these points, as well as the maximum $f(1/b) = a/be$. For example, an increase in b will shift the critical and inflection points to the left, and also lower the maximum value of f.

(c)

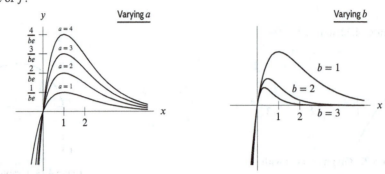

45. Cubic polynomials are all of the form $f(x) = Ax^3 + Bx^2 + Cx + D$. There is an inflection point at the origin $(0, 0)$ if $f''(0) = 0$ and $f(0) = 0$. Since $f(0) = D$, we must have $D = 0$. Since $f''(x) = 6Ax + 2B$, giving $f''(0) = 2B$, we must have $B = 0$. The family of cubic polynomials with inflection point at the origin is the two parameter family $f(x) = Ax^3 + Cx$.

49. The maximum of $y = e^{-(x-a)^2/b}$ occurs at $x = a$. (This is because the exponent $-(x-a)^2/b$ is zero when $x = a$ and negative for all other x-values. The same result can be obtained by taking derivatives.) Thus we know that $a = 2$.

Points of inflection occur where d^2y/dx^2 changes sign, that is, where $d^2y/dx^2 = 0$. Differentiating gives

$$\frac{dy}{dx} = -\frac{2(x-2)}{b}e^{-(x-2)^2/b}$$

$$\frac{d^2y}{dx^2} = -\frac{2}{b}e^{-(x-2)^2/b} + \frac{4(x-2)^2}{b^2}e^{-(x-2)^2/b} = \frac{2}{b}e^{-(x-2)^2/b}\left(-1 + \frac{2}{b}(x-2)^2\right).$$

Since $e^{-(x-2)^2/b}$ is never zero, $d^2y/dx^2 = 0$ where

$$-1 + \frac{2}{b}(x-2)^2 = 0.$$

We know $d^2y/dx^2 = 0$ at $x = 1$, so substituting $x = 1$ gives

$$-1 + \frac{2}{b}(1-2)^2 = 0.$$

Solving for b gives

$$-1 + \frac{2}{b} = 0$$

$$b = 2.$$

Since $a = 2$, the function is

$$y = e^{-(x-2)^2/2}.$$

You can check that at $x = 2$, we have

$$\frac{d^2y}{dx^2} = \frac{2}{2}e^{-0}(-1 + 0) < 0$$

so the point $x = 2$ does indeed give a maximum. See Figure 4.24.

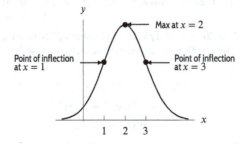

Figure 4.24: Graph of $y = e^{-(x-2)^2/2}$

53. Since the graph is symmetric about the y-axis, the polynomial must have only even powers. Also, since the y-intercept is 0, the constant term must be zero. Thus, the polynomial is of the form

$$y = ax^4 + bx^2.$$

Differentiating gives

$$\frac{dy}{dx} = 4ax^3 + 2bx = x(4ax^2 + 2b).$$

Thus $dy/dx = 0$ at $x = 0$ and when $4ax^2 + 2b = 0$. Since the maxima occur where $x = \pm 1$, we have $4a + 2b = 0$, so $b = -2a$.

We are given $y = 2$ when $x = \pm 1$, so using $y = ax^4 + bx^2$ gives

$$a + b = 2,$$
$$a - 2a = -a = 2,$$

which gives $a = -2$ and $b = 4$. Thus

$$y = -2x^4 + 4x^2.$$

To see if the points $(1, 2)$ and $(-1, 2)$ are local maxima, we take the second derivative,

$$\frac{d^2y}{dx^2} = -24x^2 + 8,$$

which is negative if $x = \pm 1$. The critical points at $x = \pm 1$ are local maxima. Since the leading coefficient of this polynomial is negative, we know the graph decreases without bound as $|x|$ approaches $\pm\infty$, so these critical points are also global maxima. Note that at $x = 0$ the second derivative is positive, so the point $(0, 0)$ is a local minimum. There is no global minimum.

57. Differentiating $y = bxe^{-ax}$ gives

$$\frac{dy}{dx} = be^{-ax} - abxe^{-ax} = be^{-ax}(1 - ax).$$

Since we have a critical point at $x = 3$, we know that $1 - 3a = 0$, so $a = 1/3$.

If $b > 0$, the first derivative goes from positive values to the left of $x = 3$ to negative values on the right of $x = 3$, so we know this critical point is a local maximum. Since the function value at this local maximum is 6, we have

$$6 = 3be^{-3/3} = \frac{3b}{e},$$

so $b = 2e$ and

$$y = 2xe^{1-x/3}.$$

61. (a) The graphs are shown in Figures 4.25–4.30.

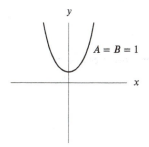

Figure 4.25: $A > 0, B > 0$

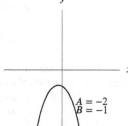

Figure 4.26: $A > 0, B < 0$

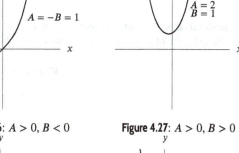

Figure 4.27: $A > 0, B > 0$

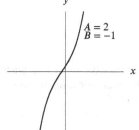

Figure 4.28: $A > 0, B < 0$

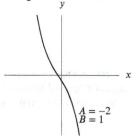

Figure 4.29: $A < 0, B < 0$ Figure 4.30: $A < 0, B > 0$

(b) If A and B have the same sign, the graph is U-shaped. If A and B are both positive, the graph opens upward. If A and B are both negative, the graph opens downward.

(c) If A and B have different signs, the graph appears to be everywhere increasing (if $A > 0, B < 0$) or decreasing (if $A < 0, B > 0$).

(d) The function appears to have a local maximum if $A < 0$ and $B < 0$, and a local minimum if $A > 0$ and $B > 0$.

To justify this, calculate the derivative

$$\frac{dy}{dx} = Ae^x - Be^{-x}.$$

Setting $dy/dx = 0$ gives

$$Ae^x - Be^{-x} = 0$$
$$Ae^x = Be^{-x}$$
$$e^{2x} = \frac{B}{A}.$$

This equation has a solution only if B/A is positive, that is, if A and B have the same sign. In that case,

$$2x = \ln\left(\frac{B}{A}\right)$$
$$x = \frac{1}{2}\ln\left(\frac{B}{A}\right).$$

This value of x gives the only critical point.

To determine whether the critical point is a local maximum or minimum, we use the first derivative test. Since

$$\frac{dy}{dx} = Ae^x - Be^{-x},$$

we see that:

If $A > 0, B > 0$, we have $dy/dx > 0$ for large positive x and $dy/dx < 0$ for large negative x, so there is a local minimum.

If $A < 0, B < 0$, we have $dy/dx < 0$ for large positive x and $dy/dx > 0$ for large negative x, so there is a local maximum.

65. (a) To find $\lim_{r \to 0^+} V(r)$, first rewrite $V(r)$ with a common denominator:

$$\lim_{r \to 0^+} V(r) = \lim_{r \to 0^+} \frac{A}{r^{12}} - \frac{B}{r^6}$$
$$= \lim_{r \to 0^+} \frac{A - Br^6}{r^{12}}$$
$$\to \frac{A}{0^+} \to +\infty.$$

As the distance between the two atoms becomes small, the potential energy diverges to $+\infty$.

(b) The critical point of $V(r)$ will occur where $V'(r) = 0$:

$$V'(r) = -\frac{12A}{r^{13}} + \frac{6B}{r^7} = 0$$
$$\frac{-12A + 6Br^6}{r^{13}} = 0$$
$$-12A + 6Br^6 = 0$$
$$r^6 = \frac{2A}{B}$$
$$r = \left(\frac{2A}{B}\right)^{1/6}$$

To determine whether this is a local maximum or minimum, i we can use the first derivative test. Since r is positive, the sign of $V'(r)$ is determined by the sign of $-12A + 6Br^6$. Notice that this is an increasing function of r for $r > 0$, so $V'(r)$ changes sign from $-$ to $+$ at $r = (2A/B)^{1/6}$. The first derivative test yields

r	$\leftarrow$	$\left(\frac{2A}{B}\right)^{1/6}$	$\rightarrow$
$V'(r)$	neg.	zero	pos.

Thus $V(r)$ goes from decreasing to increasing at the critical point $r = (2A/B)^{1/6}$, so this is a local minimum.

(c) Since $F(r) = -V'(r)$, the force is zero exactly where $V'(r) = 0$, i.e. at the critical points of V. The only critical point was the one found in part (b), so the only such point is $r = (2A/B)^{1/6}$.

(d) Since the numerator in $r = (2A/B)^{1/6}$ is proportional to $A^{1/6}$, the equilibrium size of the molecule increases when the parameter A is increased. Conversely, since B is in the denominator, when B is increased the equilibrium size of the molecules decrease.

Strengthen Your Understanding

69. Differentiating f we see that $f'(x) = -a/x^2 + b$. Solving $f'(x) = 0$ we see that $x = \pm\sqrt{a/b}$ are critical points provided a and b have the same sign. So for example, $f(x) = 1/x + x$ has two critical points. But if a and b have opposite sign, f has no critical points. The function $f(x) = 1/x - x$ has no critical points, for example.

73. Let $g(x) = ax^3 + bx^2$, where neither a nor b are allowed to be zero. Then

$$g'(x) = 3ax^2 + 2bx = x(3ax + 2b).$$

Then $g(x)$ has two distinct critical points, at $x = 0$ and at $x = -2b/3a$. Since

$$g''(x) = 6ax + 2b,$$

there is exactly one point of inflection, $x = -2b/6a = -b/3a$.

Solutions for Section 4.5

Exercises

1. The fixed costs are $5000, the marginal cost per item is $2.40, and the price per item is $4.

5. Since fixed costs are represented by the vertical intercept, they are $1.1 million. The quantity that maximizes profit is about $q = 70$, and the profit achieved is $(3.7 - 2.5) = \$1.2$ million

9. The cost function $C(q) = b + mq$ satisfies $C(0) = 5000$, so $b = 5000$, and $MC = m = 15$. So

$$C(q) = 5000 + 15q.$$

The revenue function is $R(q) = 60q$, so the profit function is

$$\pi(q) = R(q) - C(q) = 60q - 5000 - 15q = 45q - 5000.$$

13. The marginal revenue, MR, is given by differentiating the total revenue function, R. We use the chain rule so

$$MR = \frac{dR}{dq} = \frac{1}{1 + 1000q^2} \cdot \frac{d}{dq}\left(1000q^2\right) = \frac{1}{1 + 1000q^2} \cdot 2000q.$$

When $q = 10$,

$$\text{Marginal revenue} = \frac{2000 \cdot 10}{1 + 1000 \cdot 10^2} = \$0.20/\text{item}.$$

When 10 items are produced, each additional item produced gives approximately $0.20 in additional revenue.

Problems

17. Since marginal revenue is larger than marginal cost around $q = 2000$, as you produce more of the product your revenue increases faster than your costs, so profit goes up, and maximal profit will occur at a production level above 2000.

21. (a) The fixed cost is 0 because $C(0) = 0$.

(b) Profit, $\pi(q)$, is equal to money from sales, $7q$, minus total cost to produce those items, $C(q)$.

$$\pi = 7q - 0.01q^3 + 0.6q^2 - 13q$$
$$\pi' = -0.03q^2 + 1.2q - 6$$

$$\pi' = 0 \quad \text{if} \quad q = \frac{-1.2 \pm \sqrt{(1.2)^2 - 4(0.03)(6)}}{-0.06} \approx 5.9 \quad \text{or} \quad 34.1.$$

Now $\pi'' = -0.06q + 1.2$, so $\pi''(5.9) > 0$ and $\pi''(34.1) < 0$. This means $q = 5.9$ is a local min and $q = 34.1$ a local max. We now evaluate the endpoint, $\pi(0) = 0$, and the points nearest $q = 34.1$ with integer q-values:

$$\pi(35) = 7(35) - 0.01(35)^3 + 0.6(35)^2 - 13(35) = 245 - 148.75 = 96.25,$$

$$\pi(34) = 7(34) - 0.01(34)^3 + 0.6(34)^2 - 13(34) = 238 - 141.44 = 96.56.$$

So the (global) maximum profit is $\pi(34) = 96.56$. The money from sales is \$238, the cost to produce the items is \$141.44, resulting in a profit of \$96.56.

(c) The money from sales is equal to price×quantity sold. If the price is raised from \$7 by \$$x$ to \$$(7 + x)$, the result is a reduction in sales from 34 items to $(34 - 2x)$ items. So the result of raising the price by \$$x$ is to change the money from sales from $(7)(34)$ to $(7 + x)(34 - 2x)$ dollars. If the production level is fixed at 34, then the production costs are fixed at \$141.44, as found in part (b), and the profit is given by:

$$\pi(x) = (7 + x)(34 - 2x) - 141.44$$

This expression gives the profit as a function of change in price x, rather than as a function of quantity as in part (b). We set the derivative of π with respect to x equal to zero to find the change in price that maximizes the profit:

$$\frac{d\pi}{dx} = (1)(34 - 2x) + (7 + x)(-2) = 20 - 4x = 0$$

So $x = 5$, and this must give a maximum for $\pi(x)$ since the graph of π is a parabola which opens downward. The profit when the price is \$12 $(= 7 + x = 7 + 5)$ is thus $\pi(5) = (7 + 5)(34 - 2(5)) - 141.44 = \$146.56$. This is indeed higher than the profit when the price is \$7, so the smart thing to do is to raise the price by \$5.

25. (a) The value of MC is the slope of the tangent to the curve at q_0. See Figure 4.31.

(b) The line from the curve to the origin joins $(0, 0)$ and $(q_0, C(q_0))$, so its slope is $C(q_0)/q_0 = a(q_0)$.

(c) Figure 4.32 shows that the line whose slope is the minimum $a(q)$ is tangent to the curve $C(q)$. This line, therefore, also has slope MC, so $a(q) = MC$ at the q making $a(q)$ minimum.

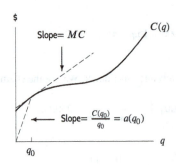

Figure 4.31

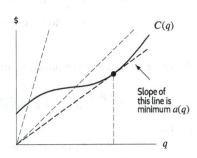

Figure 4.32

29. (a) See Figure 4.33.

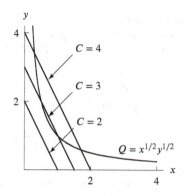

Figure 4.33

(b) Comparing the lines $C = 2$, $C = 3$, $C = 4$, we see that the cost increases as we move away from the origin. The line $C = 2$ does not cut the curve $Q = 1$; the lines $C = 3$ and $C = 4$ cut twice.

   The minimum cost occurs where a cost line is tangent to the production curve.

(c) Using implicit differentiation, the slope of $x^{1/2}y^{1/2} = 1$ is given by

$$\frac{1}{2}x^{-1/2}y^{1/2} + \frac{1}{2}x^{1/2}y^{-1/2}y' = 0$$

$$y' = \frac{-x^{-1/2}y^{1/2}}{x^{1/2}y^{-1/2}} = -\frac{y}{x}.$$

The cost lines all have slope -2. Thus, if the curve is tangent to a line, we have

$$-\frac{y}{x} = -2$$

$$y = 2x.$$

Substituting into $Q = x^{1/2}y^{1/2} = 1$ gives

$$x^{1/2}(2x)^{1/2} = 1$$

$$\sqrt{2}x = 1$$

$$x = \frac{1}{\sqrt{2}}$$

$$y = 2 \cdot \frac{1}{\sqrt{2}} = \sqrt{2}.$$

Thus the minimum cost is

$$C = 2\frac{1}{\sqrt{2}} + \sqrt{2} = 2\sqrt{2}.$$

Strengthen Your Understanding

33. We want a quantity where the slope of the cost curve, C, is greater than the slope of the revenue curve, R. For example, small quantities, such as $q = 2$, or large quantities, such as $q = 15$.

Additional Problems (online only)

37. (a) Since the company can produce more goods if it has more raw materials to use, the function $f(x)$ is increasing. Thus, we expect the derivative $f'(x)$ to be positive.

   (b) The cost to the company of acquiring x units of raw material is wx, and the revenue from the sale of $f(x)$ units of the product is $pf(x)$. The company's profit is $\pi(x) = \text{Revenue} - \text{Cost} = pf(x) - wx$.

   (c) Since profit $\pi(x)$ is maximized at $x = x^*$, we have $\pi'(x^*) = 0$. From $\pi'(x) = pf'(x) - w$, we have $pf'(x^*) - w = 0$. Thus $f'(x^*) = w/p$.

(d) Computing the second derivative of $\pi(x)$ gives $\pi''(x) = pf''(x)$. Since $\pi(x)$ has a maximum at $x = x^*$, the second derivative $\pi''(x^*) = pf''(x^*)$ is negative. Thus $f''(x^*)$ is negative.

(e) Differentiate both sides of $pf'(x^*) - w = 0$ with respect to w. The chain rule gives

$$p\frac{d}{dw}f'(x^*) - 1 = 0$$

$$pf''(x^*)\frac{dx^*}{dw} - 1 = 0$$

$$\frac{dx^*}{dw} = \frac{1}{pf''(x^*)}.$$

Since $f''(x^*) < 0$, we see dx^*/dw is negative.

(f) Since $dx^*/dw < 0$, the quantity x^* is a decreasing function of w. If the price w of the raw material goes up, the company should buy less.

Solutions for Section 4.6

Exercises

1. The rate of change of temperature is

$$\frac{dH}{dt} = 16(-0.02)e^{-0.02t} = -0.32e^{-0.02t}.$$

When $t = 0$,

$$\frac{dH}{dt} = -0.32e^0 = -0.32°C/\text{min}.$$

When $t = 10$,

$$\frac{dH}{dt} = -0.32e^{-0.02(10)} = -0.262°C/\text{min}.$$

5. (a) The rate of change of the period is given by

$$\frac{dT}{dl} = \frac{2\pi}{\sqrt{9.8}}\frac{d}{dl}(\sqrt{l}) = \frac{2\pi}{\sqrt{9.8}} \cdot \frac{1}{2}l^{-1/2} = \frac{\pi}{\sqrt{9.8}} \cdot \frac{1}{\sqrt{l}} = \frac{\pi}{\sqrt{9.8l}}.$$

(b) The rate decreases since $\sqrt{l}$ is in the denominator.

9. (a) Since $P = 25$, we have

$$25 = 30e^{-3.23\times10^{-5}h}$$

$$h = \frac{\ln(25/30)}{-3.23 \times 10^{-5}} = 5644 \text{ ft}.$$

(b) Both P and h are changing over time, and we differentiate with respect to time t, in minutes:

$$\frac{dP}{dt} = 30e^{-3.23\times10^{-5}h}\left(-3.23 \times 10^{-5}\frac{dh}{dt}\right).$$

We see in part (a) that $h = 5644$ ft when $P = 25$ inches of mercury, and we know that $dP/dt = 0.1$. Substituting gives:

$$0.1 = 30e^{-3.23\times10^{-5}\times5644.630}\left(-3.23 \times 10^{-5}\frac{dh}{dt}\right)$$

$$\frac{dh}{dt} = -124 \text{ ft/minute}.$$

The glider's altitude is decreasing at a rate of about 124 feet per minute.

13. Differentiating with respect to time t, in seconds, we have

$$\frac{d\phi}{dt} = 2\pi \left(\frac{1}{2}(x^2 + 4)^{-1/2} \left(2x\frac{dx}{dt} \right) - \frac{dx}{dt} \right).$$

Since the point is moving to the left, $dx/dt = -0.2$. Substituting $x = 3$ and $dx/dt = -0.2$, we have

$$\frac{d\phi}{dt} = 2\pi \left(\frac{1}{2}(3^2 + 4)^{-1/2}(2 \cdot 3(-0.2)) - (-0.2) \right) = 0.211.$$

The potential is changing at a rate of 0.211 units per second.

Problems

17. We begin by finding the rate of change in altitude 1 hour after takeoff. Since from about half an hour after takeoff until about two and one-half hours after takeoff the altitude is constant, we see there is no change in altitude, which means there is no change in outside air pressure.

21. Let the other side of the rectangle be x cm. Then the area is $A = 10x$, so differentiating with respect to time gives

$$\frac{dA}{dt} = 10\frac{dx}{dt}.$$

We are interested in the instant when $x = 12$ and $dx/dt = 3$, giving

$$\frac{dA}{dt} = 10\frac{dx}{dt} = 10 \cdot 3 = 30 \text{ cm}^2 \text{ per minute.}$$

25. Let x represent the distance from car A to the intersection, and y represent the distance from car B to the intersection, and z represent the distance between the two cars, as in Figure 4.34.

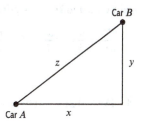

Figure 4.34

Note that all three quantities are changing over time, and they are related by the Pythagorean Theorem:

$$x^2 + y^2 = z^2.$$

Differentiating with respect to time, we have:

$$2x \cdot \frac{dx}{dt} + 2y \cdot \frac{dy}{dt} = 2z \cdot \frac{dz}{dt}.$$

We are asked to find dz/dt. Car A is moving toward the intersection at 50 mph, so $dx/dt = -50$. Car B is moving away from the intersection at 60 mph, so $dy/dt = 60$. At the point when $x = 40$ and $y = 30$, we have $z = 50$. We substitute these values and solve for dz/dt.

$$2x \cdot \frac{dx}{dt} + 2y \cdot \frac{dy}{dt} = 2z \cdot \frac{dz}{dt}$$
$$2(40)(-50) + 2(30)(60) = 2(50) \cdot \frac{dz}{dt}$$
$$-400 = 100 \cdot \frac{dz}{dt}$$
$$\frac{dz}{dt} = -4 \text{ mph.}$$

At this instant, the cars are moving closer together at a rate of 4 mph.

29. (a) (i) Differentiating thinking of r as a constant gives

$$\frac{dP}{dt} = 500e^{rt/100} \cdot \frac{r}{100} = 5re^{rt/100}.$$

Substituting $t = 0$ gives

$$\frac{dP}{dt} = 5re^{r \cdot 0/100} = 5r \text{ dollars/yr.}$$

(ii) Substituting $t = 2$ gives

$$\frac{dP}{dt} = 5re^{r \cdot 2/100} = 5re^{0.02r} \text{ dollars/yr.}$$

(b) To differentiate thinking of r as variable, think of the function as

$$P = 500e^{r(t) \cdot t/100},$$

and use the chain rule (for $e^{rt/100}$) and the product rule (for $r(t) \cdot t$):

$$\frac{dP}{dt} = 500e^{r(t) \cdot t/100} \cdot \frac{1}{100} \cdot \frac{d}{dt}(r(t) \cdot t) = 5e^{r(t) \cdot t/100}\left(r(t) \cdot 1 + r'(t) \cdot t\right).$$

Substituting $t = 2$, $r = 4$, and $r'(2) = 0.3$ gives

$$\frac{dP}{dt} = 5e^{4 \cdot 2/100}(4 + 0.3 \cdot 2) = 24.916 \text{ dollars/year}$$

Thus, the price is increasing by about \$25 per year at that time.

33. When the radius is r, the volume V of the snowball is

$$V = \frac{4}{3}\pi r^3.$$

We know that $dr/dt = -0.2$ when $r = 15$ and we want to know dV/dt at that time. Differentiating, we have

$$\frac{dV}{dt} = \frac{4}{3}\pi 3r^2 \frac{dr}{dt} = 4\pi r^2 \frac{dr}{dt}.$$

Substituting $dr/dt = -0.2$ gives

$$\left.\frac{dV}{dt}\right|_{r=15} = 4\pi(15)^2(-0.2) = -180\pi = -565.487 \text{ cm}^3/\text{hr.}$$

Thus, the volume is decreasing at 565.487 cm^3 per hour.

37. The rate of change of velocity is given by

$$\frac{dv}{dt} = -\frac{mg}{k}\left(-\frac{k}{m}e^{-kt/m}\right) = ge^{-kt/m}.$$

When $t = 0$,

$$\left.\frac{dv}{dt}\right|_{t=0} = g.$$

When $t = 1$,

$$\left.\frac{dv}{dt}\right|_{t=1} = ge^{-k/m}.$$

These answers give the acceleration at $t = 0$ and $t = 1$. The acceleration at $t = 0$ is g, the acceleration due to gravity, and at $t = 1$, the acceleration is $ge^{-k/m}$, a smaller value.

41. (a) Since the slick is circular, if its radius is r meters, its area, A, is $A = \pi r^2$. Differentiating with respect to time using the chain rule gives

$$\frac{dA}{dt} = 2\pi r \frac{dr}{dt}.$$

We know $dr/dt = 0.1$ when $r = 150$, so

$$\frac{dA}{dt} = 2\pi 150(0.1) = 30\pi = 94.248 \text{ m}^2/\text{min.}$$

(b) If the thickness of the slick is h, its volume, V, is given by

$$V = Ah.$$

Differentiating with respect to time using the product rule gives

$$\frac{dV}{dt} = \frac{dA}{dt}h + A\frac{dh}{dt}.$$

We know $h = 0.02$ and $A = \pi(150)^2$ and $dA/dt = 30\pi$. Since V is fixed, $dV/dt = 0$, so

$$0 = 0.02(30\pi) + \pi(150)^2\frac{dh}{dt}.$$

Thus

$$\frac{dh}{dt} = -\frac{0.02(30\pi)}{\pi(150)^2} = -0.0000267 \text{ m/min},$$

so the thickness is decreasing at 0.0000267 meters per minute.

45. Since the slant side of the cone makes an angle of $45°$ with the vertical, we have $r = h$, so the volume of the cone is

$$V = \frac{1}{3}\pi h^3.$$

Differentiating with respect to time gives

$$\frac{dV}{dt} = \pi h^2\frac{dh}{dt}.$$

We know $dV/dt = 2$ cubic meters/min, and we want to know dh/dt when $h = 0.5$ meters. Substituting gives

$$2 = \pi(0.5)^2\frac{dh}{dt},$$

$$\frac{dh}{dt} = \frac{2}{\pi(0.5)^2} = \frac{8}{\pi} \text{ meters/min.}$$

49. The rate of change of temperature with distance, dH/dy, at altitude 4000 ft approximated by

$$\frac{dH}{dy} \approx \frac{\Delta H}{\Delta y} = \frac{38 - 52}{6 - 4} = -7°\text{F/thousand ft.}$$

A speed of 3000 ft/min tells us $dy/dt = 3000$, so

$$\text{Rate of change of temperature with time} = \frac{dH}{dy} \cdot \frac{dy}{dt} \approx -7\frac{°\text{F}}{\text{thousand ft}} \cdot \frac{3 \text{ thousand ft}}{\text{min}} = -21°\text{F/min.}$$

Other estimates can be obtained by estimating the derivative as

$$\frac{dH}{dy} \approx \frac{\Delta H}{\Delta y} = \frac{52 - 60}{4 - 2} = -4°\text{F/thousand ft}$$

or by averaging the two estimates

$$\frac{dH}{dy} \approx \frac{-7 - 4}{2} = -5.5°\text{F/thousand ft.}$$

If the rate of change of temperature with distance is $-4°$/thousand ft, then

$$\text{Rate of change of temperature with time} = \frac{dH}{dy} \cdot \frac{dy}{dt} \approx -4\frac{°\text{F}}{\text{thousand ft}} \cdot \frac{3 \text{ thousand ft}}{\text{min}} = -12°\text{F/min.}$$

Thus, estimates for the rate at which temperature was decreasing range from 12°F/min to 21°F/min.

53. The radius r is related to the volume by the formula $V = \frac{4}{3}\pi r^3$. By implicit differentiation, we have

$$\frac{dV}{dt} = \frac{4}{3}\pi 3r^2\frac{dr}{dt} = 4\pi r^2\frac{dr}{dt}.$$

The surface area of a sphere is $4\pi r^2$, so we have

$$\frac{dV}{dt} = s \cdot \frac{dr}{dt},$$

but since $\frac{dV}{dt} = \frac{1}{3}s$ was given, we have

$$\frac{dr}{dt} = \frac{1}{3}.$$

Strengthen Your Understanding

57. The radius of a circle and the circle's diameter are related by the linear function $D = 2R$. Thus, $\dfrac{dD}{dt} = \dfrac{dD}{dR}\dfrac{dR}{dt} = 2\dfrac{dR}{dt}$. The rate of change of the diameter is twice the rate of change of the radius. For example, if the radius is increasing at a constant rate of 5 feet per second, then the circle's diameter is increasing at a constant rate of 10 feet per second.

61. True. The circumference C and radius r are related by $C = 2\pi r$, so $dC/dt = 2\pi dr/dt$. Thus if dr/dt is constant, so is dC/dt.

Additional Problems (online only)

65. The volume of a cube is $V = x^3$. So

$$\frac{dV}{dt} = 3x^2\frac{dx}{dt},$$

and

$$\frac{1}{V}\frac{dV}{dt} = \frac{3}{x}\frac{dx}{dt}.$$

The surface area of a cube is $A = 6x^2$. So

$$\frac{dA}{dt} = 12x\frac{dx}{dt},$$

and

$$\frac{1}{A}\frac{dA}{dt} = \frac{2}{x}\frac{dx}{dt}.$$

Thus the percentage rate of change of the volume of the cube, $\dfrac{1}{V}\dfrac{dV}{dt}$, is larger.

Solutions for Section 4.7

Exercises

1. Yes, l'Hopital's rule applies in this case, since the limit is in the form $0/0$.

5. No, l'Hopital's rule does not apply in this case, since the limit is not in the form $0/0$ or ∞/∞.

9. Yes, l'Hopital's rule applies in this case, since the limit is in the form $0/0$.

13. Since $\lim_{x\to 2}(x - 2) = \lim_{x\to 2}(x^2 - 4) = 0$, this is a $0/0$ form, and l'Hopital's rule applies:

$$\lim_{x\to 2}\frac{x - 2}{x^2 - 4} = \lim_{x\to 2}\frac{1}{2x} = \frac{1}{4}.$$

17. Since the limit is not in the form $0/0$ or ∞/∞, l'Hopital's rule does not apply in this case. We have

$$\lim_{x\to 0}\frac{\sin x}{e^x} = \frac{0}{1} = 0.$$

21. Since $\lim_{x\to a}(\sqrt[3]{x} - \sqrt[3]{a}) = \lim_{x\to a}(x - a) = 0$, this is a $0/0$ form, and l'Hopital's rule applies:

$$\lim_{x\to a}\frac{\sqrt[3]{x} - \sqrt[3]{a}}{x - a} = \lim_{x\to a}\frac{(1/3)x^{-2/3}}{1} = \frac{1}{3}a^{-2/3}.$$

25. Since $\lim_{x\to 1}(2x^5 - 2) = \lim_{x\to 1}(3x^4 - 3x) = 0$, this is a $0/0$ form, and l'Hopital's rule applies:

$$\lim_{x\to 1}\frac{2x^5 - 2}{3x^4 - 3x} = \lim_{x\to 1}\frac{10x^4}{12x^3 - 3} = \frac{10}{9}.$$

29. Since $\lim_{x \to 0}(\sin(5x)) = \lim_{x \to 0}(3x^2) = 0$, this is a 0/0 form, and l'Hopital's rule applies:

$$\lim_{x \to 0} \frac{\sin(5x)}{3x^2} = \lim_{x \to 0} \frac{5\cos(5x)}{6x} = \frac{5}{0}.$$

We see that the limit does not exist.

33. Since the limit is not in the form 0/0 or ∞/∞, l'Hopital's rule does not apply in this case. We have:

$$\lim_{x \to 1} \frac{3x^2 + 4}{x^2 + 3x + 5} = \frac{7}{9}.$$

37. Since $\lim_{x \to \infty}(5x + 1) = \lim_{x \to \infty}(e^x) = \infty$, this is an ∞/∞ form, and l'Hopital's rule applies:

$$\lim_{x \to \infty} \frac{5x + 1}{e^x} = \lim_{x \to \infty} \frac{5}{e^x} = 0.$$

41. The power function dominates. Using l'Hopital's rule

$$\lim_{x \to \infty} \frac{\ln(x + 3)}{x^{0.2}} = \lim_{x \to \infty} \frac{\frac{1}{(x+3)}}{0.2x^{-0.8}} = \lim_{x \to \infty} \frac{x^{0.8}}{0.2(x + 3)}.$$

Using l'Hopital's rule again gives

$$\lim_{x \to \infty} \frac{x^{0.8}}{0.2(x + 3)} = \lim_{x \to \infty} \frac{0.8x^{-0.2}}{0.2} = 0,$$

so $x^{0.2}$ dominates.

Problems

45. Since $f'(a) < 0$ and $g'(a) < 0$, l'Hopital's rule tells us that

$$\lim_{x \to a} \frac{f(x)}{g(x)} = \frac{f'(a)}{g'(a)} > 0.$$

49. The numerator goes to zero faster than the denominator, so you should expect the limit to be zero. Using l'Hopital's rule, we have

$$\lim_{x \to 0} \frac{\sin^2 x}{x} = \lim_{x \to 0} \frac{2\sin x \cos x}{1} = 0.$$

53. We have $\lim_{x \to 1} x = 1$ and $\lim_{x \to 1}(x - 1) = 0$, so l'Hopital's rule does not apply.

57. This is an ∞^0 form. With $y = \lim_{x \to \infty}(1 + x)^{1/x}$, we take logarithms to get

$$\ln y = \lim_{x \to \infty} \frac{1}{x} \ln(1 + x).$$

This limit is a $0 \cdot \infty$ form,

$$\lim_{x \to \infty} \frac{1}{x} \ln(1 + x),$$

which can be rewritten as the ∞/∞ form

$$\lim_{x \to \infty} \frac{\ln(1 + x)}{x},$$

to which l'Hopital's rule applies.

61. Let $f(x) = \ln x$ and $g(x) = 1/x$ so $f'(x) = 1/x$ and $g'(x) = -1/x^2$ and

$$\lim_{x \to 0^+} \frac{\ln x}{1/x} = \lim_{x \to 0^+} \frac{1/x}{-1/x^2} = \lim_{x \to 0^+} \frac{x}{-1} = 0.$$

65. To get this expression in a form in which l'Hopital's rule applies, we combine the fractions:

$$\frac{1}{x} - \frac{1}{\sin x} = \frac{\sin x - x}{x \sin x}.$$

Letting $f(x) = \sin x - x$ and $g(x) = x \sin x$, we have $f(0) = 0$ and $g(0) = 0$ so l'Hopital's rule can be used. Differentiating gives $f'(x) = \cos x - 1$ and $g'(x) = x \cos x + \sin x$, so $f'(0) = 0$ and $g'(0) = 0$, so $f'(0)/g'(0)$ is undefined. Therefore, to apply l'Hopital's rule we differentiate again to obtain $f''(x) = -\sin x$ and $g''(x) = 2 \cos x - x \sin x$, for which $f''(0) = 0$ and $g''(0) = 2 \neq 0$. Then

$$\lim_{x \to 0} \left(\frac{1}{x} - \frac{1}{\sin x} \right) = \lim_{x \to 0} \left(\frac{\sin x - x}{x \sin x} \right)$$
$$= \lim_{x \to 0} \left(\frac{\cos x - 1}{x \cos x + \sin x} \right)$$
$$= \lim_{x \to 0} \left(\frac{-\sin x}{2 \cos x - x \sin x} \right)$$
$$= \frac{0}{2} = 0.$$

69. Since $\lim_{t \to 0} \sin^2 At = 0$ and $\lim_{t \to 0} \cos At - 1 = 1 - 1 = 0$, this is a $0/0$ form. Applying l'Hopital's rule we get

$$\lim_{t \to 0} \frac{\sin^2 At}{\cos At - 1} = \lim_{t \to 0} \frac{2A \sin At \cos At}{-A \sin At} = \lim_{t \to 0} -2 \cos At = -2.$$

73. Let $f(x) = \cos x$ and $g(x) = x$. Observe that since $f(0) = 1$, l'Hopital's rule does not apply. But since $g(0) = 0$,

$$\lim_{x \to 0} \frac{\cos x}{x} \quad \text{does not exist.}$$

77. Let $n = 1/(kx)$, so $n \to \infty$ as $x \to 0^+$. Thus

$$\lim_{x \to 0^+} (1 + kx)^{t/x} = \lim_{n \to \infty} \left(1 + \frac{1}{n} \right)^{nkt} = \lim_{n \to \infty} \left(\left(1 + \frac{1}{n} \right)^n \right)^{kt} = e^{kt}.$$

81. This limit is of the form ∞^0 so we apply l'Hopital's rule to

$$\ln f(t) = \frac{\ln \left((3^t + 5^t)/2 \right)}{t}.$$

We have

$$\lim_{t \to +\infty} \ln f(t) = \lim_{t \to +\infty} \frac{\left((\ln 3)3^t + (\ln 5)5^t \right) / \left(3^t + 5^t \right)}{1}$$
$$= \lim_{t \to +\infty} \frac{(\ln 3)3^t + (\ln 5)5^t}{3^t + 5^t}$$
$$= \lim_{t \to +\infty} \frac{(\ln 3)(3/5)^t + \ln 5}{(3/5)^t + 1}$$
$$= \lim_{t \to +\infty} \frac{0 + \ln 5}{0 + 1} = \ln 5.$$

Thus

$$\lim_{t \to -\infty} f(t) = \lim_{t \to -\infty} e^{\ln f(t)} = e^{\lim_{t \to -\infty} \ln f(t)} = e^{\ln 5} = 5.$$

85. To evaluate, we use l'Hopital's Rule:

$$\lim_{x \to 0} \frac{1 - \cosh 5x}{x^2} = \lim_{x \to 0} \frac{-5 \sinh 5x}{2x} = \lim_{x \to 0} \frac{-25 \cosh 5x}{2} = -25/2.$$

89. Since the limit is of the form $0/0$, we can apply l'Hopital's rule. First note that

$$\frac{dx^x}{dx} = \frac{de^{x\ln x}}{dx} = e^{x\ln x}\left(x \cdot \frac{1}{x} + 1 \cdot \ln x\right) = x^x + x^x \ln x.$$

Applying l'Hopital's rule twice we have

$$\lim_{x\to 1}\frac{x^x - x}{1 - x + \ln x} = \lim_{x\to 1}\frac{x^x + x^x \ln x - 1}{-1 + 1/x}$$

$$= \lim_{x\to 1}\frac{x^x + x^x \ln x + (x^x + x^x \ln x)\ln x + x^x(1/x)}{-1/x^2} = \frac{1 + 0 + 0 + 1}{-1} = -2.$$

Strengthen Your Understanding

93. Since $\lim_{x\to\infty} \ln x = \infty$, if $\lim_{x\to\infty} f(x) = \pm\infty$ we could apply L'Hopital's rule to try to find $\lim_{x\to\infty} \frac{f(x)}{\ln x}$. One possible example is $f(x) = x$.

Solutions for Section 4.8

Exercises

1. Between times $t = 0$ and $t = 1$, x goes at a constant rate from 0 to 1 and y goes at a constant rate from 1 to 0. So the particle moves in a straight line from $(0, 1)$ to $(1, 0)$. Similarly, between times $t = 1$ and $t = 2$, it goes in a straight line to $(0, -1)$, then to $(-1, 0)$, then back to $(0, 1)$. So it traces out the diamond shown in Figure 4.35.

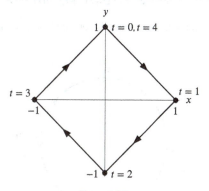

Figure 4.35

5. One possible answer is $x = 3\cos t, y = -3\sin t, 0 \le t \le 2\pi$.

9. The slope of the line is

$$m = \frac{3 - (-1)}{1 - 2} = -4.$$

The equation of the line with slope -4 through the point $(2, -1)$ is $y - (-1) = (-4)(x - 2)$, so one possible parameterization is $x = t$ and $y = -4t + 8 - 1 = -4t + 7$.

13. The particle moves clockwise: For $0 \le t \le \frac{\pi}{2}$, we have $x = \cos t$ decreasing and $y = -\sin t$ decreasing. Similarly, for the time intervals $\frac{\pi}{2} \le t \le \pi, \pi \le t \le \frac{3\pi}{2}$, and $\frac{3\pi}{2} \le t \le 2\pi$, we see that the particle moves clockwise.

17. Let $f(t) = \cos t$. Then $f'(t) = -\sin t$. The particle is moving clockwise when $f'(t) < 0$, or $-\sin t < 0$, that is, when

$$2k\pi < t < (2k + 1)\pi,$$

where k is an integer. The particle is otherwise moving counterclockwise, that is, when

$$(2k - 1)\pi < t < 2k\pi,$$

where k is an integer. Actually, the particle does not fully trace out a circle. The range of $f(t)$ is $[-1, 1]$ so the particle oscillates between the points $(\cos(-1), \sin(-1))$ and $(\cos 1, \sin 1)$.

21. We see from the parametric equations that the particle moves along a line. It suffices to plot two points: at $t = 0$, the particle is at point $(1, -4)$, and at $t = 1$, the particle is at point $(4, -3)$. Since x increases as t increases, the motion is left to right on the line as shown in Figure 4.36.

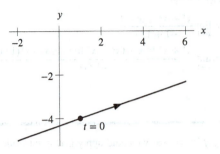

Figure 4.36

Alternately, we can solve the first equation for t, giving $t = (x - 1)/3$, and substitute this into the second equation to get

$$y = \frac{x - 1}{3} - 4 = \frac{1}{3}x - \frac{13}{3}.$$

The line is $y = \frac{1}{3}x - \frac{13}{3}$.

25. The graph is a circle centered at the origin with radius 3. The equation is

$$x^2 + y^2 = (3\cos t)^2 + (3\sin t)^2 = 9.$$

The particle is at the point $(3, 0)$ when $t = 0$, and motion is counterclockwise. See Figure 4.37.

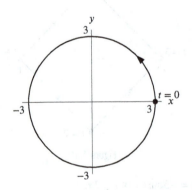

Figure 4.37

29. We have

$$\frac{dy}{dx} = \frac{dy/dt}{dx/dt} = \frac{4\cos(4t)}{3\cos(3t)}.$$

Thus when $t = \pi$, the slope of the tangent line is $-4/3$. Since $x = 0$ and $y = 0$ when $t = \pi$, the equation of the tangent line is $y = -(4/3)x$.

33. We have

$$\frac{dx}{dt} = 2t - 4, \quad \frac{dy}{dt} = 3t^2 - 12.$$

The speed is given by:

$$\text{Speed} = \sqrt{(2t - 4)^2 + (3t^2 - 12)^2}.$$

The particle stops when $2t - 4 = 0$ and $3t^2 - 12 = 0$. Since these are both satisfied only by $t = 2$, this is the only time that the particle stops.

Problems

37. (a) We get the part of the line with $x < 10$ and $y < 0$.

(b) We get the part of the line between the points $(10, 0)$ and $(11, 2)$.

41. (a) C_1 has center at the origin and radius 5, so $a = b = 0, k = 5$ or -5.

(b) C_2 has center at $(0, 5)$ and radius 5, so $a = 0, b = 5, k = 5$ or -5.

(c) C_3 has center at $(10, -10)$, so $a = 10, b = -10$. The radius of C_3 is $\sqrt{10^2 + (-10)^2} = \sqrt{200}$, so $k = \sqrt{200}$ or $k = -\sqrt{200}$.

45. (a) The chain rule gives

$$\frac{dy}{dx} = \frac{dy/dt}{dx/dt} = \frac{4e^{2t}}{e^t} = 4e^t.$$

(b) We are given $y = 2e^{2t}$ so $y = 2(e^t)^2$. Since $x = e^t$, we can substitute x for e^t. Thus $y = 2x^2$.

(c) Differentiating $y = 2x^2$ with respect to x, we get $dy/dx = 4x$. Notice that, since $x = e^t$, this is equivalent to the answer that we obtained in part (a).

49. (a) In order for the particle to stop, its velocity both dx/dt and dy/dt must be zero,

$$\frac{dx}{dt} = 3t^2 - 3 = 3(t - 1)(t + 1) = 0,$$

$$\frac{dy}{dt} = 2t - 2 = 2(t - 1) = 0.$$

The value $t = 1$ is the only solution. Therefore, the particle stops when $t = 1$ at the point $(t^3 - 3t, \ t^2 - 2t)|_{t=1} = (-2, -1)$.

(b) In order for the particle to be traveling straight up or down, the velocity in the x-direction must be 0. Thus, we solve $dx/dt = 3t^2 - 3 = 0$ and obtain $t = \pm 1$. However, at $t = 1$ the particle has no vertical motion, as we saw in part (a). Thus, the particle is moving straight up or down only when $t = -1$. The position at that time is $(t^3 - 3t, \ t^2 - 2t)|_{t=-1} = (2, 3)$.

(c) For horizontal motion we need $dy/dt = 0$. That happens when $dy/dt = 2t - 2 = 0$, and so $t = 1$. But from part (a) we also have $dx/dt = 0$ also at $t = 1$, so the particle is not moving at all when $t = 1$. Thus, there is no time when the motion is horizontal.

53. (a) To determine if the particles collide, we check whether they are ever at the same point at the same time. We first set the two x-coordinates equal to each other:

$$4t - 4 = 3t$$

$$t = 4.$$

When $t = 4$, both x-coordinates are 12. Now we check whether the y-coordinates are also equal at $t = 4$:

$$y_A(4) = 2 \cdot 4 - 5 = 3$$

$$y_B(4) = 4^2 - 2 \cdot 4 - 1 = 7.$$

Thus, the particles do not collide since they are not at the same point at the same time.

(b) For the particles to collide, we need both x- and y-coordinates to be equal. Since the x-coordinates are equal at $t = 4$, we find the k value making $y_A(4) = y_B(4)$.

Substituting $t = 4$ into $y_A(t) = 2t - k$ and $y_B(t) = t^2 - 2t - 1$, we have

$$8 - k = 16 - 8 - 1$$

$$k = 1.$$

(c) To find the speed of the particles, we differentiate.

For particle A,

$x(t) = 4t - 4$, so $x'(t) = 4$, and $x'(4) = 4$

$y(t) = 2t - 1$, so $y'(t) = 2$, and $y'(4) = 2$

$$\text{Speed}_A = \sqrt{(x'(t))^2 + (y'(t))^2} = \sqrt{4^2 + 2^2} = \sqrt{20}.$$

For particle B,

$x(t) = 3t$, so $x'(t) = 3$, and $x'(4) = 3$

$y(t) = t^2 - 2t - 1$, so $y'(t) = 2t - 2$, and $y'(4) = 6$

$$\text{Speed}_B = \sqrt{(x'(t))^2 + (y'(t))^2} = \sqrt{3^2 + 6^2} = \sqrt{45}.$$

Thus, when $t = 4$, particle B is moving faster.

57. (a) The particle touches the x-axis when $y = 0$. Since $y = \cos(2t) = 0$ for the first time when $2t = \pi/2$, we have $t = \pi/4$. To find the speed of the particle at that time, we use the formula

$$\text{Speed} = \sqrt{\left(\frac{dx}{dt}\right)^2 + \left(\frac{dy}{dt}\right)^2} = \sqrt{(\cos t)^2 + (-2\sin(2t))^2}.$$

When $t = \pi/4$,

$$\text{Speed} = \sqrt{(\cos(\pi/4))^2 + (-2\sin(\pi/2))^2} = \sqrt{(\sqrt{2}/2)^2 + (-2 \cdot 1)^2} = \sqrt{9/2}.$$

(b) The particle is at rest when its speed is zero. Since $\sqrt{(\cos t)^2 + (-2\sin(2t))^2} \geq 0$, the speed is zero when

$$\cos t = 0 \quad \text{and} \quad -2\sin(2t) = 0.$$

Now $\cos t = 0$ when $t = \pi/2$ or $t = 3\pi/2$. Since $-2\sin(2t) = -4\sin t \cos t$, we see that this expression also equals zero when $t = \pi/2$ or $t = 3\pi/2$.

(c) We need to find d^2y/dx^2. First, we must determine dy/dx. We know

$$\frac{dy}{dx} = \frac{dy/dt}{dx/dt} = \frac{-2\sin 2t}{\cos t} = \frac{-4\sin t \cos t}{\cos t} = -4\sin t.$$

Since $dy/dx = -4\sin t$, we can now use the formula:

$$\frac{d^2y}{dx^2} = \frac{dw/dt}{dx/dt} \quad \text{where} \quad w = \frac{dy}{dx}$$

$$\frac{d^2y}{dx^2} = \frac{-4\cos t}{\cos t} = -4.$$

Since d^2y/dx^2 is always negative, our graph is concave down everywhere.

Using the identity $y = \cos(2t) = 1 - 2\sin^2 t$, we can eliminate the parameter and write the original equation as $y = 1 - 2x^2$, which is a parabola that is concave down everywhere.

Strengthen Your Understanding

61. One possible choice is $x = 2\cos t$, $y = 2\sin t$, $0 \leq t \leq \frac{\pi}{2}$. There are many other possibilities.

Additional Problems (online only)

65. Let

$$w = \frac{dy}{dx} = \frac{dy/dt}{dx/dt}.$$

We want to find

$$\frac{d^2y}{dx^2} = \frac{dw}{dx} = \frac{dw/dt}{dx}dt.$$

To find dw/dt, we use the quotient rule:

$$\frac{dw}{dt} = \frac{(dx/dt)(d^2y/dt^2) - (dy/dt)(d^2x/dt^2)}{(dx/dt)^2}.$$

We then divide this by dx/dt again to get the required formula, since

$$\frac{d^2y}{dx^2} = \frac{dw}{dx} = \frac{dw/dt}{dx/dt}.$$

69. This curve never closes on itself. The plot for $0 \le t \le 8\pi$ is in Figure 4.38.

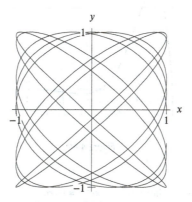

Figure 4.38

Solutions for Chapter 4 Review

Exercises

1. See Figure 4.39.

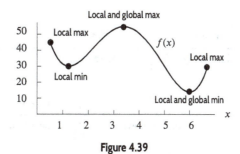

Figure 4.39

5. (a) First we find f' and f'':

$$f'(x) = -e^{-x}\sin x + e^{-x}\cos x$$
$$f''(x) = e^{-x}\sin x - e^{-x}\cos x$$
$$\qquad\qquad -e^{-x}\cos x - e^{-x}\sin x$$
$$= -2e^{-x}\cos x$$

(b) The critical points are $x = \pi/4, 5\pi/4$, since $f'(x) = 0$ here.

(c) The inflection points are $x = \pi/2, 3\pi/2$, since f'' changes sign at these points.

(d) At the endpoints, $f(0) = 0$, $f(2\pi) = 0$. So we have $f(\pi/4) = (e^{-\pi/4})(\sqrt{2}/2)$ as the global maximum; $f(5\pi/4) = -e^{-5\pi/4}(\sqrt{2}/2)$ as the global minimum.

(e) See Figure 4.40.

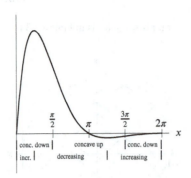

Figure 4.40

9. As $x \to -\infty$, $e^{-x} \to \infty$, so $xe^{-x} \to -\infty$. Thus $\lim_{x \to -\infty} xe^{-x} = -\infty$.

As $x \to \infty$, $\frac{x}{e^x} \to 0$, since e^x grows much more quickly than x. Thus $\lim_{x \to \infty} xe^{-x} = 0$.

Using the product rule,

$$f'(x) = e^{-x} - xe^{-x} = (1 - x)e^{-x},$$

which is zero when $x = 1$, negative when $x > 1$, and positive when $x < 1$. Thus $f(1) = 1/e^1 = 1/e$ is a local maximum.

Again, using the product rule,

$$f''(x) = -e^{-x} - e^{-x} + xe^{-x}$$
$$= xe^{-x} - 2e^{-x}$$
$$= (x - 2)e^{-x},$$

which is zero when $x = 2$, positive when $x > 2$, and negative when $x < 2$, giving an inflection point at $(2, \frac{2}{e^2})$. With the above, we have the following diagram:

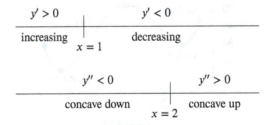

The graph of f is shown in Figure 4.41.

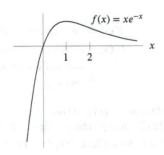

Figure 4.41

and $f(x)$ has one global maximum at $1/e$ and no local or global minima.

13. Since $f(x) = e^x + \cos x$ is continuous on the closed interval $0 \leq x \leq \pi$, there must be a global maximum and minimum. The possible candidates are critical points in the interval and endpoints. Since there are no points in the interval where $f'(x)$ is undefined, we solve $f'(x) = 0$ to find the critical points:

$$f'(x) = e^x - \sin x = 0.$$

Since $e^x > 1$ for all $x > 0$ and $\sin x \leq 1$ for all x, the only possibility is $x = 0$, but $e^0 - \sin 0 = 1$. Thus there are no critical points in the interval. We then compare the values of f at the endpoints:

$$f(0) = 2, \quad f(\pi) = e^\pi + \cos(\pi) = 22.141.$$

Thus, the global maximum is 22.141 at $x = \pi$, and the global minimum is 2 at $x = 0$.

17. Since $g(t)$ is always decreasing for $t \geq 0$, we expect it to a global maximum at $t = 0$ but no global minimum. At $t = 0$, we have $g(0) = 1$, and as $t \to \infty$, we have $g(t) \to 0$.

Alternatively, rewriting as $g(t) = (t^3 + 1)^{-1}$ and differentiating using the chain rule gives

$$g'(t) = -(t^3 + 1)^{-2} \cdot 3t^2.$$

Since $3t^2 = 0$ when $t = 0$, there is a critical point at $t = 0$, and g decreases for all $t > 0$. See Figure 4.42.

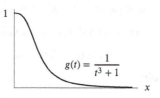

Figure 4.42

21. $\lim_{x \to +\infty} f(x) = +\infty$, and $\lim_{x \to 0^+} f(x) = +\infty$.

Hence, $x = 0$ is a vertical asymptote.

$f'(x) = 1 - \dfrac{2}{x} = \dfrac{x - 2}{x}$, so $x = 2$ is the only critical point.

$f''(x) = \dfrac{2}{x^2}$, which can never be zero. So there are no inflection points.

x		2	
f'	$-$	0	$+$
f''	$+$	$+$	$+$
f	$\searrow$		$\nearrow$

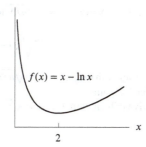

Thus, $f(2)$ is a local and global minimum.

25. For $0 \leq r \leq a$, the speed at which air is expelled is given by

$$v(r) = k(a - r)r^2 = kar^2 - kr^3.$$

Thus, the derivative is defined for all r and given by

$$v'(r) = 2kar - 3kr^2 = 2kr\left(a - \frac{3}{2}r\right).$$

The derivative is zero if $r = \frac{2}{3}a$ or $r = 0$. These are the critical points of v. To decide if the critical points give global maxima or minima, we evaluate v at the critical point:

$$\begin{aligned} v\left(\frac{2}{3}a\right) &= k\left(a - \frac{2}{3}a\right)\left(\frac{2}{3}a\right)^2 \\ &= k\left(\frac{a}{3}\right)\frac{4a^2}{9} \\ &= \frac{4ka^3}{27}, \end{aligned}$$

and we evaluate v at the endpoints:

$$v(0) = v(a) = 0.$$

Thus, v has a global maximum at $r = \frac{2}{3}a$. The global minimum of $v = 0$ occurs at both endpoints $r = 0$ and $r = a$.

29. (a) We set the derivative equal to zero and solve for x to find critical points:

$$\begin{aligned} f'(x) = 4x^3 - 4ax &= 0 \\ 4x(x^2 - a) &= 0. \end{aligned}$$

We see that there are three critical points:

$$\text{Critical points:} \quad x = 0, \quad x = \sqrt{a}, \quad x = -\sqrt{a}.$$

To find possible inflection points, we set the second derivative equal to zero and solve for x:

$$f''(x) = 12x^2 - 4a = 0.$$

There are two possible inflection points:

$$\text{Possible inflection points:} \quad x = \sqrt{\frac{a}{3}}, \quad x = -\sqrt{\frac{a}{3}}.$$

To see if these are inflection points, we determine whether concavity changes by evaluating f'' at values on either side of each of the potential inflection points. We see that

$$f''\left(-2\sqrt{\frac{a}{3}}\right) = 12\left(4\frac{a}{3}\right) - 4a = 16a - 4a = 12a > 0,$$

so f is concave up to the left of $x = -\sqrt{a/3}$. Also,

$$f''(0) = -4a < 0,$$

so f is concave down between $x = -\sqrt{a/3}$ and $x = \sqrt{a/3}$. Finally, we see that

$$f''\left(2\sqrt{\frac{a}{3}}\right) = 12\left(4\frac{a}{3}\right) - 4a = 16a - 4a = 12a > 0,$$

so f is concave up to the right of $x = \sqrt{a/3}$. Since $f(x)$ changes concavity at $x = \sqrt{a/3}$ and $x = -\sqrt{a/3}$, both points are inflection points.

(b) The only positive critical point is at $x = \sqrt{a}$, so to have a critical point at $x = 2$, we substitute:

$$\begin{aligned} x &= \sqrt{a} \\ 2 &= \sqrt{a} \\ a &= 4. \end{aligned}$$

Since the critical point is at the point $(2, 5)$, we have

$$\begin{aligned} f(2) &= 5 \\ 2^4 - 2(4)2^2 + b &= 5 \\ 16 - 32 + b &= 5 \\ b &= 21. \end{aligned}$$

The function is $f(x) = x^4 - 8x^2 + 21$.

(c) We have seen that $a = 4$, so the inflection points are at $x = \sqrt{4/3}$ and $x = -\sqrt{4/3}$.

33. Let the numbers be x, y, z and let $y = 2x$. Then

$$xyz = 2x^2z = 192, \quad \text{so} \quad z = \frac{192}{2x^2} = \frac{96}{x^2}.$$

Since all the numbers are positive, we restrict to $x > 0$.
    The sum is

$$S = x + y + z = x + 2x + \frac{96}{x^2} = 3x + \frac{96}{x^2}.$$

Differentiating to find the minimum,

$$\frac{dS}{dx} = 3 - 2\frac{96}{x^3} = 0$$

$$3 = \frac{192}{x^3}$$

$$x^3 = \frac{192}{3} = 64 \quad \text{so} \quad x = 4.$$

There is only one critical point at $x = 4$. We find

$$\frac{d^2S}{dx^2} = 3\frac{192}{x^4}.$$

Since $d^2S/dx^2 > 0$ for all x, there is a local minimum at $x = 4$. The derivative dS/dx is negative for $0 < x < 4$ and dS/dx is positive for $x > 4$. Thus, $x = 4$ gives the global minimum for $x > 0$.
    The minimum value of the sum obtained from the three numbers 4, 8, and 6 is

$$S = 3 \cdot 4 + \frac{96}{4^2} = 18.$$

37. We have

$$\frac{dM}{dt} = (3x^2 + 0.4x^3)\frac{dx}{dt}.$$

If $x = 5$, then

$$\frac{dM}{dt} = [3(5^2) + 0.4(5^3)](0.02) = 2.5 \text{ gm/hr}.$$

Problems

41. (a) The function f is a local maximum where $f'(x) = 0$ and $f' > 0$ to the left, $f' < 0$ to the right. This occurs at the point x_3.
  (b) The function f is a local minimum where $f'(x) = 0$ and $f' < 0$ to the left, $f' > 0$ to the right. This occurs at the points x_1 and x_5.
  (c) The graph of f is climbing fastest where f' is a maximum, which is at the point x_2.
  (d) The graph of f is falling most steeply where f' is the most negative, which is at the point 0.

45. Since the x^3 term has coefficient of 1, the cubic polynomial is of the form $y = x^3 + ax^2 + bx + c$. We now find a, b, and c. Differentiating gives

$$\frac{dy}{dx} = 3x^2 + 2ax + b.$$

The derivative is 0 at local maxima and minima, so

$$\frac{dy}{dx}\bigg|_{x=1} = 3(1)^2 + 2a(1) + b = 3 + 2a + b = 0$$

$$\frac{dy}{dx}\bigg|_{x=3} = 3(3)^2 + 2a(3) + b = 27 + 6a + b = 0$$

Subtracting the first equation from the second and solving for a and b gives

$$24 + 4a = 0 \quad \text{so} \quad a = -6$$

$$b = -3 - 2(-6) = 9.$$

Since the y-intercept is 5, the cubic is

$$y = x^3 - 6x^2 + 9x + 5.$$

Since the coefficient of x^3 is positive, $x = 1$ is the maximum and $x = 3$ is the minimum. See Figure 4.43. To confirm that $x = 1$ gives a maximum and $x = 3$ gives a minimum, we calculate

$$\frac{d^2y}{dx^2} = 6x + 2a = 6x - 12.$$

At $x = 1$, $\dfrac{d^2y}{dx^2} = -6 < 0$, so we have a maximum.

At $x = 3$, $\dfrac{d^2y}{dx^2} = 6 > 0$, so we have a minimum.

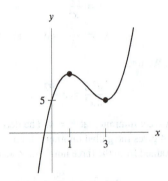

Figure 4.43: Graph of $y = x^3 - 6x^2 + 9x + 5$

49. First notice that since this function approaches 0 as x approaches either plus or minus infinity, any local extrema that we find are also global extrema.

Differentiating $y = axe^{-bx^2}$ gives

$$\frac{dy}{dx} = ae^{-bx^2} - 2abx^2e^{-bx^2} = ae^{-bx^2}(1 - 2bx^2).$$

Since we have a critical points at $x = 1$ and $x = -1$, we know $1 - 2b = 0$, so $b = 1/2$.

The global maximum is 2 at $x = 1$, so we have $2 = ae^{-1/2}$ which gives $a = 2e^{1/2}$. Notice that this value of a also gives the global minimum at $x = -1$.

Thus,

$$y = 2xe^{\left(\frac{1-x^2}{2}\right)}.$$

53. We want to maximize the volume $V = \pi r^2 h$ of the cylinder, shown in Figure 4.44. The cylinder has 2 pieces: the end disk, of area πr^2 and the tube, which has area $2\pi rh$. Thus $8 = \pi r^2 + 2\pi rh$. Solving for h gives

$$h = \frac{8 - \pi r^2}{2\pi r}.$$

Substituting this expression in for h in the formula for V gives

$$V = \pi r^2 \cdot \frac{8 - \pi r^2}{2\pi r} = \frac{1}{2}(8r - \pi r^3).$$

Differentiating gives

$$\frac{dV}{dr} = \frac{1}{2}(8 - 3\pi r^2).$$

To maximize V we look for critical points, so we solve $0 = (8 - 3\pi r^2)/2$, thus $r = \pm\sqrt{8/(3\pi)}$. We discard the negative solution, since r is a positive length. Substituting this value in for r in the formula for h gives

$$h = \frac{8 - \pi\left(\frac{8}{3\pi}\right)}{2\pi\sqrt{\frac{8}{3\pi}}} = \frac{\frac{16}{3}}{2\pi\sqrt{\frac{8}{3\pi}}} = \frac{\frac{8}{3\pi}}{\sqrt{\frac{8}{3\pi}}} = \sqrt{\frac{8}{3\pi}}.$$

We can check that this critical point is a maximum of V by checking the sign of

$$\frac{d^2V}{dr^2} = -3\pi r$$

which is negative when $r > 0$. So V is concave down at the critical point and therefore $r = \sqrt{8/(3\pi)}$ is a maximum.

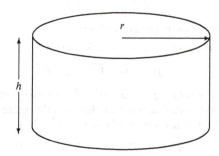

Figure 4.44

57. We first solve for P

$$P = -6jm^2 + 4jk - 5km,$$

and find the derivative

$$\frac{dP}{dm} = -12jm - 5k.$$

Since the derivative is defined for all m, we find the critical points by solving $dP/dm = 0$:

$$\frac{dP}{dm} = -12jm - 5k = 0,$$

$$m = -\frac{5k}{12j}.$$

There is one critical point at $m = -5k/(12j)$. Since P is a quadratic function of m with a negative leading coefficient $-6j$, the critical point gives the global maximum of P. There is no global minimum because $P \to -\infty$ as $m \to \pm\infty$.

61.

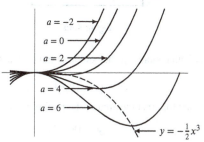

To solve for the critical points, we set $\frac{dy}{dx} = 0$. Since $\frac{d}{dx}\left(x^3 - ax^2\right) = 3x^2 - 2ax$, we want $3x^2 - 2ax = 0$, so $x = 0$ or $x = \frac{2}{3}a$. At $x = 0$, we have $y = 0$. This first critical point is independent of a and lies on the curve $y = -\frac{1}{2}x^3$. At $x = \frac{2}{3}a$, we calculate $y = -\frac{4}{27}a^3 = -\frac{1}{2}\left(\frac{2}{3}a\right)^3$. Thus the second critical point also lies on the curve $y = -\frac{1}{2}x^3$.

65. Since $I(t)$ is a periodic function with period $2\pi/w$, it is enough to consider $I(t)$ for $0 \le wt \le 2\pi$. Differentiating, we find

$$\frac{dI}{dt} = -w\sin(wt) + \sqrt{3}w\cos(wt).$$

At a critical point

$$-w\sin(wt) + \sqrt{3}w\cos(wt) = 0$$

$$\sin(wt) = \sqrt{3}\cos(wt)$$

$$\tan(wt) = \sqrt{3}.$$

So $wt = \pi/3$ or $4\pi/3$, or these values plus multiples of 2π. Substituting into I, we see

$$\text{At } wt = \frac{\pi}{3}: \quad I = \cos\left(\frac{\pi}{3}\right) + \sqrt{3}\sin\left(\frac{\pi}{3}\right) = \frac{1}{2} + \sqrt{3} \cdot \left(\frac{\sqrt{3}}{2}\right) = 2.$$

$$\text{At } wt = \frac{4\pi}{3}: \quad I = \cos\left(\frac{4\pi}{3}\right) + \sqrt{3}\sin\left(\frac{4\pi}{3}\right) = -\frac{1}{2} - \sqrt{3} \cdot \left(\frac{\sqrt{3}}{2}\right) = -2.$$

Thus, the maximum value is 2 amps and the minimum is -2 amps.

69. The distance from a given point on the parabola (x, x^2) to $(1, 0)$ is given by

$$D = \sqrt{(x-1)^2 + (x^2 - 0)^2}.$$

Minimizing this is equivalent to minimizing $d = (x-1)^2 + x^4$. (We can ignore the square root if we are only interested in minimizing because the square root is smallest when the thing it is the square root of is smallest.) To minimize d, we find its critical points by solving $d' = 0$. Since $d = (x-1)^2 + x^4 = x^2 - 2x + 1 + x^4$,

$$d' = 2x - 2 + 4x^3 = 2(2x^3 + x - 1).$$

By graphing $d' = 2(2x^3 + 2x - 1)$ on a calculator, we see that it has only 1 root, $x \approx 0.59$. This must give a minimum because $d \to \infty$ as $x \to -\infty$ and as $x \to +\infty$, and d has only one critical point. This is confirmed by the second derivative test: $d'' = 12x^2 + 2 = 2(6x^2 + 1)$, which is always positive. Thus the point $(0.59, 0.59^2) \approx (0.59, 0.35)$ is approximately the closest point of $y = x^2$ to $(1, 0)$.

73. Figure 4.45 shows the the pool has dimensions x by y and the deck extends 5 feet at either side and 10 feet at the ends of the pool.

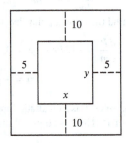

Figure 4.45

The dimensions of the plot of land containing the pool are then $(x + 5 + 5)$ by $(y + 10 + 10)$. The area of the land is then

$$A = (x + 10)(y + 20),$$

which is to be minimized. We also are told that the area of the pool is $xy = 1800$, so

$$y = 1800/x$$

and

$$A = (x + 10)\left(\frac{1800}{x} + 20\right)$$

$$= 1800 + 20x + \frac{18000}{x} + 200.$$

We find dA/dx and set it to zero to get

$$\frac{dA}{dx} = 20 - \frac{18000}{x^2} = 0$$
$$20x^2 = 18000$$
$$x^2 = 900$$
$$x = 30 \text{ feet.}$$

Since $A \to \infty$ as $x \to 0^+$ and as $x \to \infty$, this critical point must be a global minimum. Also, $y = 1800/30 = 60$ feet. The plot of land is therefore $(30 + 10) = 40$ by $(60 + 20) = 80$ feet.

77.

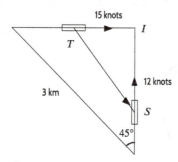

Figure 4.46: Position of the tanker
and ship

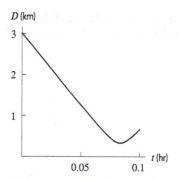

Figure 4.47: Distance between the
ship at S and the tanker at T

Suppose t is the time, in hours, since the ships were 3 km apart. Then $\overline{TI} = \frac{3\sqrt{2}}{2} - (15)(1.85)t$ and $\overline{SI} = \frac{3\sqrt{2}}{2} - (12)(1.85)t$. So the distance, $D(t)$, in km, between the ships at time t is

$$D(t) = \sqrt{\left(\frac{3\sqrt{2}}{2} - 27.75t\right)^2 + \left(\frac{3\sqrt{2}}{2} - 22.2t\right)^2}.$$

Differentiating gives

$$\frac{dD}{dt} = \frac{-55.5\left(\frac{3}{\sqrt{2}} - 27.75\,t\right) - 44.4\left(\frac{3}{\sqrt{2}} - 22.2\,t\right)}{2\sqrt{\left(\frac{3}{\sqrt{2}} - 27.75\,t\right)^2 + \left(\frac{3}{\sqrt{2}} - 22.2\,t\right)^2}}.$$

Solving $dD/dt = 0$ gives a critical point at $t = 0.0839$ hours when the ships will be approximately 331 meters apart. So the ships do not need to change course. Alternatively, tracing along the curve in Figure 4.47 gives the same result. Note that this is after the eastbound ship crosses the path of the northbound ship.

81. (a) Since the volume of water in the container is proportional to its depth, and the volume is increasing at a constant rate,

$$d(t) = \text{Depth at time } t = Kt,$$

where K is some positive constant. So the graph is linear, as shown in Figure 4.48. Since initially no water is in the container, we have $d(0) = 0$, and the graph starts from the origin.

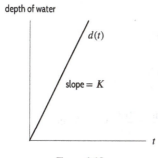

Figure 4.48

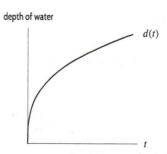

Figure 4.49

(b) As time increases, the additional volume needed to raise the water level by a fixed amount increases. Thus, although the depth, $d(t)$, of water in the cone at time t, continues to increase, it does so more and more slowly. This means $d'(t)$ is positive but decreasing, i.e., $d(t)$ is concave down. See Figure 4.49.

85. We have $\lim_{x \to 1} \sin \pi x = \sin \pi = 0$, and $\lim_{x \to 1} x - 1 = 0$, so this is a 0/0 form and l'Hopital's rule applies directly.

89. If $f(x) = x - \sinh x$ and $g(x) = x^3$, then $f(0) = g(0) = 0$. However, $f'(0) = g'(0) = f''(0) = g''(0) = 0$ also, so we use l'Hopital's Rule three times. Since $f'''(x) = -\cosh x$ and $g'''(x) = 6$:

$$\lim_{x \to 0} \frac{x - \sinh x}{x^3} = \lim_{x \to 0} \frac{1 - \cosh x}{3x^2} = \lim_{x \to 0} \frac{-\sinh x}{6x} = \lim_{x \to 0} \frac{-\cosh x}{6} = -\frac{1}{6}.$$

93. (a) Since $d\theta/dt$ represents the rate of change of θ with time, $d\theta/dt$ represents the angular velocity of the disk.

(b) Suppose P is the point on the rim shown in Figure 4.50.

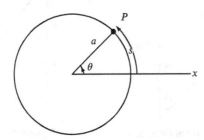

Figure 4.50

Any other point on the rim is moving at the same speed, though in a different direction. We know that since θ is in radians,

$$s = a\theta.$$

Since a is a constant, we know

$$\frac{ds}{dt} = a\frac{d\theta}{dt}.$$

But $ds/dt = v$, the speed of the point on the rim, so

$$v = a\frac{d\theta}{dt}.$$

97. The volume, V, of a cone of height h and radius r is

$$V = \frac{1}{3}\pi r^2 h.$$

Since the angle of the cone is $\pi/6$, so $r = h\tan(\pi/6) = h/\sqrt{3}$

$$V = \frac{1}{3}\pi\left(\frac{h}{\sqrt{3}}\right)^2 h = \frac{1}{9}\pi h^3.$$

Differentiating gives

$$\frac{dV}{dh} = \frac{1}{3}\pi h^2.$$

To find dh/dt, use the chain rule to obtain

$$\frac{dV}{dt} = \frac{dV}{dh}\frac{dh}{dt}.$$

So,

$$\frac{dh}{dt} = \frac{dV/dt}{dV/dh} = \frac{0.1\,\text{meters/hour}}{\pi h^2/3} = \frac{0.3}{\pi h^2}\ \text{meters/hour}.$$

Since $r = h\tan(\pi/6) = h/\sqrt{3}$, we have

$$\frac{dr}{dt} = \frac{dh}{dt}\frac{1}{\sqrt{3}} = \frac{1}{\sqrt{3}}\frac{0.3}{\pi h^2}\ \text{meters/hour}.$$

101. From Figure 4.51, Pythagoras' Theorem shows that the ground distance, d, between the train and the point, B, vertically below the plane is given by

$$d^2 = x^2 + y^2.$$

Figure 4.52 shows that

$$z^2 = d^2 + 4^2$$

so

$$z^2 = x^2 + y^2 + 4^2.$$

We know that when $x = 1$, $dx/dt = 80$, $y = 5$, $dy/dt = 500$, and we want to know dz/dt. First, we find z:

$$z^2 = 1^2 + 5^2 + 4^2 = 42, \text{ so } z = \sqrt{42}.$$

Differentiating $z^2 = x^2 + y^2 + 4^2$ gives

$$2z\frac{dz}{dt} = 2x\frac{dx}{dt} + 2y\frac{dy}{dt}.$$

Canceling 2s and substituting gives

$$\sqrt{42}\frac{dz}{dt} = 1(80) + 5(500)$$

$$\frac{dz}{dt} = \frac{2580}{\sqrt{42}} = 398.103 \text{ mph.}$$

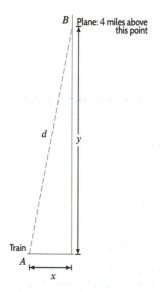

Figure 4.51: View from air

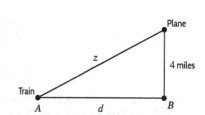

Figure 4.52: Vertical view

CAS Challenge Problems

105. (a) The graph has a jump discontinuity whose position depends on a. The function is increasing, and the slope at a given x-value seems to be the same for all values of a. See Figure 4.53.

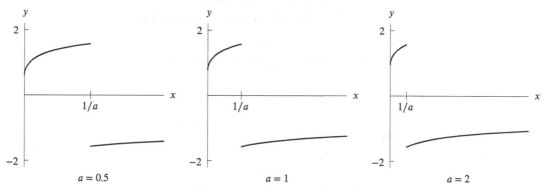

Figure 4.53

(b) Most computer algebra systems will give a fairly complicated answer for the derivative. Here is one example; others may be different.

$$\frac{dy}{dx} = \frac{\sqrt{x} + \sqrt{a}\,\sqrt{ax}}{2x\left(1 + a + 2\sqrt{a}\,\sqrt{x} + x + ax - 2\sqrt{ax}\right)}.$$

When we graph the derivative, it appears that we get the same graph for all values of a. See Figure 4.54.

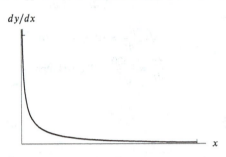

Figure 4.54

(c) Since a and x are positive, we have $\sqrt{ax} = \sqrt{a}\sqrt{x}$. We can use this to simplify the expression we found for the derivative:

$$\frac{dy}{dx} = \frac{\sqrt{x} + \sqrt{a}\sqrt{ax}}{2x\left(1 + a + 2\sqrt{a}\sqrt{x} + x + ax - 2\sqrt{ax}\right)}$$

$$= \frac{\sqrt{x} + \sqrt{a}\sqrt{a}\sqrt{x}}{2x\left(1 + a + 2\sqrt{a}\sqrt{x} + x + ax - 2\sqrt{a}\sqrt{x}\right)}$$

$$= \frac{\sqrt{x} + a\sqrt{x}}{2x\,(1 + a + x + ax)} = \frac{(1+a)\sqrt{x}}{2x(1+a)(1+x)} = \frac{\sqrt{x}}{2x(1+x)}.$$

Since a has canceled out, the derivative is independent of a. This explains why all the graphs look the same in part (b). (In fact they are not exactly the same, because $f'(x)$ is undefined where $f(x)$ has its jump discontinuity. The point at which this happens changes with a.)

109. (a) Different CASs give different answers. (In fact, their answers could be more complicated than what you get by hand.) One possible answer is

$$\frac{dy}{dx} = \frac{\tan\left(\frac{x}{2}\right)}{2\sqrt{\frac{1-\cos x}{1+\cos x}}}.$$

(b) The graph in Figure 4.55 is a step function:

$$f(x) = \begin{cases} 1/2 & 2n\pi < x < (2n+1)\pi \\ -1/2 & (2n+1)\pi < x < (2n+2)\pi. \end{cases}$$

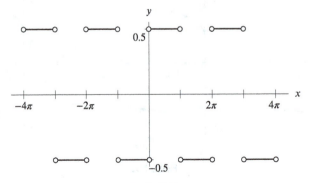

Figure 4.55

Figure 4.55, which shows the graph in disconnected line segments, is correct. However, unless you select certain graphing options in your CAS, it may join up the segments. Use the double angle formula $\cos(x) = \cos^2(x/2) - \sin^2(x/2)$ to simplify the answer in part (a). We find

$$\frac{dy}{dx} = \frac{\tan(x/2)}{2\sqrt{\frac{1-\cos x}{1+\cos x}}} = \frac{\tan(x/2)}{2\sqrt{\frac{1-\cos(2\cdot(x/2))}{1+\cos(2\cdot(x/2))}}} = \frac{\tan(x/2)}{2\sqrt{\frac{1-\cos^2(x/2)+\sin^2(x/2)}{1+\cos^2(x/2)-\sin^2(x/2)}}}$$

$$= \frac{\tan(x/2)}{2\sqrt{\frac{2\sin^2(x/2)}{2\cos^2(x/2)}}} = \frac{\tan(x/2)}{2\sqrt{\tan^2(x/2)}} = \frac{\tan(x/2)}{2\,|\tan(x/2)|}$$

Thus, $dy/dx = 1/2$ when $\tan(x/2) > 0$, i.e. when $0 < x < \pi$ (more generally, when $2n\pi < x < (2n+1)\pi$), and $dy/dx = -1/2$ when $\tan(x/2) < 0$, i.e., when $\pi < x < 2\pi$ (more generally, when $(2n+1)\pi < x < (2n+2)\pi$, where n is any integer).

CHAPTER FIVE

Solutions for Section 5.1

Exercises

1. (a) Left sum
   (b) Upper estimate
   (c) 6
   (d) $\Delta t = 2$
   (e) Upper estimate is approximately $4 \cdot 2 + 2.9 \cdot 2 + 2 \cdot 2 + 1.5 \cdot 2 + 1 \cdot 2 + 0.8 \cdot 2 = 24.4$.

5. The distance traveled is represented by area under the velocity curve. We can approximate the area using left- and right-hand sums. Alternatively, counting the squares (each of which has area 10), and allowing for the broken squares, we can see that the area under the curve from 0 to 6 is between 140 and 150. Hence the distance traveled is between 140 and 150 meters.

9. (a) Since the car starts at a velocity of 90 ft/sec, $b = 90$. Since the car takes 12 seconds to reach its minimum velocity of 20 ft/sec, $a = 20$ and $c = 12$. See Figure 5.1.
   (b) The distance traveled is the area under the velocity graph, which can be viewed as a rectangle with base 12 and height 20 and a triangle with base 12 and height 70. So

$$\text{Distance traveled } = 12 \cdot 20 + \frac{1}{2} \cdot 12 \cdot 70 = 660 \text{ feet.}$$

Figure 5.1

13. Lower estimate: $2.7 \cdot 2 + 2.7 \cdot 2 + 4 \cdot 2 + 6.3 \cdot 2 + 8.5 \cdot 2 + 11.6 \cdot 2 + 13.4 \cdot 2 + 17.4 \cdot 2 + 21.9 \cdot 2 + 29.1 \cdot 2 = 235.2$ meters
   Upper estimate: $2.7 \cdot 2 + 4 \cdot 2 + 6.3 \cdot 2 + 8.5 \cdot 2 + 11.6 \cdot 2 + 13.4 \cdot 2 + 17.4 \cdot 2 + 21.9 \cdot 2 + 29.1 \cdot 2 + 32.6 \cdot 2 = 295.0$ meters

17. From $t = 0$ to $t = 3$ the velocity is constant and positive, so the change in position is $2 \cdot 3$ cm, that is, 6 cm to the right. From $t = 3$ to $t = 5$, the velocity is negative and constant, so the change in position is $-3 \cdot 2$ cm, that is, 6 cm to the left. Thus, the total change in position is 0. The particle moves 6 cm to the right, followed by 6 cm to the left, and returns to where it started.

For total distance, we add the distance traveled in each direction, giving $6 + 6 = 12$ cm.

21. (a) Since $v(t)$ is increasing, we use a left sum to get a lower estimate with $\Delta t = 2$. Evaluating $v(t)$ at $t = 0, 2, 4$:

$$\text{Lower estimate } = v(0) \cdot 2 + v(2) \cdot 2 + v(4) \cdot 2 = 0 \cdot 2 + 20 \cdot 2 + 40 \cdot 2 = 120.$$

To get an upper estimate, we use a right sum. Evaluating $v(t)$ at $t = 2, 4, 6$:

$$\text{Upper estimate } = v(2) \cdot 2 + v(4) \cdot 2 + v(6) \cdot 2 = 20 \cdot 2 + 40 \cdot 2 + 60 \cdot 2 = 240$$

Thus, the lower estimate for the distance traveled is 120 ft, the upper estimate is 240 ft, and the average of the two is 180 ft.
   (b) The distance traveled is the area between the t-axis and the graph of $v(t) = 10t$ for $0 \leq t \leq 6$. This is a triangle with base 6 seconds and height 60 feet/sec, so the distance traveled is

$$\text{Area of the triangle } = \frac{1}{2} \cdot 6 \cdot 60 = 180 \text{ ft.}$$

This is the same as the average of the upper and lower estimates in part (a).

Problems

25. (a) Note that 15 minutes equals 0.25 hours. Lower estimate $= 11(0.25) + 10(0.25) = 5.25$ miles. Upper estimate $=$ $12(0.25) + 11(0.25) = 5.75$ miles.

(b) Lower estimate $= 11(0.25) + 10(0.25) + 10(0.25) + 8(0.25) + 7(0.25) + 0(0.25) = 11.5$ miles. Upper estimate $= 12(0.25) + 11(0.25) + 10(0.25) + 10(0.25) + 8(0.25) + 7(0.25) = 14.5$ miles.

(c) The difference between Roger's pace at the beginning and the end of his run is 12 mph. If the time between the measurements is h, then the difference between the upper and lower estimates is $12h$. We want $12h < 0.1$, so

$$h < \frac{0.1}{12} \approx 0.0083 \text{ hours} = 30 \text{ seconds}$$

Thus Jeff would have to measure Roger's pace every 30 seconds.

29. Since f is increasing, the right-hand sum is the upper estimate and the left-hand sum is the lower estimate. We have $f(0) = 0$, $f(\pi/2) = 1$ and $\Delta t = (b - a)/n = \pi/200$. Thus,

$$|\text{Difference in estimates}| = |f(b) - f(a)|\Delta t$$
$$= |1 - 0|\frac{\pi}{200} = 0.0157.$$

33. From $t = 0$ to $t = 3$, you are moving away from home ($v > 0$); thereafter you move back toward home. So you are the farthest from home at $t = 3$. To find how far you are then, we can measure the area under the v curve as about 9 squares, or $9 \cdot 10 \text{ km/hr} \cdot 1 \text{ hr} = 90$ km. To find how far away from home you are at $t = 5$, we measure the area from $t = 3$ to $t = 5$ as about 25 km, except that this distance is directed toward home, giving a total distance from home during the trip of $90 - 25 = 65$ km.

37. (a) Since car B starts at $t = 2$, the tick marks on the horizontal axis (which we assume are equally spaced) are 2 hours apart. Thus car B stops at $t = 6$ and travels for 4 hours.
Car A starts at $t = 0$ and stops at $t = 8$, so it travels for 8 hours.

(b) Car A's maximum velocity is approximately twice that of car B, that is 100 km/hr.

(c) The distance traveled is given by the area under the velocity graph. Using the formula for the area of a triangle, the distances are given approximately by

$$\text{Car } A \text{ travels} = \frac{1}{2} \cdot \text{Base} \cdot \text{Height} = \frac{1}{2} \cdot 8 \cdot 100 = 400 \text{ km}$$

$$\text{Car } B \text{ travels} = \frac{1}{2} \cdot \text{Base} \cdot \text{Height} = \frac{1}{2} \cdot 4 \cdot 50 = 100 \text{ km}.$$

41. The graph of her velocity against time is a straight line from 0 mph to 60 mph; see Figure 5.2. Since the distance traveled is the area under the curve, we have

$$\text{Shaded area} = \frac{1}{2} \cdot t \cdot 60 = 10 \text{ miles}$$

Solving for t gives

$$t = \frac{1}{3}\text{hr} = 20 \text{ minutes} .$$

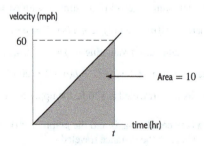

Figure 5.2

Strengthen Your Understanding

45. If f is decreasing on the interval then the right sum is less than the left sum, so $f(x) = x^2$, $[a, b] = [-2, -1]$ is an example.

49. False. For example, for a constant function, the difference does not get smaller, since it is always 0. Another example is the velocity function $f(x) = x^2$ on the interval $-1 \leq x \leq 1$. By the symmetry of the graph, for an even number of subdivisions the difference between the left and right sums is always 0.

Solutions for Section 5.2

Exercises

1. (a) Right sum
   (b) Upper estimate
   (c) 3
   (d) $\Delta x = 2$

5.

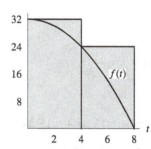

Figure 5.3: Left Sum, $\Delta t = 4$

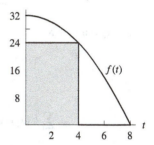

Figure 5.4: Right Sum, $\Delta t = 4$

(a) Left-hand sum $= 32 \cdot 4 + 24 \cdot 4 = 224$.
(b) Right-hand sum $= 24 \cdot 4 + 0 \cdot 4 = 96$.

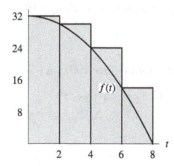

Figure 5.5: Left Sum, $\Delta t = 2$

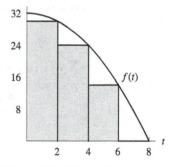

Figure 5.6: Right Sum, $\Delta t = 2$

(c) Left-hand sum $= 32 \cdot 2 + 30 \cdot 2 + 24 \cdot 2 + 14 \cdot 2 = 200$.
(d) Right-hand sum $= 30 \cdot 2 + 24 \cdot 2 + 14 \cdot 2 + 0 \cdot 2 = 136$.

9. Since we have 5 subdivisions,

$$\Delta x = \frac{b - a}{n} = \frac{7 - 3}{5} = 0.8.$$

The interval begins at $x = 3$ and ends at $x = 7$. Table 5.1 gives the value of $f(x)$ at the pertinent points.

Table 5.1

x	3.0	3.8	4.6	5.4	6.2	7.0
$f(x)$	$\dfrac{1}{1+3.0}$	$\dfrac{1}{1+3.8}$	$\dfrac{1}{1+4.6}$	$\dfrac{1}{1+5.4}$	$\dfrac{1}{1+6.2}$	$\dfrac{1}{1+7.0}$

So a right-hand sum is

$$\frac{1}{1+3.8}(0.8) + \frac{1}{1+4.6}(0.8) + \cdots + \frac{1}{1+7.0}(0.8).$$

13. We use a calculator or computer to see that $\int_1^4 (x^2 + x)\,dx = 28.5$.

17. We use a calculator or computer to see that $\int_0^1 \sin(t^2)dt = 0.3103$.

21. (a) Since $\Delta x = 2$ and $x_0 = 2$ we have

$$x_0 = 2, x_1 = 4, x_2 = 6, x_3 = 8, x_4 = 10.$$

Thus,

$$\sum_{i=0}^{4} h\left(x_i\right)\Delta x = h(2)\Delta x + h(4)\Delta x + h(6)\Delta x + h(8)\Delta x + h(10)\Delta x$$

$$= 6 \cdot 2 + 7 \cdot 2 + 8 \cdot 2 + 9 \cdot 2 + 10 \cdot 2$$

$$= 80$$

(b) Since $\Delta x = 3$ and $x_0 = 0$ we have

$$x_0 = 0, x_1 = 3, x_2 = 6, x_3 = 9, x_4 = 12, x_5 = 15.$$

Thus,

$$\sum_{i=2}^{5} h\left(x_i\right)\Delta x = h(6)\Delta x + h(9) + h(12)\Delta x + h(15)\Delta x$$

$$= 8 \cdot 3 + 9.5 \cdot 3 + 11 \cdot 3 + 12.5 \cdot 3$$

$$= 123$$

(c) Since $\Delta x = 2$ and $x_0 = 1$ we have

$$x_0 = 1, x_1 = 3, x_2 = 5, x_3 = 7, x_4 = 9, x_5 = 11, x_6 = 13, x_7 = 15$$

Thus,

$$\sum_{i=4}^{7} h\left(x_i\right)\Delta x = h(9)\Delta x + h(11)\Delta x + h(13)\Delta x + h(15)\Delta x$$

$$= 9.5 \cdot 2 + 10.5 \cdot 2 + 11.5 \cdot 2 + 12.5 \cdot 2$$

$$= 88$$

Problems

25. The graph of $y = 4 - x^2$ crosses the x-axis at $x = 2$ since solving $y = 4 - x^2 = 0$ gives $x = \pm 2$. See Figure 5.7. To find the total area, we find the area above the axis and the area below the axis separately. We have

$$\int_0^2 (4 - x^2)dx = 5.3333 \quad \text{and} \quad \int_2^3 (4 - x^2)dx = -2.3333.$$

As expected, the integral from 2 to 3 is negative. The area above the axis is 5.3333 and the area below the axis is 2.3333, so

$$\text{Total area} = 5.3333 + 2.3333 = 7.6666.$$

Thus the area is 7.667.

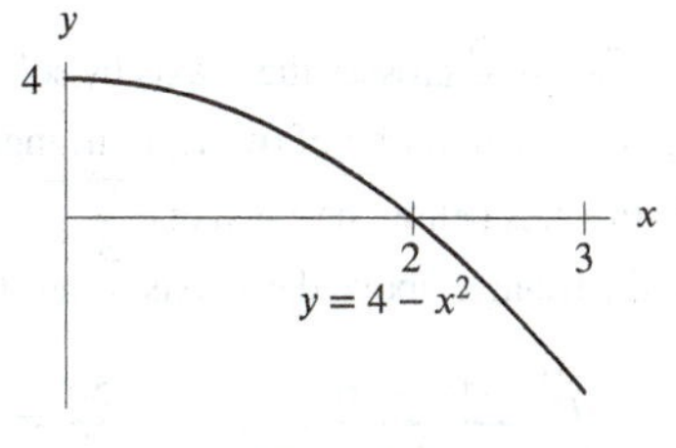

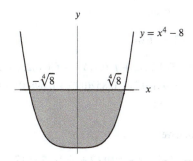

Figure 5.7

29. A graph of $y = \ln x$ shows that this function is non-negative on the interval $x = 1$ to $x = 4$. Thus,

$$\text{Area} = \int_1^4 \ln x \, dx = 2.545.$$

The integral was evaluated on a calculator.

33. The graph of $y = x^4 - 8$ has intercepts $x = \pm\sqrt[4]{8}$. See Figure 5.8. Since the region is below the x-axis, the integral is negative, so

$$\text{Area} = -\int_{-\sqrt[4]{8}}^{\sqrt[4]{8}} (x^4 - 8) \, dx = 21.527.$$

The integral was evaluated on a calculator.

y

$-\sqrt[4]{8}$ $\sqrt[4]{8}$ $y = x^4 - 8$

x

Figure 5.8

37. We can compute each integral in this problem by finding the difference between the area that lies above the x-axis and the area that lies below the x-axis on the given interval.

  (a) For $\int_0^2 f(x) \, dx$, on $0 \le x \le 1$ the area under the graph is 1; on $1 \le x \le 2$ the areas above and below the x-axis are equal and cancel each other out. Therefore, $\int_0^2 f(x) \, dx = 1$.

  (b) On $3 \le x \le 7$ the graph of $f(x)$ is the upper half circle of radius 2 centered at $(5, 0)$. The integral is equal to the area between the graph of $f(x)$ and the x-axis, which is the area of a semicircle of radius 2. This area is 2π, and so

$$\int_3^7 f(x) \, dx = \frac{\pi 2^2}{2} = 2\pi.$$

  (c) On $2 \le x \le 7$ we are looking at two areas: We already know that the area of the semicircle on $3 \le x \le 7$ is 2π. On $2 \le x \le 3$, the graph lies below the x-axis and the area of the triangle is $\frac{1}{2}$. Therefore,

$$\int_2^7 f(x) \, dx = -\frac{1}{2} + 2\pi.$$

  (d) For this portion of the problem, we can split the region between the graph and the x-axis into a quarter circle on $5 \le x \le 7$ and a trapezoid on $7 \le x \le 8$ below the x-axis. The semicircle has area π, the trapezoid has area $3/2$. Therefore,

$$\int_5^8 f(x) \, dx = \pi - \frac{3}{2}.$$

41. We first find where the graph of $f(x) = \dfrac{-4x - 16}{3}$ crosses the x-axis by solving the equation $\dfrac{-4x - 16}{3} = 0$, so at $x = -4$. Making a sketch, we see that the graph is composed of two right triangles. The area of the triangle that lies above the x-axis is 24 and the area of the triangle that lies below the x-axis is $\dfrac{50}{3}$.

The integral is equal to the area of the triangle above the x-axis minus the area of the triangle below the x-axis. Thus,

$$\int_{-10}^{1} \frac{-4x - 16}{3}\, dx = 24 - \frac{50}{3} = \frac{22}{3}.$$

45. Since e^{-x^2} is decreasing between $x = 0$ and $x = 1$, the left sum is an overestimate and the right sum is an underestimate of the integral. Letting $f(x) = e^{-x^2}$, we divide the interval $0 \le x \le 1$ into $n = 5$ sub-intervals to create Table 5.2.

Table 5.2

x	0.0	0.2	0.4	0.6	0.8	1.0
$f(x)$	1.000	0.961	0.852	0.698	0.527	0.368

(a) Letting $\Delta x = (1 - 0)/5 = 0.2$, we have:

$$\text{Left-hand sum} = f(0)\Delta x + f(0.2)\Delta x + f(0.4)\Delta x + f(0.6)\Delta x + f(0.8)\Delta x$$
$$= 1(0.2) + 0.961(0.2) + 0.852(0.2) + 0.698(0.2) + 0.527(0.2)$$
$$= 0.808.$$

(b) Again letting $\Delta x = (1 - 0)/5 = 0.2$, we have:

$$\text{Right-hand sum} = f(0.2)\Delta x + f(0.4)\Delta x + f(0.6)\Delta x + f(0.8)\Delta x + f(1)\Delta x$$
$$= 0.961(0.2) + 0.852(0.2) + 0.698(0.2) + 0.527(0.2) + 0.368(0.2)$$
$$= 0.681.$$

49. Using $n = 4$, for the left-hand sum we have

$$\text{Left-hand sum} = f(-4) \cdot 2 + f(-2) \cdot 2 + f(0) \cdot 2 + f(2) \cdot 2 = 17 \cdot 2 + 5 \cdot 2 + 2 \cdot 2 + 1.25 \cdot 2 = 50.5.$$

Using $n = 4$, for the right-hand sum we have

$$\text{Right-hand sum} = f(-2) \cdot 2 + f(0) \cdot 2 + f(2) \cdot 2 + f(4) \cdot 2 = 5 \cdot 2 + 2 \cdot 2 + 1.25 \cdot 2 + 1.0625 \cdot 2 = 18.625.$$

53. Using $n = 3$, for the left-hand sum we have

$$\text{Left-hand sum} = f(1) \cdot 1 + f(2) \cdot 1 + f(3) \cdot 1 = 1 \cdot 1 + \sqrt{2} \cdot 1 + \sqrt{3} \cdot 1 = 1 + \sqrt{2} + \sqrt{3}.$$

Using $n = 3$, for the right-hand sum we have

$$\text{Right-hand sum} = f(2) \cdot 1 + f(3) \cdot 1 + f(4) \cdot 1 = \sqrt{2} \cdot 1 + \sqrt{3} \cdot 1 + 2 \cdot 1 = \sqrt{2} + \sqrt{3} + 2.$$

57. (a) See Figure 5.9.

(b) Since each of the triangular regions in Figure 5.9 have area $1/2$, we have

$$\int_{0}^{2} f(x)\, dx = \frac{1}{2} + \frac{1}{2} = 1.$$

(c) Using $\Delta x = 1/2$ in the 4-term Riemann sum shown in Figure 5.10, we have

$$\text{Left hand sum} = f(0)\Delta x + f(0.5)\Delta x + f(1)\Delta x + f(1.5)\Delta x$$
$$= 1\left(\frac{1}{2}\right) + \frac{1}{2}\left(\frac{1}{2}\right) + 0\left(\frac{1}{2}\right) + \frac{1}{2}\left(\frac{1}{2}\right) = 1.$$

We notice that in this case the approximation is exactly equal to the exact value of the integral.

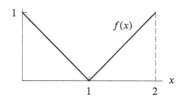

Figure 5.9

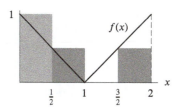

Figure 5.10

61. The lower limit is $i = 0$ so the Riemann sum is a left-hand sum of the form $\sum_{i=0}^{n-1} f(t_i)\Delta t$.

Since $t_i = 3i/n$ we have $t_0 = 0/n$, $t_1 = 3/n$, $t_2 = 6/n$, $t_3 = 9/n$ and so on, so the width of each subinterval is $\Delta t = 3/n$. This means $f(t_i) = 2(3i/n) = 2t_i$, so the Riemann sum is for $f(t) = 2t$.

To find the limits of integration we observe that $t_0 = 0/n = 0$ and $t_n = 3n/n = 3$, so the integral goes from $t = 0$ to $t = 3$. Thus:

$$\lim_{n \to \infty} \sum_{i=0}^{n-1} 2\left(\frac{3i}{n}\right)\left(\frac{3}{n}\right) = \int_0^3 2t \, dt.$$

65. The lower limit is $i = 1$ so the Riemann sum is a right-hand sum of the form $\sum_{i=1}^{n} f(t_i)\Delta t$.

Since $t_i = 1 + 3i/n$ we have $t_1 = 1 + 3/n$, $t_2 = 1 + 6/n$, $t_3 = 1 + 9/n$ and so on, so the width of each subinterval is $\Delta t = 3/n$. Therefore,

$$f(t_i) = \sqrt{\left(1 + \frac{3i}{n}\right)^2 + \left(1 + \frac{3i}{n}\right)} = \sqrt{t_i^2 + t_i},$$

so the Riemann sum is for $f(t) = \sqrt{t^2 + t}$.

To find the limits of integration, we observe that $t_0 = 1 + 0/n = 1$ and $t_n = 1 + 3n/n = 4$, which means the integral goes from from $t = 1$ to $t = 4$. Thus:

$$\lim_{n \to \infty} \sum_{i=1}^{n} \frac{3}{n}\sqrt{\left(1 + \frac{3i}{n}\right)^2 + \left(1 + \frac{3i}{n}\right)} = \int_1^4 \sqrt{t^2 + t} \, dt.$$

69. Since the lower limit is $i = 1$, the Riemann sum can be interpreted as a right-hand sum of the form $\sum_{i=1}^{n} f(t_i)\Delta t$.

Choosing $t_i = 1 + (i/n)$, we have $t_0 = 1 + 0/n$, $t_1 = 1 + 1/n$, $t_2 = 1 + 2/n$, $t_3 = 1 + 3/n$ and so on, so the width of each subinterval is $\Delta t = 1/n$. Therefore,

$$f(t_i) = 8\left(1 + \frac{i}{n}\right) - 8 = 8t_i - 8,$$

so the Riemann sum is for $f(t) = 8t - 8$.

To find the limits of integration, we observe that $t_0 = 1 + (0/n) = 1$ and $t_n = 1 + (n/n) = 2$, which means that the integral goes from $t = 1$ to $t = 2$. Thus:

$$\lim_{n \to \infty} \sum_{i=1}^{n} \left(8\left(1 + \frac{i}{n}\right) - 8\right) \cdot \frac{1}{n} = \int_1^2 (8t - 8) \, dt.$$

Observe that the integral $\int_1^2 (8t - 8) \, dt$ is the area of the shaded triangular region in Figure 5.11, so

$$\int_1^2 (8t - 8) \, dt = \frac{1}{2} \cdot 1 \cdot 8 = 4.$$

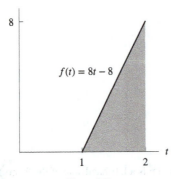

Figure 5.11

73. Since
$$\int_a^b f(x)\,dx = \quad \text{Area above } x\text{-axis} \quad - \quad \text{Area below } x\text{-axis}$$
between $x = a$ and $x = b$, we need to choose a and b so that the total area above the x-axis minus the total area below the x-axis is smallest.

For a fixed a, we consider what happens to the value of $\int_a^b f(x)\,dx$ as we move the upper limit b to the right starting from $b = a$.

If we move b to the right over an interval where $f(x) > 0$ then more area above the x-axis is included between the limits of integration, so $\int_a^b f(x)\,dx$ gets larger.

If we move b to the right over an interval where $f(x) < 0$ then more area below the x-axis is included between the limits of integration, so $\int_a^b f(x)\,dx$ gets smaller.

Therefore, to make $\int_a^b f(x)\,dx$ smallest, we choose b at a point where $\int_a^b f(x)\,dx$ stops getting smaller. From Figure 5.12, this gives either $b = 3$ or $b = 5$.

By similar reasoning, for a fixed b, if we move a to the left from $a = b$ over an interval where $f(x) > 0$, then $\int_a^b f(x)\,dx$ gets larger and if we move a to the left over an interval where $f(x) < 0$, then $\int_a^b f(x)\,dx$ gets smaller.

Therefore, to make $\int_a^b f(x)\,dx$ smallest, we choose a at a point where $\int_a^b f(x)\,dx$ stops getting smaller. From Figure 5.12, this gives either $a = 1$ or $a = 4$.

Since $a < b$, this gives three possible integrals:
$$\int_1^3 f(x)\,dx, \int_1^5 f(x)\,dx, \quad \text{or} \quad \int_4^5 f(x)\,dx.$$

From Figure 5.12, the area below the x-axis between $x = 4$ and $x = 5$ is less than that between $x = 1$ and $x = 3$, so $\int_4^5 f(x)\,dx$ cannot be smallest.

We know
$$\int_1^3 f(x)\,dx = - \text{ Area between 1 and 3}$$
and
$$\int_1^5 f(x)\,dx = \quad \text{Area between 3 and 4} \quad - \quad \text{Area between 1 and 3} \quad - \quad \text{Area between 4 and 5}.$$

From Figure 5.12 we see the area between $x = 4$ and $x = 5$ is larger than the area between $x = 3$ and $x = 4$, so the area under the x-axis is larger in $\int_1^5 f(x)\,dx$ than in $\int_1^3 f(x)\,dx$. Therefore, $\int_1^5 f(x)\,dx$ is smallest, so $a = 1$ and $b = 5$.

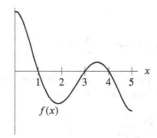

Figure 5.12

Strengthen Your Understanding

77. Any function which is negative on the whole interval will do, for example $f(x) = -1$ and $[a, b] = [0, 1]$. There are also examples like $f(x) = -1 + x$ with $0 \leq x \leq 1.1$.

81. False. A counterexample is given by the functions f and g in Figure 5.13. The function f is decreasing, g is increasing, and we have

$$\int_1^2 f(x)\,dx = \int_1^2 g(x)\,dx,$$

because both integrals equal 1/2, the area of of the same sized triangle.

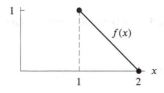

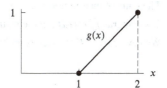

Figure 5.13

85. An example is graphed in Figure 5.14.

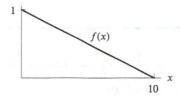

Figure 5.14

Additional Problems (online only)

89. (a) If the interval $1 \leq t \leq 2$ is divided into n equal subintervals of length $\Delta t = 1/n$, the subintervals are given by

$$1 \leq t \leq 1 + \frac{1}{n},\ 1 + \frac{1}{n} \leq t \leq 1 + \frac{2}{n},\ \dots,\ 1 + \frac{n-1}{n} \leq t \leq 2.$$

The left-hand sum is given by

$$\text{Left sum} = \sum_{r=0}^{n-1} f\left(1 + \frac{r}{n}\right)\frac{1}{n} = \sum_{r=0}^{n-1} \frac{1}{1 + r/n} \cdot \frac{1}{n} = \sum_{r=0}^{n-1} \frac{1}{n + r}.$$

and the right-hand sum is given by

$$\text{Right sum} = \sum_{r=1}^{n} f\left(1 + \frac{r}{n}\right)\frac{1}{n} = \sum_{r=1}^{n} \frac{1}{n+r}.$$

Since $f(t) = 1/t$ is decreasing in the interval $1 \le t \le 2$, we know that the right-hand sum is less than $\int_{1}^{2} 1/t \, dt$ and the left-hand sum is larger than this integral. Thus we have

$$\sum_{r=1}^{n} \frac{1}{n+r} \;<\; \int_{1}^{t} \frac{1}{t} \, dt \;<\; \sum_{r=0}^{n-1} \frac{1}{n+r}.$$

(b) Subtracting the sums gives

$$\sum_{r=0}^{n-1} \frac{1}{n+r} - \sum_{r=1}^{n} \frac{1}{n+r} = \frac{1}{n} - \frac{1}{2n} = \frac{1}{2n}.$$

(c) Here we need to find n such that

$$\frac{1}{2n} \le 5 \times 10^{-6}, \quad \text{so} \quad n \ge \frac{1}{10} \times 10^{6} = 10^{5}.$$

Solutions for Section 5.3

Exercises

1. The units of measurement are dollars.

5. The integral $\int_{0}^{6} a(t) \, dt$ represents the change in velocity between times $t = 0$ and $t = 6$ hours; it is measured in km/hr.

9. We have $f(t) = F'(t) = 2t$, so by the Fundamental Theorem of Calculus,

$$\int_{1}^{3} 2t \, dt = F(3) - F(1) = 9 - 1 = 8.$$

13. We have $f(t) = F'(t) = 7\ln(4) \cdot 4^{t}$, so by the Fundamental Theorem of Calculus,

$$\int_{2}^{3} 7\ln(4) \cdot 4^{t} \, dt = F(3) - F(2) = 448 - 112 = 336.$$

Problems

17. (a) By the chain rule,

$$\frac{d}{dx}\left(\frac{1}{2}\sin^{2} t\right) = \frac{1}{2} \cdot 2\sin t \cos t = \sin t \cos t.$$

(b) (i) Using a calculator, $\int_{0.2}^{0.4} \sin t \cos t \, dt = 0.056$

(ii) The Fundamental Theorem of Calculus tells us that the integral is

$$\int_{0.2}^{0.4} \sin t \cos t \, dt = F(0.4) - F(0.2) = \frac{1}{2}\left(\sin^{2}(0.4) - \sin^{2}(0.2)\right) = 0.05609.$$

21. (a) On day 12 pollution is removed from the lake at a rate of 500 kg/day.

(b) The limits of the integral are $t = 5$ and $t = 15$. Since t is time in days, the units of the 5 and 15 are days. The units of the integral are obtained by multiplying the units of $f(t)$, kg/day, by the units of dt, day. Thus the units of the integral are

$$\frac{\text{kg}}{\text{day}} \times \text{day} = \text{kg}.$$

The 4000 has units of kilograms.

(c) The integral of a rate gives the total change. Here $f(t)$ is the rate of change of the quantity of pollution that has been removed from the lake. The integral gives the change in the quantity of pollution that has been removed during the time interval; in other words, the total quantity removed during that time period. During the 10 days from day 5 to day 15, a total of 4000 kg were removed from the lake.

25. (a) Quantity used $= \int_0^5 f(t)\,dt$.
(b) Using a left sum, our approximation is

$$32e^{0.05(0)} + 32e^{0.05(1)} + 32e^{0.05(2)} + 32e^{0.05(3)} + 32e^{0.05(4)} = 177.270.$$

Since f is an increasing function, this represents an underestimate.
(c) Each term is a lower estimate of one year's consumption of oil.

29. We use left- and right-hand sums to estimate the total amount of coal produced during this period:

$$\text{Left sum} = (1.090)2 + (1.094)2 + (1.121)2 + (1.072)2 + (1.132)2 + (1.147)2 = 13.312.$$

$$\text{Right sum} = (1.094)2 + (1.121)2 + (1.072)2 + (1.132)2 + (1.147)2 + (1.073)2 = 13.278.$$

We see that

$$\text{Total amount of coal produced} \approx \frac{13.312 + 13.278}{2} = 13.295 \text{ billion tons.}$$

The total amount of coal produced is the definite integral of the rate of coal production $r = f(t)$ given in the table. Since t is in years since 1997, the limits of integration are $t = 0$ and $t = 12$. We have

$$\text{Total amount of coal produced} = \int_0^{12} f(t)\,dt \text{ billion tons.}$$

33. The units of the horizontal axis are hours and the units of the vertical axis are w/m^2. We use the graph to find the energy produced by 1 square meter solar panel, in watt-hours, by estimating the area under the curve using a Riemann Sum.

There is no measurable solar radiation between midnight and 6 am, or after 6 pm, so we only consider values between 6 am and 6 pm. Between 6 am and 12 noon, a left-hand sum gives an underestimate of the energy produced since solar radiation is increasing then. Between 12 noon and 6 pm, a right-hand sum gives an underestimate of the energy produced since solar radiation is decreasing. (Overestimates are obtained similarly.)

For example, during the hour from 9 to 10 am, if 100% of the radiation were absorbed, an underestimate is

$$\text{Energy generated by 1 square meter of solar panels} = (174 \text{ watts})(1 \text{ hour}) = 174 \text{ watt-hours.}$$

Since only 18% of the radiation is absorbed, we estimate the energy generated during this hour to be $174(0.18) = 31.32$ watt-hours.

Over the entire day an underestimate of the energy generated is

$$\text{Energy} = (0+1+32+174+291+378+412+350+263+134+19+0)(0.18) = 2054(0.18) = 369.72 \text{ watt-hours/m}^2.$$

Similarly, an overestimate of the energy generated is

$$\text{Energy} = (1+32+174+291+378+420+420+412+350+263+134+19)(0.18) = 2894(0.18) = 520.92 \text{ watt-hours/m}^2.$$

The average between the over- and the underestimate is 445.32 watt-hours/m^2. Therefore, a 20-square-meter-sized solar array generates approximately $20 \cdot 445.32 = 8906.4$ watt-hours or 8.9064 kwh.

37. (a) The statement means that 1 hour after the drug is administered the concentration of the medicine in the plasma is increasing at a rate of 50 mg/ml per hour.
(b) The quantity $\int_0^3 h(t)\,dt = 480$ represents the increase in the concentration of the drug in the plasma during the first three hours after it is administered. Since there is an initial concentration of 250 mg/ml when the drug is administered, at the end of three hours we have

$$\text{Concentration} = 250 + \int_0^3 h(t)\,dt = 250 + 480 = 730 \text{ mg/ml in the plasma.}$$

41. According to the Fundamental Theorem,

$$f(2) - \underbrace{f(1)}_{7} = \int_1^2 f'(t)\,dt$$

$$f(2) = 7 + \int_1^2 e^{-t^2}\,dt, \quad \text{since } f'(t) = e^{-t^2}.$$

We estimate the integral using left and right sums. Since $f'(t) = e^{-t^2}$ is decreasing between $t = 1$ and $t = 2$, the left sum overestimates and the right sum underestimates the integral.

To find left- and right-hand sums of 5 rectangles, we let $\Delta t = (2 - 1)/5 = 0.2$. The table gives values of $f'(t)$.

t	1.0	1.2	1.4	1.6	1.8	2.0
$f'(t)$	0.368	0.237	0.141	0.077	0.039	0.018

We have the following estimates:

$$LHS = \Delta t \left(f'(1.0) + f'(1.2) + f'(1.4) + f'(1.6) + f'(1.8) \right) = 0.2\,(0.862) = 0.1724$$
$$RHS = \Delta t \left(f'(1.2) + f'(1.4) + f'(1.6) + f'(1.8) + f'(2.0) \right) = 0.2(0.512) = 0.1024.$$

So

$$0.1024 < \int_1^2 e^{-t^2}\, dt < 0.1734.$$

Adding 7, we estimate that

$$7.1024 < f(2) < 7.1724.$$

45. The value of this integral tells us how much oil is pumped from the well between day $t = 0$ and day $t = t_0$.

49. (a) The water stored in Lake Sonoma on March 1, 2014, is $S(0) = 182{,}566$ acre-feet.

Since $t = 3$ is June 1, we find the water stored in Lake Sonoma then by starting with the water stored in March and adding the change per month for the next three months:

$$S(3) = 182{,}566 + 3003 - 5631 - 8168 = 171{,}770 \text{ acre-feet.}$$

(b) We estimate the maxima and minima of $S(t)$ by examining where the changes go from positive to negative values or vice versa.

We can think of the change in water stored as $S'(t)$. When the data are positive more water is flowing into the reservoir than flowing out; when the data are negative more water is flowing out than in.

Therefore, we have a local maximum at the beginning of April at $t = 1$ and a local minimum at the beginning of November at $t = 8$.

(c) To locate an inflection point, we compute the rate of change of the change in water stored per month, that is $S''(t)$. See Table 5.3. To estimate $S''(t)$ we find the difference between successive months of $S'(t)$.

Since $S''(t)$ changes sign only once, from negative in June to positive in July 2014, an inflection point is located somewhere in that time. Until June, the water stored was decreasing more and more each month. Starting in July it was still decreasing but more slowly.

Table 5.3

Month	Mar. 2014	Apr. 2014	May 2014	June 2014	July 2014
$S'(t)$	3003	−5631	−8168	−8620	−8270
$S''(t)$		−8634	−2537	−452	350

Month	Aug. 2014	Sept. 2014	Oct. 2014	Nov. 2014	
$S'(t)$	−7489	−6245	−4593	54,743	
$S''(t)$	781	1244	1652	59,336	

53. Since C is an antiderivative of c, we know by the Fundamental Theorem that

$$\int_{15}^{24} c(n)\, dn = C(24) - C(15) = 13 - 8 = 5.$$

Since $C(24)$ is the cost to plow a 24-inch snowfall, and $C(15)$ is the cost to plow a 15-inch snowfall, this tells us that it costs \$5 million more to plow a 24-inch snowfall than a 15-inch snowfall.

57. The expression $\int_0^4 r(t)\,dt$ represents the amount of water that leaked from the ruptured pipe during the first four hours. The expression $r(4)$ represents the rate water was leaking from the pipe at time $t = 4$, or four hours after the leak began. Had water leaked at this rate during the whole four-hour time period,

$$\underbrace{\text{Total amount of water leaked}}_{\text{at a constant rate of } r(4)} = \underbrace{\text{Rate}}_{r(4)} \times \underbrace{\text{Time elapsed}}_{4} = 4r(4) \text{ gallons.}$$

According to the article, the leak "worsened throughout the afternoon," so the rate that water leaked out initially was less than $r(4)$. Thus, $4r(4)$ overestimates the total amount of water leaked during the first four hours.

We conclude that

$$\int_0^4 r(t)\,dt < 4r(4).$$

61. Expression (I) and expression (II) represent the area under the graph of $r(t)$ over different three-hour intervals. Since $R'(T) = r(t)$, expression (III) also represents area under the graph of $r(t)$ over a three-hour interval. We have that $r(t) > 0$ and $r'(t) < 0$, so $r(t)$ is a positive, decreasing function. Thus, the areas over any three-hour interval get smaller as t increases. We have that (I) is the earliest interval, (II) is the next earliest, and (III) is the latest. Thus, the order from smallest to greatest is (III), (II), (I).

Strengthen Your Understanding

65. If the car travels at a constant velocity of 50 miles per hour, it travels 200 miles in 4 hours as shown in Figure 5.15.

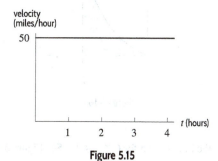

Figure 5.15

Solutions for Section 5.4

Exercises

1. Note that $\int_a^b g(x)\,dx = \int_a^b g(t)\,dt$. Thus, we have

$$\int_a^b (f(x) + g(x))\,dx = \int_a^b f(x)\,dx + \int_a^b g(x)\,dx = 8 + 2 = 10.$$

5. We write

$$\int_a^b \left(c_1 g(x) + (c_2 f(x))^2\right)\,dx = \int_a^b \left(c_1 g(x) + c_2^2 (f(x))^2\right)\,dx$$

$$= \int_a^b c_1 g(x)\,dx + \int_a^b c_2^2 (f(x))^2\,dx$$

$$= c_1 \int_a^b g(x)\,dx + c_2^2 \int_a^b (f(x))^2\,dx$$

$$= c_1(2) + c_2^2(12) = 2c_1 + 12c_2^2.$$

9. We have

$$\text{Average value} = \frac{1}{b-a}\int_a^b f(x)\,dx$$

$$= \frac{1}{b-a}\int_a^b 2\,dx = \frac{1}{b-a}\cdot\left(\begin{array}{c}\text{Area of rectangle}\\\text{of height 2 and base }b-a\end{array}\right)$$

$$= \frac{1}{b-a}[2(b-a)] = 2.$$

13. The graph of $y = e^x$ is above the line $y = 1$ for $0 \le x \le 2$. See Figure 5.16. Therefore

$$\text{Area} = \int_0^2 (e^x - 1)\,dx = 4.389.$$

The integral was evaluated on a calculator.

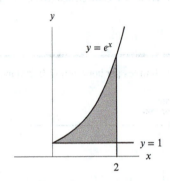

Figure 5.16

17. The graph of $y = \sin x + 2$ is above the line $y = 0.5$ for $6 \le x \le 10$. See Figure 5.17. Therefore

$$\text{Area} = \int_6^{10} \sin x + 2 - 0.5\,dx = 7.799.$$

The integral was evaluated on a calculator.

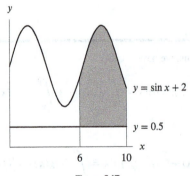

Figure 5.17

21. For $0 \le x \le 1$, we know that x^2 is smaller than x, so $\int_0^1 x^2\,dx$ is smaller.

Problems

25. (a) The integral is the difference between the two right-hand areas in Figure 5.18, with the one above the x-axis counted positively and the one below the axis counted negatively. So

$$\int_0^4 f(x)\,dx = 8 - 4 = 4.$$

(b) The regions under the curve on the left of the vertical axis have the same area as those on the right; see Figure 5.18. Thus,

$$\int_{-2}^2 f(x)\,dx = \int_{-2}^0 f(x)\,dx + \int_0^2 f(x)\,dx = 8 + 8 = 16.$$

(c) Breaking up the area we have

$$\int_{-4}^2 f(x)\,dx = \int_{-4}^{-2} f(x)\,dx + \int_{-2}^0 f(x)\,dx + \int_0^2 f(x)\,dx = -4 + 8 + 8 = 12.$$

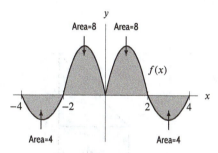

Figure 5.18

29. (a) A property of definite integrals tells us that:

$$\int_1^9 f(x)\,dx = \int_1^5 f(x)\,dx + \int_5^9 f(x)\,dx$$

so $$\int_5^9 f(x)\,dx = \underbrace{\int_1^9 f(x)\,dx}_{3} - \underbrace{\int_1^5 f(x)\,dx}_{5} = -2.$$

(b) A property of definite integrals tells us that:

$$\int_3^9 f(x)\,dx = \int_3^5 f(x)\,dx + \int_5^9 f(x)\,dx$$

so $$\int_3^5 f(x)\,dx = \int_3^9 f(x)\,dx - \int_5^9 f(x)\,dx.$$

We are given that $\int_3^9 f(x)\,dx = 4$. From part (a), we have $\int_5^9 f(x)\,dx = -2$

Thus, $$\int_3^5 f(x)\,dx = \underbrace{\int_3^9 f(x)\,dx}_{4} - \underbrace{\int_5^9 f(x)\,dx}_{-2} = 6.$$

(c) A property of definite integrals tells us that $\int_3^1 f(x)\,dx = -\int_1^3 f(x)\,dx.$
Another property tells us that

$$\int_1^9 f(x)\,dx = \int_1^3 f(x)\,dx + \int_3^9 f(x)\,dx$$

so $\displaystyle\int_1^3 f(x)\,dx = \underbrace{\int_1^9 f(x)\,dx}_{3} - \underbrace{\int_3^9 f(x)\,dx}_{4} = -1.$

Thus, $\displaystyle\int_3^1 f(x)\,dx = -\underbrace{\int_1^3 f(x)\,dx}_{-1} = 1.$

33. (a) The integral represents the area above the x-axis and below the line $y = x$ between $x = a$ and $x = b$. See Figure 5.19. This area is

$$A_1 + A_2 = a(b-a) + \frac{1}{2}(b-a)^2 = \left(a + \frac{b-a}{2}\right)(b-a) = \frac{b+a}{2}(b-a) = \frac{b^2 - a^2}{2}.$$

The formula holds similarly for negative values.

(b) (i) $\int_2^5 x\,dx = 21/2.$

(ii) $\int_{-3}^8 x\,dx = 55/2.$

(iii) $\int_1^3 5x\,dx = 5\int_1^3 x\,dx = 20.$

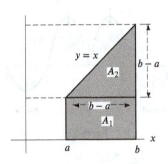

Figure 5.19

37. On the interval $2 \le x \le 5$,

$$\text{Average value of } f = \frac{1}{5-2}\int_2^5 f(x)\,dx = 4,$$

so

$$\int_2^5 f(x)\,dx = 12.$$

Thus

$$\int_2^5 (3f(x) + 2)\,dx = 3\int_2^5 f(x)\,dx + 2\int_2^5 1\,dx = 3(12) + 2(5-2) = 42.$$

41. (a) At the end of one hour $t = 60$, and $H = 22°C$.

(b)

$$\text{Average temperature} = \frac{1}{60}\int_0^{60}(20 + 980e^{-0.1t})\,dt$$

$$= \frac{1}{60}(10976) = 183°C.$$

(c) Average temperature at beginning and end of hour $= (1000 + 22)/2 = 511°C$. The average found in part (b) is smaller than the average of these two temperatures because the bar cools quickly at first and so spends less time at high temperatures. Alternatively, the graph of H against t is concave up.

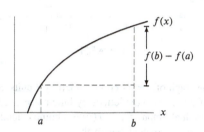

The graph shows $H(°C)$ versus t (mins), with values 1000°, 511°, 22°, and 60 marked. The curve is labeled $H = 20 + 980e^{-0.1t}$.

45. Since $f(x) = \sqrt{25 - x^2}$ is decreasing for $0 \le x \le 3$, we have

$$f(3) = 4 \le f(x) \le f(0) = 5.$$

Thus, a lower bound is $m(b - a) = 4(3 - 0) = 12$ and an upper bound is $M(b - a) = 5(3 - 0) = 15$.

49. See Figure 5.20.

The graph shows $f(x)$ as an increasing concave-down curve, with a and b marked on the x-axis and $f(b) - f(a)$ indicated.

Figure 5.20

53. (a) We know that $\displaystyle\int_3^8 f(x)\,dx = \int_3^6 f(x)\,dx + \int_6^8 f(x)\,dx$.

From the graph, we see that the area corresponding to $\displaystyle\int_3^6 f(x)\,dx$ is larger than the area corresponding to $\displaystyle\int_6^8 f(x)\,dx$.

However, the first area lies below the x-axis, so it is counted negatively. This means:

$$\int_3^8 f(x)\,dx = \underbrace{\int_3^6 f(x)\,dx}_{\text{large negative}} + \underbrace{\int_6^8 f(x)\,dx}_{\text{small positive}} = \text{A negative number.}$$

We conclude that the integral is negative.

(b) We know that $\displaystyle\int_6^0 f(x)\,dx = -\int_0^6 f(x)\,dx$.

We also know that $\displaystyle\int_0^6 f(x)\,dx = \int_0^3 f(x)\,dx + \int_3^6 f(x)\,dx$.

From the graph, we see that the area corresponding to $\displaystyle\int_0^3 f(x)\,dx$ is larger than the area corresponding to $\displaystyle\int_3^6 f(x)\,dx$.

The first area lies above the x-axis, whereas the second area lies below it, so it is counted negatively. This means:

$$\int_0^6 f(x)\,dx = \underbrace{\int_0^3 f(x)\,dx}_{\text{large positive}} + \underbrace{\int_3^6 f(x)\,dx}_{\text{small negative}} = \text{A positive number.}$$

Therefore, $\displaystyle\int_6^0 f(x)\,dx = -\underbrace{\int_0^6 f(x)\,dx}_{\text{positive}} = \text{A negative number. We conclude that the integral is negative.}$

(c) The graph of $y = f(x) + 6$ is the graph of $y = f(x)$ shifted vertically up by 6 units. See Figure 5.21.

Thus, the value of $\int_8^{12} (f(x) + 6)\, dx$ corresponds to the area bounded by this vertically shifted graph. The area bounded by the graph of $y = f(x)$ from $x = 8$ to $x = 12$ lies entirely below the x-axis. However, the area bounded by the shifted graph lies *above* the x-axis. Thus, the value of the integral is positive.

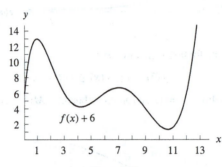

Figure 5.21

(d) The graph of $y = f(x + 3)$ is the graph of $y = f(x)$ shifted left by 3 units. See Figure 5.22.

When we shift the graph of $y = f(x)$ to the left, everything moves, including the "bumps."

The area bounded by the original graph of $y = f(x)$ is a bump extending from $x = 3$ to $x = 6$. This bump lies entirely below the x-axis, so the area is counted negatively.

The area bounded by the horizontally shifted graph, from $x = 0$ to $x = 3$, still lies below the x-axis and is therefore still counted negatively.

Thus, the value of $\int_0^3 f(x + 3)\, dx$ is negative.

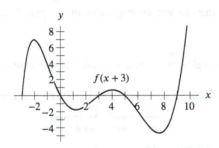

Figure 5.22

57. (a) Splitting the integral in order to make use of the values in the table gives:

$$\frac{1}{\sqrt{2\pi}} \int_1^3 e^{-x^2/2}\, dx = \frac{1}{\sqrt{2\pi}} \int_0^3 e^{-x^2/2}\, dx - \frac{1}{\sqrt{2\pi}} \int_0^1 e^{-x^2/2}\, dx = 0.4987 - 0.3413 = 0.1574.$$

(b) Using the symmetry of $e^{x^2/2}$, we have

$$\frac{1}{\sqrt{2\pi}} \int_{-2}^3 e^{-x^2/2}\, dx = \frac{1}{\sqrt{2\pi}} \int_{-2}^0 e^{-x^2/2}\, dx + \frac{1}{\sqrt{2\pi}} \int_0^3 e^{-x^2/2}\, dx$$

$$= \frac{1}{\sqrt{2\pi}} \int_0^2 e^{-x^2/2}\, dx + \frac{1}{\sqrt{2\pi}} \int_0^3 e^{-x^2/2}\, dx$$

$$= 0.4772 + 0.4987 = 0.9759.$$

61. There is not enough information to decide. In terms of area, knowing the area under the graph of f on $0 \le x \le 7$ does not help us find the area on $0 \le x \le 3.5$. It depends on the shape of the graph and where it lies above and below the x-axis.

65. See Figure 5.23.

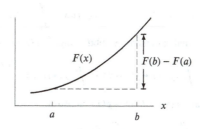

Figure 5.23

Strengthen Your Understanding

69. The time needs to be expressed in days in the definite integral, since the population is a function of time t in days. If the six-month period contains 181 days, the correct integral is

$$\frac{1}{181} \int_0^{181} f(t)\, dt.$$

For a different six-month period, the number 181 may be different.

73. True, since $\int_0^2 (f(x) + g(x))\, dx = \int_0^2 f(x)\, dx + \int_0^2 g(x)\, dx.$

77. False. This would be true if $h(x) = 5f(x)$. However, we cannot assume that $f(5x) = 5f(x)$, so for many functions this statement is false. For example, if f is the constant function $f(x) = 3$, then $h(x) = 3$ as well, so $\int_0^2 f(x)\, dx = \int_0^2 h(x)\, dx = 6$.

81. False. If the graph of f is symmetric about the y-axis, this is true, but otherwise it is usually not true. For example, if $f(x) = x + 1$ the area under the graph of f for $-1 \le x \le 0$ is less than the area under the graph of f for $0 \le x \le 1$, so $\int_{-1}^1 f(x)\, dx < 2 \int_0^1 f(x)\, dx$. See Figure 5.24.

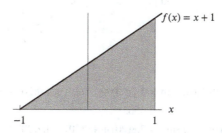

Figure 5.24

85. True. We have by the properties of integrals in Theorem 5.3,

$$\int_1^9 f(x)\, dx = \int_1^4 f(x)\, dx + \int_4^9 f(x)\, dx.$$

Since $(1/(4 - 1)) \int_1^4 f(x)\, dx = A$ and $(1/(9 - 4)) \int_4^9 f(x)\, dx = B$, we have

$$\int_1^9 f(x)\, dx = 3A + 5B.$$

Dividing this equation through by 8, we get that the average value of f on the interval $[1, 9]$ is $(3/8)A + (5/8)B$.

89. (a) Does not follow; the statement implies that

$$\int_a^b f(x)\, dx + \int_a^b g(x)\, dx = 5 + 7 = 12,$$

but the fact that the two integrals add to 12 does not tell us what the integrals are individually. For example, we could have $\int_a^b f(x)\, dx = 10$ and $\int_a^b g(x)\, dx = 2$.

(b) This follows:

$$\int_a^b (f(x) + g(x)) \, dx = \int_a^b f(x) \, dx + \int_a^b g(x) \, dx = 7 + 7 = 14.$$

(c) This follows: rearranging the original statement by subtracting $\int_a^b g(x) \, dx$ from both sides gives

$$\int_a^b (f(x) + g(x)) \, dx - \int_a^b g(x) \, dx = \int_a^b f(x) \, dx.$$

Since $f(x) + g(x) = h(x)$, we have $f(x) = h(x) - g(x)$. Substituting for $f(x)$, we get

$$\int_a^b h(x) \, dx - \int_a^b g(x) \, dx = \int_a^b (h(x) - g(x)) \, dx.$$

Solutions for Chapter 5 Review

Exercises

1. (a) Lower estimate $= (45)(2) + (16)(2) + (0)(2) = 122$ feet.
Upper estimate $= (88)(2) + (45)(2) + (16)(2) = 298$ feet.

(b)

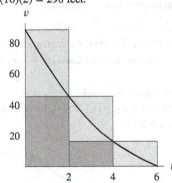

5. We take $\Delta t = 20$. Then:

$$\text{Left-hand sum} = 1.2(20) + 2.8(20) + 4.0(20) + 4.7(20) + 5.1(20)$$
$$= 356.$$
$$\text{Right-hand sum} = 2.8(20) + 4.0(20) + 4.7(20) + 5.1(20) + 5.2(20)$$
$$= 436.$$
$$\int_0^{100} f(t) \, dt \approx \text{Average} = \frac{356 + 436}{2} = 396.$$

9. We have $f(t) = F'(t) = 12t^3 - 15t^2 + 5$, so by the Fundamental Theorem of Calculus,

$$\int_{-2}^1 (12t^3 - 15t^2 + 5) \, dt = F(1) - F(-2) = 3 - 78 = -75.$$

13. Since the θ intercepts of $y = \sin \theta$ are
$\theta = 0, \pi, 2\pi, \ldots,$

$$\text{Area} = \int_0^\pi (1 - \sin \theta) \, d\theta = 1.142.$$

The integral was evaluated on a calculator.

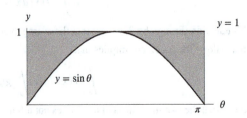

Problems

17. This integral represents the area of two triangles, each of base 1 and height 1. See Figure 5.25. Therefore:

$$\int_{-1}^{1} |x|\, dx = \frac{1}{2} \cdot 1 \cdot 1 + \frac{1}{2} \cdot 1 \cdot 1 = 1.$$

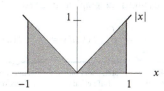

Figure 5.25

21. Distance traveled $= \displaystyle\int_{0}^{1.1} \sin(t^2)\, dt = 0.399$ miles.

25. (a) An overestimate is 7 tons. An underestimate is 5 tons.

(b) An overestimate is $7 + 8 + 10 + 13 + 16 + 20 = 74$ tons. An underestimate is $5 + 7 + 8 + 10 + 13 + 16 = 59$ tons.

(c) If measurements are made every Δt months, then the error is $|f(6) - f(0)| \cdot \Delta t$. So for this to be less than 1 ton, we need $(20 - 5) \cdot \Delta t < 1$, or $\Delta t < 1/15$. So measurements every 2 days or so will guarantee an error in over- and underestimates of less than 1 ton.

29. (a) A graph of $f'(x) = \sin(x^2)$ is shown in Figure 5.26. Since the derivative $f'(x)$ is positive between $x = 0$ and $x = 1$, the change in $f(x)$ is positive, so $f(1)$ is larger than $f(0)$. Between $x = 2$ and $x = 2.5$, we see that $f'(x)$ is negative, so the change in $f(x)$ is negative; thus, $f(2)$ is greater than $f(2.5)$.

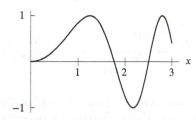

Figure 5.26: Graph of $f'(x) = \sin(x^2)$

(b) The change in $f(x)$ between $x = 0$ and $x = 1$ is given by the Fundamental Theorem of Calculus:

$$f(1) - f(0) = \int_{0}^{1} \sin(x^2)dx = 0.310.$$

Since $f(0) = 2$, we have

$$f(1) = 2 + 0.310 = 2.310.$$

Similarly, since

$$f(2) - f(0) = \int_{0}^{2} \sin(x^2)dx = 0.805,$$

we have

$$f(2) = 2 + 0.805 = 2.805.$$

Since

$$f(3) - f(0) = \int_{0}^{3} \sin(x^2)dx = 0.774,$$

we have

$$f(3) = 2 + 0.774 = 2.774.$$

The results are shown in the table.

x	0	1	2	3
$f(x)$	2	2.310	2.805	2.774

33. Since f is even, $\int_0^2 f(x)\,dx = (1/2)6 = 3$ and $\int_0^5 f(x)\,dx = (1/2)14 = 7$. Therefore

$$\int_2^5 f(x)\,dx = \int_0^5 f(x)\,dx - \int_0^2 f(x)\,dx = 7 - 3 = 4.$$

37. Let $A = \int_0^p f(x)\,dx$. We are told $A > 0$. We can interpret A as the area under f on this interval. Since the graph of f repeats periodically every p units, the area under the graph also repeats. This means

$$\underbrace{\int_0^p f(x)\,dx}_{\substack{\text{area under one} \\ \text{full period}}} = \underbrace{\int_p^{2p} f(x)\,dx}_{\substack{\text{area under one} \\ \text{full period}}} = \underbrace{\int_{2p}^{3p} f(x)\,dx}_{\substack{\text{area under one} \\ \text{full period}}} = \cdots = A.$$

We have that

$$\int_0^{2p} f(x)\,dx = \underbrace{\int_0^p f(x)\,dx}_{A} + \underbrace{\int_p^{2p} f(x)\,dx}_{A} = 2A$$

$$\int_p^{5p} f(x)\,dx = \underbrace{\int_p^{2p} f(x)\,dx}_{A} + \cdots + \underbrace{\int_{4p}^{5p} f(x)\,dx}_{A} = 4A$$

$$\int_{5p}^{7p} f(x)\,dx = \underbrace{\int_{5p}^{6p} f(x)\,dx}_{A} + \underbrace{\int_{6p}^{7p} f(x)\,dx}_{A} = 2A,$$

so

$$\frac{\displaystyle\int_0^{2p} f(x)\,dx + \int_p^{5p} f(x)\,dx}{\displaystyle\int_{5p}^{7p} f(x)\,dx} = \frac{2A + 4A}{2A} = 3.$$

41. (a) The model indicates that wind capacity was 318,128 in 2013 and $318{,}000e^{0.11 \cdot 7} = 711{,}556.784$ in 2020.
(b) Since the continuous growth rate was 0.115, the wind energy generating capacity was increasing by 11.5% per year.
(c) We use the formula for average value between $t = 0$ and $t = 7$:

$$\text{Average value} = \frac{1}{7-0} \int_0^7 318{,}128e^{0.115t}\,dt = \frac{1}{7}(3{,}421{,}111) = 488{,}730 \text{ megawatts.}$$

45. We know that T_h is the time at which half the fuel has burned, so $\int_0^{T_h} r(t)\,dt$ equals half the fuel burned.

We know that T is the time at which all the fuel has burned, so $0.5T$ is halfway through the entire time period. This means $\int_0^{0.5T} r(t)\,dt$ is the amount of fuel burned halfway through the total time period.

Since $r'(t) < 0$, we know the rate fuel burns is going down. This means more than half the fuel is burned during the first half of the total time period than the second half.

We conclude that more than half the fuel is burned during the first half of the time period, so $\int_0^{T_h} r(t)\,dt <$

$$\int_0^{0.5T} r(t)\,dt$$

49. In Figure 5.27 the area A_1 is largest, A_2 is next, and A_3 is smallest. We have

$$\text{I} = \int_a^b f(x)\,dx = A_1, \quad \text{II} = \int_a^c f(x)\,dx = A_1 - A_2, \quad \text{III} = \int_a^e f(x)\,dx = A_1 - A_2 + A_3,$$

$$\text{IV} = \int_b^e f(x)\,dx = -A_2 + A_3, \quad \text{V} = \int_b^c f(x)\,dx = -A_2.$$

The relative sizes of A_1, A_2, and A_3 mean that I is positive and largest, III is next largest (since $-A_2 + A_3$ is negative, but less negative than $-A_2$), II is next largest, but still positive (since A_1 is larger than A_2). The integrals IV and V are both negative, but V is more negative. Thus

$$\text{V} < \text{IV} < 0 < \text{II} < \text{III} < \text{I}.$$

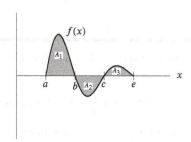

Figure 5.27

53. (a) For $-2 \leq x \leq 2$, f is symmetrical about the y-axis, so $\int_{-2}^{0} f(x)\,dx = \int_{0}^{2} f(x)\,dx$ and $\int_{-2}^{2} f(x)\,dx = 2\int_{0}^{2} f(x)\,dx$.

  (b) For any function f, $\int_{0}^{2} f(x)\,dx = \int_{0}^{5} f(x)\,dx - \int_{2}^{5} f(x)\,dx$.

  (c) Note that $\int_{-2}^{0} f(x)\,dx = \frac{1}{2}\int_{-2}^{2} f(x)\,dx$, so $\int_{0}^{5} f(x)\,dx = \int_{-2}^{5} f(x)\,dx - \int_{-2}^{0} f(x)\,dx = \int_{-2}^{5} f(x)\,dx - \frac{1}{2}\int_{-2}^{2} f(x)\,dx$.

57. (a) The acceleration is positive for $0 \leq t < 40$ and for a tiny period before $t = 60$, since the slope is positive over these intervals. Just to the left of $t = 40$, it looks like the acceleration is approaching 0. Between $t = 40$ and a moment just before $t = 60$, the acceleration is negative.

  (b) The maximum altitude was about 500 feet, when t was a little greater than 40 (here we are estimating the area under the graph for $0 \leq t \leq 42$).

  (c) The acceleration is greatest when the slope of the velocity is most positive. This happens just before $t = 60$, where the magnitude of the velocity is plunging and the direction of the acceleration is positive, or up.

  (d) The deceleration is greatest when the slope of the velocity is most negative. This happens just after $t = 40$.

  (e) After the Montgolfier Brothers hit their top climbing speed (at $t = 40$), they suddenly stopped climbing and started to fall. This suggests some kind of catastrophe—the flame going out, the balloon ripping, etc. (In actual fact, in their first flight in 1783, the material covering their balloon, held together by buttons, ripped and the balloon landed in flames.)

  (f) The total change in altitude for the Montgolfiers and their balloon is the definite integral of their velocity, or the total area under the given graph (counting the part after $t = 42$ as negative, of course). As mentioned before, the total area of the graph for $0 \leq t \leq 42$ is about 500. The area for $t > 42$ is about 220. So subtracting, we see that the balloon finished 280 feet or so higher than where it began.

61. From the Fundamental Theorem of Calculus, we see that

$$f(3) + \int_{3}^{3.1} f'(t)\,dt = f(3) + f(3.1) - f(3) = f(3.1),$$

so this value is exact.

65. (a) The slope of a line is the tangent of the angle the line makes with the horizontal. Thus the slope, $f'(x)$, of the tangent line at x is the tangent of the angle θ.

  (b) Using the identity

$$1 + \tan^2\theta = \frac{1}{\cos^2\theta}$$

and part (a) we have

$$1 + (f'(x))^2 = \frac{1}{\cos^2\theta}.$$

  (c) Using the chain rule, we have

$$\frac{d}{dx}(\tan\theta) = \frac{1}{\cos^2\theta}\frac{d\theta}{dx}.$$

Therefore, using parts (a) and (b), we have

$$\frac{d\theta}{dx} = \cos^2\theta\frac{d}{dx}f'(x)$$

$$= \frac{1}{1 + (f'(x))^2}f''(x).$$

(d) By the Fundamental Theorem of Calculus we have

$$\theta(b) - \theta(a) = \int_a^b \theta'(x)\,dx = \int_a^b \frac{f''(x)}{1 + (f'(x))^2}\,dx.$$

CAS Challenge Problems

69. (a) Since the length of the interval of integration is $2 - 1 = 1$, the width of each subdivision is $\Delta t = 1/n$. Thus the endpoints of the subdivision are

$$t_0 = 1, \quad t_1 = 1 + \Delta t = 1 + \frac{1}{n}, \quad t_2 = 1 + 2\Delta t = 1 + \frac{2}{n}, \dots,$$

$$t_i = 1 + i\Delta t = 1 + \frac{i}{n}, \dots, \quad t_{n-1} = 1 + (n-1)\Delta t = 1 + \frac{n-1}{n}.$$

Thus, since the integrand is $f(t) = t^2$,

$$\text{Left-hand sum} = \sum_{i=0}^{n-1} f(t_i)\Delta t = \sum_{i=0}^{n-1} t_i^2 \Delta t = \sum_{i=0}^{n-1}\left(1 + \frac{i}{n}\right)^2 \frac{1}{n} = \sum_{i=0}^{n-1} \frac{(n+i)^2}{n^3}.$$

(b) Using a CAS to find the sum, we get

$$\sum_{i=0}^{n-1} \frac{(n+i)^2}{n^3} = \frac{(-1+2\,n)\,(-1+7\,n)}{6\,n^2} = \frac{7}{3} + \frac{1}{6\,n^2} - \frac{3}{2\,n}.$$

(c) Taking the limit as $n \to \infty$

$$\lim_{n\to\infty}\left(\frac{7}{3} + \frac{1}{6\,n^2} - \frac{3}{2\,n}\right) = \lim_{n\to\infty}\frac{7}{3} + \lim_{n\to\infty}\frac{1}{6\,n^2} - \lim_{n\to\infty}\frac{3}{2\,n} = \frac{7}{3} + 0 + 0 = \frac{7}{3}.$$

(d) We have calculated $\int_1^2 t^2\,dt$ using Riemann sums. Since t^2 is above the t-axis between $t = 1$ and $t = 2$, this integral is the area; so the area is 7/3.

CHAPTER SIX

Solutions for Section 6.1

Exercises

1. (a) If $f(x)$ is positive over an interval, then $F(x)$ is increasing over the interval.
  (b) If $f(x)$ is increasing over an interval, then $F(x)$ is concave up over the interval.

5. See Figure 6.1.

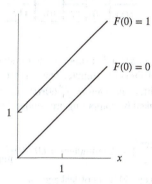

Figure 6.1

9. See Figure 6.2.

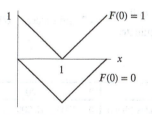

Figure 6.2

13. The Fundamental Theorem tells us that

$$\text{Change in } F = F(5) - F(0) = \int_0^5 f(x)\,dx = 12.$$

Since $F(0) = 50$, we have

$$F(5) = 50 + 12 = 62.$$

Problems

17. (a) The integral is the area above the x-axis counted positively plus the area below the x-axis counted negatively, so the value of the integral is negative. The area below the x-axis is a triangle with base 3 and height 12; the area above the x-axis is a triangle with base 1 and height 4. So

$$\int_0^4 f(x)\,dx = -\frac{1}{2} \cdot 3 \cdot 12 + \frac{1}{2} \cdot 1 \cdot 4 = -18 + 2 = -16.$$

(b) The Fundamental Theorem tells us that

$$\text{Change in } F = F(4) - F(0) = \int_0^4 f(x)\,dx = -16.$$

Since $F(0) = 100$, we have

$$F(4) = 100 + (-16) = 84.$$

21. The change in $f(x)$ between 0 and 2 is equal to $\int_0^2 f'(x)\,dx$. A left-hand estimate for this integral is $(17)(2) = 34$ and a right hand estimate is $(15)(2) = 30$. Our best estimate is the average, 32. The change in $f(x)$ between 0 and 2 is $+32$. Since $f(0) = 50$, we have $f(2) = 82$. We find the other values similarly. The results are shown in Table 6.1.

Table 6.1

x	0	2	4	6
$f(x)$	50	82	107	119

25. (a) Let $r(t)$ be the leakage rate in liters per second at time t minutes, shown in the graph. Since time is in minutes, it is helpful to express leakage in units of liters per minute. The leakage rate in liters per minute is $60r(t)$. The quantity leaked during the first b minutes, in liters, is given by $\int_0^b 60r(t)\,dt$.

We can evaluate the quantity leaked by computing area under the rate graph, by counting grid squares. Each grid square contributes area

$$1 \text{ grid square} = \left(10\,\frac{\text{liter}}{\text{sec}}\right) \times (10 \text{ minutes}) = 60 \cdot 10\frac{\text{liters}}{\text{minute}} \times (10 \text{ minutes}) = 6000 \text{ liters.}$$

Thus each grid square represents 6000 liters of leakage. So

$$\begin{array}{l}\text{Total spill over} \\ \text{first 10 minutes}\end{array} = \frac{1}{2} \text{ grid square} \cdot 6000 = 3000 \text{ liters.}$$

$$\begin{array}{l}\text{Total spill over} \\ \text{20 minutes}\end{array} = 2 \text{ grid squares} \cdot 6000 = 12{,}000 \text{ liters.}$$

Continuing, we have

Time t (minutes)	0	10	20	30	40	50
Total spill, over t minutes (liters)	0	3000	12,000	21,000	27,000	30,000

(b) Plotting the values from part (a) and connecting with a smooth curve gives Figure 6.3.

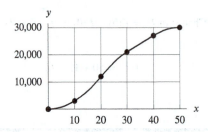

Figure 6.3

29. See Figure 6.4. Note that since $f(x_1) = 0$ and $f'(x_1) > 0$, $F(x_1)$ is a local minimum; since $f(x_3) = 0$ and $f'(x_3) < 0$, $F(x_3)$ is a local maximum. Also, since $f'(x_2) = 0$ and f changes from decreasing to increasing about $x = x_2$, F has an inflection point at $x = x_2$.

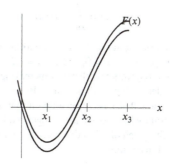

Figure 6.4

33. (a) We know that $\int_0^3 f'(x)dx = f(3) - f(0)$ from the Fundamental Theorem of Calculus. From the graph of f' we can see that $\int_0^3 f'(x)dx = 2-1 = 1$ by subtracting areas between f' and the x-axis. Since $f(0) = 0$, we find that $f(3) = 1$. Similar reasoning gives $f(7) = \int_0^7 f'(x)dx = 2 - 1 + 2 - 4 + 1 = 0$.

   (b) We have $f(0) = 0$, $f(2) = 2$, $f(3) = 1$, $f(4) = 3$, $f(6) = -1$, and $f(7) = 0$. So the graph, beginning at $x = 0$, starts at zero, increases to 2 at $x = 2$, decreases to 1 at $x = 3$, increases to 3 at $x = 4$, then passes through a zero as it decreases to -1 at $x = 6$, and finally increases to 0 at 7. Thus, there are three zeroes: $x = 0$, $x = 5.5$, and $x = 7$.

   (c)

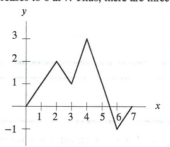

Strengthen Your Understanding

37. The statement has $f(x)$ and $F(x)$ reversed. Namely if an antiderivative $F(x)$ is increasing, then $F'(x) = f(x) \geq 0$.

   We can see the statement given is not always true by looking for a counterexample. The function $f(x) = 2x$ is always increasing, but it has antiderivatives that are less than 0. For example, the antiderivative $F(x)$ with $F(0) = -1$ is negative at 0. See Figure 6.5.

   A correct statement is: If $f(x) > 0$ everywhere, then $F(x)$ is increasing everywhere.

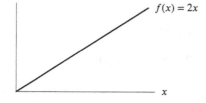

Figure 6.5: $f(x) = 2x$ increasing and its antiderivative $F(x)$ is not always positive

41. True. A function can have only one derivative.

Additional Problems (online only)

45. Let $y'(t) = dy/dt$. Then y is the antiderivative of y' such that $y(0) = 0$. We know that

$$y(x) = \int_0^x y'(t)\,dt.$$

Thus $y(x)$ is the area under the graph of dy/dt from $t = 0$ to $t = x$, with regions below the t-axis contributing negatively to the integral. We see that $y(t_1) = -2$, $y(t_3) = -2 + 2 = 0$, and $y(t_5) = -2$. See Figure 6.6.

Since y' is positive on the interval (t_1, t_3), we know that y is increasing on that interval. Since y' is negative on the intervals $(0, t_1)$ and (t_3, ∞), we know y is decreasing on those intervals.

Since y' is increasing on $(0, t_2)$, we know that y is concave up on that interval. Since y' is decreasing on (t_2, t_4), we know that y is concave down there. The point where concavity changes, t_2, is an inflection point. In addition, since y' is a negative constant on the interval (t_4, ∞), the graph of y is a line with negative slope on this interval. The value $y(t_1) = -2$ is a local minimum and $y(t_3) = 0$ is a local maximum.

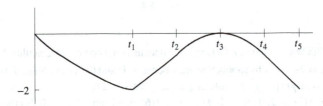

Figure 6.6

Solutions for Section 6.2

Exercises

1. Since the derivative of $p(x)$ is $q(x)$, we know that an antiderivative of $q(x)$ is $p(x)$. The indefinite integral is the most general antiderivative, so we have

$$\int q(x)\,dx = p(x) + C.$$

5. Recall that $F(x)$ is an antiderivative of $f(x)$ if $F'(x) = f(x)$, so to determine which are antiderivatives, we differentiate each and check whether the derivative is equal to $f(x)$. We have

$$\frac{d}{dx}(-2\sin x \cos x) = -2\ \sin x(-\sin x) + (-2\cos x)(\cos x) = 2\sin^2 x - 2\cos^2 x \neq f(x)$$

$$\frac{d}{dx}(2\cos^2 x - 2\sin^2 x) = -4\cos x \sin x - 4\sin x \cos x = -8\sin x \cos x \neq f(x)$$

$$\frac{d}{dx}(\sin^2 x) = 2\sin x \cos x = f(x)$$

$$\frac{d}{dx}(-\cos^2 x) = -2\cos x(-\sin x) = 2\sin x \cos x = f(x)$$

$$\frac{d}{dx}(2\sin^2 x + \cos^2 x) = 4\sin x \cos x + 2\cos x(-\sin x) = 2\sin x \cos x = f(x).$$

Therefore, (III), (IV) and (V) are antiderivatives.

9. $\frac{1}{3}t^3 + \frac{1}{2}t^2$

13. $\sin t$

17. $-\cos t$

21. $\dfrac{t^4}{4} - \dfrac{t^3}{6} - \dfrac{t^2}{2}$

25. $G(t) = \int \sqrt{t}\,dt = \frac{2}{3}t^{3/2} + C$

29. $H(x) = \int (4x^3 - 7)\,dx = x^4 - 7x + C$

33. $H(t) = \int \frac{7}{\cos^2 t}\,dt = 7\tan t + C$

37. $f(x) = 2 + 4x + 5x^2$, so $F(x) = 2x + 2x^2 + \frac{5}{3}x^3 + C$. $F(0) = 0$ implies that $C = 0$. Thus $F(x) = 2x + 2x^2 + \frac{5}{3}x^3$ is the only possibility.

41. $f(x) = \sin x$, so $F(x) = -\cos x + C$. $F(0) = 0$ implies that $-\cos 0 + C = 0$, so $C = 1$. Thus $F(x) = -\cos x + 1$ is the only possibility.

45. $\int 7e^x\,dx = 7e^x + C$

49. $\int (x+3)^2\,dx = \int (x^2 + 6x + 9)\,dx = \frac{x^3}{3} + 3x^2 + 9x + C$

53. Expand the integrand and then integrate

$$\int t^3(t^2 + 1)\,dt = \int (t^5 + t^3)\,dt = \frac{1}{6}t^6 + \frac{1}{4}t^4 + C.$$

57. $\int_1^3 \frac{1}{t}\,dt = \ln|t|\Big|_1^3 = \ln|3| - \ln|1| = \ln 3 \approx 1.0986.$

61. $\int_2^5 (x^3 - \pi x^2)\,dx = \left(\frac{x^4}{4} - \frac{\pi x^3}{3} \right)\Big|_2^5 = \frac{609}{4} - 39\pi \approx 29.728.$

65. $\int_0^{\pi/4} (\sin t + \cos t)\,dt = (-\cos t + \sin t)\Big|_0^{\pi/4} = \left(-\frac{\sqrt{2}}{2} + \frac{\sqrt{2}}{2} \right) - (-1 + 0) = 1.$

69. We differentiate the right-hand side:

$$\frac{d}{dx}\left(-\frac{1}{x} + C \right) = \frac{1}{x^2} = x^{-2},$$

which is the integrand of the left side of the given statement.
    Therefore, the integration is correct.

73. We differentiate the right-hand side:

$$\frac{d}{dx}(3\sin x + C) = 3\cos x,$$

which is the integrand of the left side of the given statement.
    Therefore, the integration is correct.

Problems

77. Since $y = x^3 - x = x(x-1)(x+1)$, the graph crosses the axis at the three points shown in Figure 6.7. The two regions have the same area (by symmetry). Since the graph is below the axis for $0 < x < 1$, we have

$$\text{Area} = 2\left(-\int_0^1 (x^3 - x)\,dx \right)$$

$$= -2\left[\frac{x^4}{4} - \frac{x^2}{2} \right]_0^1 = -2\left(\frac{1}{4} - \frac{1}{2} \right) = \frac{1}{2}.$$

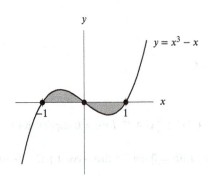

Figure 6.7

81. The graph of $y = e^x - 2$ is below the x-axis at $x = 0$ and above the x-axis at $x = 2$. The graph crosses the axis where

$$e^x - 2 = 0$$
$$x = \ln 2.$$

See Figure 6.8. Thus we find the area by dividing the region at $x = \ln 2$:

$$\text{Area} = -\int_0^{\ln 2} (e^x - 2)\, dx + \int_{\ln 2}^2 (e^x - 2)\, dx$$

$$= (-e^x + 2x)\Big|_0^{\ln 2} + (e^x - 2x)\Big|_{\ln 2}^2$$

$$= -e^{\ln 2} + 2\ln 2 + e^0 + e^2 - 4 - (e^{\ln 2} - 2\ln 2)$$

$$= -2e^{\ln 2} + 4\ln 2 - 3 + e^2$$

$$= -2 \cdot 2 + 4\ln 2 - 3 + e^2 = e^2 + 4\ln 2 - 7.$$

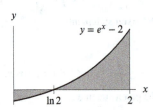

Figure 6.8

85. The area under $f(x) = 8x$ between $x = 1$ and $x = b$ is given by $\int_1^b (8x)dx$. Using the Fundamental Theorem to evaluate the integral:

$$\text{Area} = 4x^2\Big|_1^b = 4b^2 - 4.$$

Since the area is 192, we have

$$4b^2 - 4 = 192$$
$$4b^2 = 196$$
$$b^2 = 49$$
$$b = \pm 7.$$

Since b is larger than 1, we have $b = 7$.

89. (a) The average value of $f(t) = \sin t$ over $0 \leq t \leq 2\pi$ is given by the formula

$$\text{Average} = \frac{1}{2\pi - 0}\int_0^{2\pi} \sin t\, dt$$

$$= \frac{1}{2\pi}(-\cos t)\Big|_0^{2\pi}$$

$$= \frac{1}{2\pi}(-\cos 2\pi - (-\cos 0)) = 0.$$

We can check this answer by looking at the graph of $\sin t$ in Figure 6.9. The area below the curve and above the t-axis over the interval $0 \leq t \leq \pi, A_1$, is the same as the area above the curve but below the t-axis over the interval $\pi \leq t \leq 2\pi, A_2$. When we take the integral of $\sin t$ over the entire interval $0 \leq t \leq 2\pi$, we get $A_1 - A_2 = 0$.

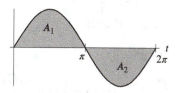

Figure 6.9

(b) Since

$$\int_0^{\pi} \sin t\, dt = -\cos t\Big|_0^{\pi} = -\cos \pi - (-\cos 0) = -(-1) - (-1) = 2,$$

the average value of $\sin t$ on $0 \leq t \leq \pi$ is given by

$$\text{Average value} = \frac{1}{\pi}\int_0^{\pi} \sin t\, dt = \frac{2}{\pi}.$$

93. Since $C'(x) = 4000 + 10x$ we want to evaluate the indefinite integral

$$\int (4000 + 10x)\, dx = 4000x + 5x^2 + K$$

where K is a constant. Thus $C(x) = 5x^2 + 4000x + K$, and the fixed cost of 1,000,000 riyal means that $C(0) = 1,000,000 = K$. Therefore, the total cost is

$$C(x) = 5x^2 + 4000x + 1,000,000 \text{ riyals.}$$

Since $C(x)$ depends on x^2, the square of the depth drilled, costs will increase dramatically when x grows large.

97. The function $\displaystyle\int g(2x)\, dx$ is an antiderivative of $g(2x)$, so we need to determine which, if any, of (I)–(III) is an antiderivative of $g(2x)$. In order to do this, we differentiate each and check whether it is equal to $g(2x)$ using the fact that the derivative of C is 0 and $G'(x) = g(x)$. Applying the chain rule, we have

$$\frac{d}{dx}(0.5G(0.5x) + C) = 0.5G'(0.5x) \cdot \frac{d}{dx}(0.5x) = 0.5g(0.5x) \cdot 0.5 = 0.25g(0.5x) \neq g(2x),$$

$$\frac{d}{dx}(0.5G(2x) + C) = 0.5G'(2x) \cdot \frac{d}{dx}(2x) = 0.5g(2x) \cdot 2 = g(2x),$$

$$\frac{d}{dx}(2G(0.5x) + C) = 2G'(0.5x) \cdot \frac{d}{dx}(0.5x) = 2g(0.5x) \cdot 0.5 = g(0.5x) \neq g(2x).$$

Thus, (II) is equal to $\displaystyle\int g(2x)\, dx$.

Strengthen Your Understanding

101. The statement is true for all $n \neq -1$, since $\int x^{-1}\, dx = \ln|x| + C$.

105. True. Any antiderivative of $3x^2$ is obtained by adding a constant to x^3.

109. True. Adding a constant to an antiderivative gives another antiderivative.

Additional Problems (online only)

113. We have

$$f\left(x^{-1}\right) = 4\left(x^{-1}\right)^{-3} \qquad \text{because } f(x) = 4x^{-3}$$
$$= 4x^3$$

so $$\int_1^3 f\left(x^{-1}\right) dx = \int_1^3 4x^3 \, dx$$
$$= 4\left(\frac{1}{4}\right)x^4 \Big|_1^3$$
$$= 3^4 - 1^4 = 80.$$

117. The area beneath the curve in Figure 6.10 is given by

$$\int_0^a y \, dx = \int_0^a (\sqrt{a} - \sqrt{x})^2 dx = \left[ax - \frac{4\sqrt{a}\,x^{3/2}}{3} + \frac{x^2}{2}\right]_0^a = \frac{a^2}{6}.$$

The area of the square is a^2 so the area above the curve is $5a^2/6$. Thus, the ratio of the areas is 5 to 1.

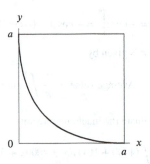

Figure 6.10: The curve
$\sqrt{x} + \sqrt{y} = \sqrt{a}$

Solutions for Section 6.3

Exercises

1. We differentiate $y = xe^{-x} + 2$ using the product rule to obtain

$$\frac{dy}{dx} = x\left(e^{-x}(-1)\right) + (1)e^{-x} + 0$$
$$= -xe^{-x} + e^{-x}$$
$$= (1-x)e^{-x},$$

and so $y = xe^{-x} + 2$ satisfies the differential equation. We now check that $y(0) = 2$:

$$y = xe^{-x} + 2$$
$$y(0) = 0e^0 + 2 = 2.$$

5. We need to find a function whose derivative is $x^3 + 5x^4$, so one antiderivative is $x^4/4 + x^5$. The general solution is

$$y = \frac{x^4}{4} + x^5 + C.$$

To check, we differentiate to get

$$\frac{dy}{dx} = \frac{d}{dx}\left(\frac{x^4}{4} + x^5 + C\right) = x^3 + 5x^4,$$

as required.

9. Integrating gives

$$\int \frac{dy}{dx}\, dx = \int (3x^2)\, dx = x^3 + C.$$

If $y = 5$ when $x = 0$, then $0^3 + C = 5$ so $C = 5$. Thus $y = x^3 + 5$.

Problems

13. The acceleration is $a(t) = -32$, so the velocity is $v(t) = -32t + C$. We find C using $v(0) = -10$, (negative velocity is downward) so

$$v(t) = -32t - 10.$$

Then, the height of the rock above the water is

$$s(t) = -16t^2 - 10t + D.$$

We find D using $s(0) = 100$, so

$$s(t) = -16t^2 - 10t + 100.$$

Now we can find when the rock hits the water by solving the quadratic equation

$$0 = -16t^2 - 10t + 100.$$

There are two solutions: $t = 2.207$ seconds and $t = -2.832$ seconds. We discard the negative solution. So, at $t = 2.207$ the rock is traveling with a velocity of

$$v(2.207) = -32(2.207) - 10 = -80.624 \text{ ft/sec}.$$

Thus the rock is traveling with a speed of 80.624 ft/sec downward when it hits the water.

17. (a) Since

$$R'(p) = 25 - 2p$$
$$R(p) = \int (25 - 2p)\, dp = 25p - p^2 + C.$$

We assume that the revenue is 0 when the price is 0. Substituting gives

$$0 = 25 \cdot 0 - 0^2 + C$$
$$C = 0.$$

Thus

$$R(p) = 25p - p^2.$$

(b) The revenue increases with price if $R'(p) > 0$, that is $25 - 2p > 0$, so $p < 12.5$ dollars. The revenue decreases with price if $R'(p) < 0$, that is $25 - 2p < 0$, so $p > 12.5$.

21. (a) The velocity is decreasing at 32 ft/sec^2, the acceleration due to gravity.

(b) The graph is a line because the velocity is decreasing at a constant rate.

(c) The highest point is reached when the velocity is 0, which occurs when

$$\text{Time} = \frac{160}{32} = 5 \text{ sec}.$$

(d) The object hits the ground at $t = 10$ seconds, since by symmetry if the object takes 5 seconds to go up, it takes 5 seconds to come back down.

(e) See Figure 6.11.

(f) The maximum height is the distance traveled when going up, which is represented by the area A of the triangle above the time axis.

$$\text{Area} = \frac{1}{2}(160 \text{ ft/sec})(5 \text{ sec}) = 400 \text{ feet}.$$

(g) The slope of the line is -32 so

$$v(t) = -32t + 160.$$

Antidifferentiating, we get

$$s(t) = -16t^2 + 160t + s_0.$$

Since the object starts on the ground, $s_0 = 0$, so

$$s(t) = -16t^2 + 160t.$$

At $t = 5$, we have

$$s(t) = -400 + 800 = 400 \text{ ft}.$$

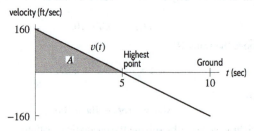

Figure 6.11

25. The first thing we should do is convert our units. We'll bring everything into feet and seconds. Thus, the initial speed of the car is

$$\frac{70 \text{ miles}}{\text{hour}} \left(\frac{1 \text{ hour}}{3600 \text{ sec}} \right) \left(\frac{5280 \text{ feet}}{1 \text{ mile}} \right) \approx 102.7 \text{ ft/sec}.$$

We assume that the acceleration is constant as the car comes to a stop. A graph of its velocity versus time is given in Figure 6.12. We know that the area under the curve represents the distance that the car travels before it comes to a stop, 157 feet. But this area is a triangle, so it is easy to find t_0, the time the car comes to rest. We solve

$$\frac{1}{2}(102.7)t_0 = 157,$$

which gives

$$t_0 \approx 3.06 \text{ sec}.$$

Since acceleration is the rate of change of velocity, the car's acceleration is given by the slope of the line in Figure 6.12. Thus, the acceleration, k, is given by

$$k = \frac{102.7 - 0}{0 - 3.06} \approx -33.56 \text{ ft/sec}^2.$$

Notice that k is negative because the car is slowing down.

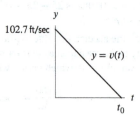

Figure 6.12: Graph of velocity versus time

29. (a) $s = v_0 t - 16t^2$, where $v_0 = $ initial velocity, and $v = s' = v_0 - 32t$. At the maximum height, $v = 0$, so $v_0 = 32t_{\max}$. Plugging into the distance equation yields $100 = 32t_{\max}^2 - 16t_{\max}^2 = 16t_{\max}^2$, so $t_{\max} = \frac{5}{2}$ seconds, from which we get

$$v_0 = 32\left(\frac{5}{2}\right) = 80 \text{ ft/sec}.$$

(b) This time $g = 5$ ft/sec^2, so $s = v_0 t - 2.5t^2 = 80t - 2.5t^2$, and $v = s' = 80 - 5t$. At the highest point, $v = 0$, so $t_{\max} = \frac{80}{5} = 16$ seconds. Plugging into the distance equation yields $s = 80(16) - 2.5(16)^2 = 640$ ft.

33. Let time $t = 0$ be the moment when the astronaut jumps up. If acceleration due to gravity is 5 ft/sec^2 and initial velocity is 10 ft/sec, then the velocity of the astronaut is described by

$$v(t) = 10 - 5t.$$

Suppose $y(t)$ describes his distance from the surface of the moon. By the Fundamental Theorem,

$$y(t) - y(0) = \int_0^t (10 - 5x)\, dx$$

$$y(t) = 10t - \frac{1}{2}5t^2.$$

since $y(0) = 0$ (assuming the astronaut jumps off the surface of the moon).

The astronaut reaches the maximum height when his velocity is 0, i.e. when

$$\frac{dy}{dt} = v(t) = 10 - 5t = 0.$$

Solving for t, we get $t = 2$ sec as the time at which he reaches the maximum height from the surface of the moon. At this time his height is

$$y(2) = 10(2) - \frac{1}{2}5(2)^2 = 10 \text{ ft}.$$

When the astronaut is at height $y = 0$, he either just landed or is about to jump. To find how long it is before he comes back down, we find when he is at height $y = 0$. Set $y(t) = 0$ to get

$$0 = 10t - \frac{1}{2}5t^2$$
$$0 = 20t - 5t^2$$
$$0 = 4t - t^2$$
$$0 = t(t - 4).$$

So we have $t = 0$ sec (when he jumps off) and $t = 4$ sec (when he lands, which gives the time he spent in the air).

Strengthen Your Understanding

37. If $y = \cos(t^2)$, then $dy/dt = -2t\sin(t^2)$. Thus, $y = \cos(t^2)$ does not satisfy the differential equation.

41. If $y = \cos(5x)$, we have $dy/dx = -5\sin(5x)$. The equation $dy/dx = -5\sin(5x)$ is therefore a differential equation that has $y = \cos(5x)$ as a solution.

45. True. If acceleration is $a(t) = k$ for some constant k, $k \neq 0$, then we have

$$\text{Velocity} = v(t) = \int a(t)\, dt = \int k\, dt = kt + C_1,$$

for some constant C_1. We integrate again to find position as a function of time:

$$\text{Position} = s(t) = \int v(t)\, dt = \int (kt + C_1)\, dt = \frac{kt^2}{2} + C_1 t + C_2,$$

for some constant C_2. Since $k \neq 0$, this is a quadratic polynomial.

49. True. Two solutions $y = F(x)$ and $y = G(x)$ of the same differential equation $dy/dx = f(x)$ are both antiderivatives of $f(x)$ and hence they differ by a constant: $F(x) - G(x) = C$ for all x. Since $F(3) \neq G(3)$ we have $C \neq 0$.

Additional Problems (online only)

53. (a) $t = \dfrac{s}{\frac{1}{2}v_{max}}$, where t is the time it takes for an object to travel the distance s, starting from rest with uniform acceleration

      a. v_{max} is the highest velocity the object reaches. Since its initial velocity is 0, the mean of its highest velocity and initial velocity is $\frac{1}{2}v_{max}$.

  (b) By Problem 52, $s = \frac{1}{2}gt^2$, where g is the acceleration due to gravity, so it takes $\sqrt{200/32} = 5/2$ seconds for the body to hit the ground. Since $v = gt, v_{max} = 32(\frac{5}{2}) = 80$ ft/sec. Galileo's statement predicts (100 ft)/(40 ft/sec) = 5/2 seconds, and so Galileo's result is verified.

  (c) If the acceleration is a constant a, then $s = \frac{1}{2}at^2$, and $v_{max} = at$. Thus

$$\frac{s}{\frac{1}{2}v_{max}} = \frac{\frac{1}{2}at^2}{\frac{1}{2}at} = t.$$

Solutions for Section 6.4

Exercises

1.

Table 6.2

x	0	0.5	1	1.5	2
$I(x)$	0	0.50	1.09	2.03	3.65

5. If $f'(x) = \dfrac{\sin x}{x}$, then $f(x)$ is of the form

$$f(x) = C + \int_a^x \frac{\sin t}{t}\, dt.$$

Since $f(1) = 5$, we take $a = 1$ and $C = 5$, giving

$$f(x) = 5 + \int_1^x \frac{\sin t}{t}\, dt.$$

9. Since f is always positive, F is always increasing. F has an inflection point where $f' = 0$. Since $F(0) = \int_0^0 f(t)dt = 0$, F goes through the origin. See Figure 6.13.

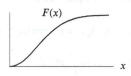

Figure 6.13

13. $(1 + x)^{200}$.

Problems

17. We need to find where $F''(x)$ is positive or negative. First, we compute

$$F'(x) = e^{-x^2},$$

then
$$F''(x) = -2xe^{-x^2}.$$

Since $e^{-x^2} > 0$ for all x, we see that
$$F''(x) > 0 \text{ for } x < 0$$
and
$$F''(x) < 0 \text{ for } x > 0.$$

Thus the graph of $F(x)$ is concave up for $x < 0$ and concave down for $x > 0$.

21. Using the solution to Problem 20, we see that $F(x)$ is increasing for $x < \sqrt{\pi}$ and $F(x)$ is decreasing for $x > \sqrt{\pi}$. Thus $F(x)$ has its maximum value when $x = \sqrt{\pi}$.

By the Fundamental Theorem of Calculus,
$$F(\sqrt{\pi}) - F(0) = \int_0^{\sqrt{\pi}} f(t)\, dt = \int_0^{\sqrt{\pi}} \sin(t^2)\, dt.$$

Since $F(0) = 0$, calculating Riemann sums gives
$$F(\sqrt{\pi}) = \int_0^{\sqrt{\pi}} \sin(t^2)\, dt = 0.895.$$

25. (a) The definition of g gives $g(0) = \int_0^0 f(t)\, dt = 0$.
    (b) The Fundamental Theorem gives $g'(1) = f(1) = -2$.
    (c) The function g is concave upward where g'' is positive. Since $g'' = f'$, we see that g is concave up where f is increasing. This occurs on the interval $1 \le x \le 6$.
    (d) The function g decreases from $x = 0$ to $x = 3$ and increases for $3 < x \le 8$, and the magnitude of the increase is more than the magnitude of the decrease. Thus g takes its maximum value at $x = 8$.

29. Since $G'(x) = \cos(x^2)$ and $G(0) = -3$, we have
$$G(x) = G(0) + \int_0^x \cos(t^2)\, dt = -3 + \int_0^x \cos(t^2)\, dt.$$

Substituting $x = -1$ and evaluating the integral numerically gives
$$G(-1) = -3 + \int_0^{-1} \cos(t^2)\, dt = -3.905.$$

33. Since $v(t) = \int_0^t q'(x)\, dx$, we know from the Construction Theorem that v is an antiderivative of q'. This means:
$$v'(t) = q'(t)$$
$$\text{so} \quad v'(0.4) = q'(0.4)$$
$$\approx \frac{q(0.5) - q(0.4)}{0.1} \quad \text{average rate of change of } q$$
$$= \frac{3.1 - 3.9}{0.1} = -8.$$

Since q' and q'' are both negative, we know q is decreasing and its graph is concave-down. This tells us that the secant line from $x = 0.4$ to $x = 0.5$ is steeper (its slope is "more negative") than the tangent line, making -8 an underestimate.

Alternatively we can write
$$v'(0.4) = q'(0.4) \approx \frac{q(0.4) - q(0.3)}{0.1} \quad \text{average rate of change of } q$$
$$= \frac{3.9 - 4.5}{0.1} = -6.$$

Here, the secant line from $x = 0.3$ to $x = 0.4$ is less steep (its slope is "less negative") than the tangent line, making -6 an overestimate.

37. If we split the integral at $x = 0$, we have

$$\int_{-x^2}^{x^2} e^{t^2} \, dt = \int_{-x^2}^{0} e^{t^2} \, dt + \int_{0}^{x^2} e^{t^2} \, dt = -\int_{0}^{-x^2} e^{t^2} \, dt + \int_{0}^{x^2} e^{t^2} \, dt.$$

If we let

$$F(x) = \int_{0}^{x} e^{t^2} \, dt,$$

using the chain rule on each part separately gives

$$\frac{d}{dx}[-F(-x^2) + F(x^2)] = -(-2x)F'(-x^2) + (2x)F'(x^2) = (2x)e^{(-x^2)^2} + (2x)e^{(x^2)^2} = 4xe^{x^4}.$$

41. We split the integral $\int_{x}^{x^3} e^{-t^2} \, dt$ into two pieces, say at $t = 1$ (though it could be at any other point):

$$\int_{x}^{x^3} e^{-t^2} \, dt = \int_{1}^{x^3} e^{-t^2} \, dt + \int_{x}^{1} e^{-t^2} \, dt = \int_{1}^{x^3} e^{-t^2} \, dt - \int_{1}^{x} e^{-t^2} \, dt.$$

We have used the fact that $\int_{x}^{1} e^{-t^2} \, dt = -\int_{1}^{x} e^{-t^2} \, dt$. Differentiating gives

$$\frac{d}{dx}\left(\int_{x}^{x^3} e^{-t^2} \, dt\right) = \frac{d}{dx}\left(\int_{1}^{x^3} e^{-t^2} \, dt\right) - \frac{d}{dx}\left(\int_{1}^{x} e^{-t^2} \, dt\right)$$

For the first integral, we use the chain rule with $g(x) = x^3$ as the inside function, so the final answer is

$$\frac{d}{dx}\left(\int_{x}^{x^3} e^{-t^2} \, dt\right) = e^{-(x^3)^2} \cdot 3x^2 - e^{-x^2} = 3x^2 e^{-x^6} - e^{-x^2}.$$

Strengthen Your Understanding

45. $f(x) = x^2$ has a minimum at $x = 0$. However, $F(x)$ is non-decreasing for all x since $F'(x) = x^2 \geq 0$ everywhere.

49. True. The Construction Theorem for Antiderivatives gives a method for building an antiderivative with a definite integral.

53. True. Since F and G are both antiderivatives of f, they must differ by a constant. In fact, we can see that the constant C is equal to $\int_{0}^{2} f(t) \, dt$ since

$$F(x) = \int_{0}^{x} f(t) \, dt = \int_{2}^{x} f(t) \, dt + \int_{0}^{2} f(t) \, dt = G(x) + C.$$

Additional Problems (online only)

57. (a) The definition of R gives

$$R(0) = \int_{0}^{0} \sqrt{1 + t^2} \, dt = 0$$

and

$$R(-x) = \int_{0}^{-x} \sqrt{1 + t^2} \, dt.$$

Changing the variable of integration by letting $t = -z$ gives

$$\int_{0}^{-x} \sqrt{1 + t^2} \, dt = \int_{0}^{x} \sqrt{1 + (-z)^2}(-dz) = -\int_{0}^{x} \sqrt{1 + z^2} \, dz.$$

Thus R is an odd function.

(b) Using the Second Fundamental Theorem gives $R'(x) = \sqrt{1 + x^2}$, which is always positive, so R is increasing everywhere.

(c) Since

$$R''(x) = \frac{x}{\sqrt{1 + x^2}},$$

then R is concave up if $x > 0$ and concave down if $x < 0$.

(d) See Figure 6.14.

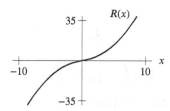

Figure 6.14

(e) We have

$$\lim_{x \to \infty} \frac{R(x)}{x^2} = \lim_{x \to \infty} \frac{\int_0^x \sqrt{1 + t^2}\, dt}{x^2}.$$

Using l'Hopital's rule gives

$$\lim_{x \to \infty} \frac{\sqrt{1 + x^2}}{2x} = \lim_{x \to \infty} \frac{\sqrt{1/x^2 + 1}}{2} = \frac{1}{2}.$$

Thus the limit exists; its value is 1/2.

Solutions for Chapter 6 Review

Exercises

1. We find the changes in $f(x)$ between any two values of x by counting the area between the curve of $f'(x)$ and the x-axis. Since $f'(x)$ is linear throughout, this is quite easy to do. From $x = 0$ to $x = 1$, we see that $f'(x)$ outlines a triangle of area 1/2 below the x-axis (the base is 1 and the height is 1). By the Fundamental Theorem,

$$\int_0^1 f'(x)\, dx = f(1) - f(0),$$

so

$$f(0) + \int_0^1 f'(x)\, dx = f(1)$$

$$f(1) = 2 - \frac{1}{2} = \frac{3}{2}$$

Similarly, between $x = 1$ and $x = 2$ we can see that $f'(x)$ outlines a rectangle below the x-axis with area 1, so $f(2) = 3/2 - 1 = 1/2$. Continuing with this procedure (note that at $x = 4$, $f'(x)$ becomes positive), we get the table below.

x	0	1	2	3	4	5	6
$f(x)$	2	3/2	1/2	−1/2	−1	−1/2	1/2

5. (a) The value of the integral is negative since the area below the x-axis is greater than the area above the x-axis. We count boxes: The area below the x-axis includes approximately 11.5 boxes and each box has area $(2)(1) = 2$, so

$$\int_0^5 f(x)dx \approx -23.$$

The area above the x-axis includes approximately 2 boxes, each of area 2, so

$$\int_5^7 f(x)dx \approx 4.$$

So we have

$$\int_0^7 f(x)dx = \int_0^5 f(x)dx + \int_5^7 f(x)dx \approx -23 + 4 = -19.$$

(b) By the Fundamental Theorem of Calculus, we have

$$F(7) - F(0) = \int_0^7 f(x)dx$$

so,

$$F(7) = F(0) + \int_0^7 f(x)dx = 25 + (-19) = 6.$$

9. $\displaystyle\int (x^3 - 2)\,dx = \frac{x^4}{4} - 2x + C$

13. $\displaystyle\int 4\sqrt{w}\,dw = \frac{8}{3}w^{3/2} + C$

17. $\dfrac{x^2}{2} + 2x^{1/2} + C$

21. $\tan x + C$

25. $\dfrac{1}{\ln 2}2^x + C$, since $\dfrac{d}{dx}(2^x) = (\ln 2) \cdot 2^x$

29. We have

$$\int_{-\pi/2}^{\pi/2} 2\cos\phi\,d\phi = 2\left(\sin\phi\Big|_{-\pi/2}^{\pi/2}\right) = 2\left(\sin\frac{\pi}{2} - \sin\frac{-\pi}{2}\right) = 2\left(1 - (-1)\right). = 4.$$

33. $F(x) = \displaystyle\int e^x\,dx = e^x + C$. If $F(0) = 4$, then $F(0) = 1 + C = 4$ and thus $C = 3$. So $F(x) = e^x + 3$.

37. Differentiating y with respect to x gives

$$y' = nx^{n-1}$$

for all values of A.

41. $r = \displaystyle\int 3\sin p\,dp = -3\cos p + C$

45. Integrating gives

$$\int \frac{dq}{dz}\,dz = \int (2 + \sin z)\,dz = 2z - \cos z + C.$$

If $q = 5$ when $z = 0$, then $2(0) - \cos(0) + C = 5$ so $C = 6$. Thus $q = 2z - \cos z + 6$.

Problems

49. Between time $t = 0$ and time $t = B$, the velocity of the cork is always positive, which means the cork is moving upward. At time $t = B$, the velocity is zero, and so the cork has stopped moving altogether. Since shortly thereafter the velocity of the cork becomes negative, the cork will next begin to move downward. Thus when $t = B$ the cork has risen as far as it ever will, and is riding on top of the crest of the wave.

From time $t = B$ to time $t = D$, the velocity of the cork is negative, which means it is falling. When $t = D$, the velocity is again zero, and the cork has ceased to fall. Thus when $t = D$ the cork is riding on the bottom of the trough of the wave.

Since the cork is on the crest at time B and in the trough at time D, it is probably midway between crest and trough when the time is midway between B and D. Thus at time $t = C$ the cork is moving through the equilibrium position on its way down. (The equilibrium position is where the cork would be if the water were absolutely calm.) By symmetry, $t = A$ is the time when the cork is moving through the equilibrium position on the way up.

Since acceleration is the derivative of velocity, points where the acceleration is zero would be critical points of the velocity function. Since point A (a maximum) and point C (a minimum) are critical points, the acceleration is zero there.

A possible graph of the height of the cork is shown in Figure 6.15. The horizontal axis represents a height equal to the average depth of the ocean at that point (the equilibrium position of the cork).

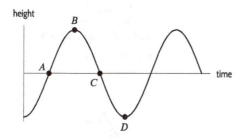

Figure 6.15

53. (a) Starting at $x = 3$, we are given that $f(3) = 0$. Moving to the left on the interval $2 < x < 3$, we have $f'(x) = -1$, so $f(2) = f(3) - (1)(-1) = 1$. On the interval $0 < x < 2$, we have $f'(x) = 1$, so

$$f(0) = f(2) + 1(-2) = -1.$$

Moving to the right from $x = 3$, we know that $f'(x) = 2$ on $3 < x < 4$. So $f(4) = f(3) + 2 = 2$. On the interval $4 < x < 6$, $f'(x) = -2$ so

$$f(6) = f(4) + 2(-2) = -2.$$

On the interval $6 < x < 7$, we have $f'(x) = 1$, so

$$f(7) = f(6) + 1 = -2 + 1 = -1.$$

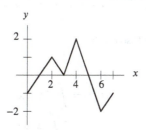

(b) In part (a) we found that $f(0) = -1$ and $f(7) = -1$.

(c) The integral $\int_0^7 f'(x)\,dx$ is given by the sum

$$\int_0^7 f'(x)\,dx = (1)(2) + (-1)(1) + (2)(1) + (-2)(2) + (1)(1) = 0.$$

Alternatively, knowing $f(7)$ and $f(0)$ and using the Fundamental Theorem of Calculus, we have

$$\int_0^7 f'(x)\,dx = f(7) - f(0) = -1 - (-1) = 0.$$

57. Since $y = 0$ only when $x = 0$ and $x = 1$, the area lies between these limits and is given by

$$\text{Area} = \int_0^1 x^2(1-x)^2 dx = \int_0^1 x^2(1 - 2x + x^2)\,dx = \int_0^1 (x^2 - 2x^3 + x^4)\,dx$$

$$= \frac{x^3}{3} - \frac{2}{4}x^4 + \frac{x^5}{5}\Big|_0^1 = \frac{1}{30}.$$

61. Since the graph of $y = e^x$ is above the graph of $y = \cos x$ (see Figure 6.16) we have

$$\text{Area} = \int_0^1 (e^x - \cos x)\, dx$$

$$= \int_0^1 e^x\, dx - \int_0^1 \cos x\, dx$$

$$= e^x \Big|_0^1 - \sin x \Big|_0^1$$

$$= e^1 - e^0 - \sin 1 + \sin 0$$

$$= e - 1 - \sin 1.$$

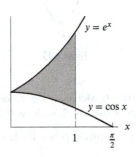

Figure 6.16

65. The average value of $v(x)$ on the interval $1 \le x \le c$ is

$$\frac{1}{c-1} \int_1^c \frac{6}{x^2}\, dx = \frac{1}{c-1} \left(-\frac{6}{x} \right) \Big|_1^c = \frac{1}{c-1} \left(\frac{-6}{c} + 6 \right) = \frac{6}{c}.$$

Since $\dfrac{1}{c-1} \displaystyle\int_1^c \dfrac{6}{x^2}\, dx = 1$, we have $\dfrac{6}{c} = 1$, so $c = 6$.

69. We have:

$$f(2) = \int_0^2 2x^2\, dx$$

$$= 2 \int_0^2 x^2\, dx$$

$$= 2 \cdot \frac{1}{3} x^3 \Big|_0^2$$

$$= \frac{16}{3}.$$

73. If we split the integral at $x = 0$, we have

$$\int_{-x}^x e^{-t^4}\, dt = \int_{-x}^0 e^{-t^4}\, dt + \int_0^x e^{-t^4}\, dt = -\int_0^{-x} e^{-t^4}\, dt + \int_0^x e^{-t^4}\, dt.$$

If we let

$$F(x) = \int_0^x e^{-t^4}\, dt,$$

using the chain rule on each part separately gives

$$\frac{d}{dx}[-F(-x) + F(x)] = -(-1)F'(-x) + (1)F'(x) = e^{-(-x)^4} + e^{-x^4} = 2e^{-x^4}.$$

77. (a) Since $f'(t)$ is positive on the interval $0 < t < 2$ and negative on the interval $2 < t < 5$, the function $f(t)$ is increasing on $0 < t < 2$ and decreasing on $2 < t < 5$. Thus $f(t)$ attains its maximum at $t = 2$. Since the area under the t-axis is greater than the area above the t-axis, the function $f(t)$ decreases more than it increases. Thus, the minimum is at $t = 5$.

(b) To estimate the value of f at $t = 2$, we see that the area under $f'(t)$ between $t = 0$ and $t = 2$ is about 1 box, which has area 5. Thus,

$$f(2) = f(0) + \int_0^2 f'(t)dt \approx 50 + 5 = 55.$$

The maximum value attained by the function is $f(2) \approx 55$.

The area between $f'(t)$ and the t-axis between $t = 2$ and $t = 5$ is about 3 boxes, each of which has an area of 5. Thus

$$f(5) = f(2) + \int_2^5 f'(t)dt \approx 55 + (-15) = 40.$$

The minimum value attained by the function is $f(5) = 40$.

(c) Using part (b), we have $f(5) - f(0) = 40 - 50 = -10$. Alternately, we can use the Fundamental Theorem:

$$f(5) - f(0) = \int_0^5 f'(t)dt \approx 5 - 15 = -10.$$

81. Since the car's acceleration is constant, a graph of its velocity against time t is linear, as shown below.

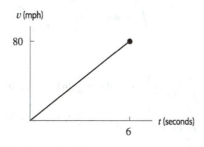

The acceleration is just the slope of this line:

$$\frac{dv}{dt} = \frac{80 - 0 \text{ mph}}{6 \text{ sec}} = \frac{40}{3} = 13.33 \frac{\text{mph}}{\text{sec}}.$$

To convert our units into ft/sec²,

$$\frac{40}{3} \cdot \frac{\text{mph}}{\text{sec}} \cdot \frac{5280 \text{ ft}}{1 \text{ mile}} \cdot \frac{1 \text{ hour}}{3600 \text{ sec}} = 19.55 \frac{\text{ft}}{\text{sec}^2}$$

85. Since $V'(r) = S(r)$ and $S(r) = 4\pi r^2$, we have

$$V'(r) = 4\pi r^2.$$

Thus, we have, for some arbitrary constant K:

$$V(r) = \int 4\pi r^2 \, dr = 4\pi \int r^2 \, dr = 4\pi \frac{r^3}{3} + K = \frac{4}{3}\pi r^3 + K.$$

Since a sphere of radius $r = 0$ has volume $= 0$, we substitute to find K:

$$0 = \frac{4}{3}\pi 0^3 + K$$
$$K = 0.$$

Thus

$$V(r) = \frac{4}{3}\pi r^3.$$

89. By the Second Fundamental Theorem, we know that $N'(t) = r(t)$. Since $r(t) > 0$, this means $N'(t) > 0$, so N is an increasing function. Since $r'(t) < 0$, this means $N''(t) < 0$, so the graph of N is concave down.

CAS Challenge Problems

93. (a) We have $\Delta x = \dfrac{(b-a)}{n}$ and $x_i = a + i(\Delta x) = a + i\left(\dfrac{b-a}{n}\right)$, so, since $f(x_i) = x_i^3$,

$$\text{Riemann sum} = \sum_{i=1}^{n} f(x_i)\Delta x = \sum_{i=1}^{n} \left[a + i\left(\frac{b-a}{n}\right)\right]^3 \left(\frac{b-a}{n}\right).$$

(b) A CAS gives

$$\sum_{i=1}^{n} \left[a + \frac{i(b-a)}{n}\right]^3 \frac{(b-a)}{n} = -\frac{(a-b)(a^3(n-1)^2 + (a^2b + ab^2)(n^2-1) + b^3(n+1)^3)}{4n^2}.$$

Taking the limit as $n \to \infty$ gives

$$\lim_{n \to \infty} \sum_{i=1}^{n} \left[a + i\left(\frac{b-a}{n}\right)\right]^3 \left(\frac{b-a}{n}\right) = -\frac{(a+b)(a-b)(a^2+b^2)}{4}.$$

(c) The answer to part (b) simplifies to $\dfrac{b^4}{4} - \dfrac{a^4}{4}$. Since $\dfrac{d}{dx}\left(\dfrac{x^4}{4}\right) = x^3$, the Fundamental Theorem of Calculus says that

$$\int_a^b x^3 dx = \frac{x^4}{4}\Big|_a^b = \frac{b^4}{4} - \frac{a^4}{4}.$$

97. (a) A CAS gives

$$\int \frac{1}{(x-1)(x-3)}\,dx = \frac{1}{2}(\ln|x-3| - \ln|x-1|)$$

$$\int \frac{1}{(x-1)(x-4)}\,dx = \frac{1}{3}(\ln|x-4| - \ln|x-1|)$$

$$\int \frac{1}{(x-1)(x+3)}\,dx = \frac{1}{4}(\ln|x+3| - \ln|x-1|).$$

Although the absolute values are needed in the answer, some CASs may not include them.

(b) The three integrals in part (a) obey the rule

$$\int \frac{1}{(x-a)(x-b)}\,dx = \frac{1}{b-a}(\ln|x-b| - \ln|x-a|).$$

(c) Checking the formula by calculating the derivative

$$\frac{d}{dx}\left(\frac{1}{b-a}(\ln|x-b| - \ln|x-a|)\right) = \frac{1}{b-a}\left(\frac{1}{x-b} - \frac{1}{x-a}\right)$$

$$= \frac{1}{b-a}\left(\frac{(x-a) - (x-b)}{(x-a)(x-b)}\right)$$

$$= \frac{1}{b-a}\left(\frac{b-a}{(x-a)(x-b)}\right) = \frac{1}{(x-a)(x-b)}.$$

CHAPTER SEVEN

Solutions for Section 7.1

Exercises

1. (a) We substitute $w = 1 + x^2$, $dw = 2x\,dx$.

$$\int_{x=0}^{x=1} \frac{x}{1+x^2}\,dx = \frac{1}{2}\int_{w=1}^{w=2} \frac{1}{w}\,dw = \frac{1}{2}\ln|w|\Big|_1^2 = \frac{1}{2}\ln 2.$$

(b) We substitute $w = \cos x$, $dw = -\sin x\,dx$.

$$\int_{x=0}^{x=\frac{\pi}{4}} \frac{\sin x}{\cos x}\,dx = -\int_{w=1}^{w=\sqrt{2}/2} \frac{1}{w}\,dw$$

$$= -\ln|w|\Big|_1^{\sqrt{2}/2} = -\ln\frac{\sqrt{2}}{2} = \frac{1}{2}\ln 2.$$

5. We use the substitution $w = -x$, $dw = -dx$.

$$\int e^{-x}dx = -\int e^w dw = -e^w + C = -e^{-x} + C.$$

Check: $\frac{d}{dx}(-e^{-x} + C) = -(-e^{-x}) = e^{-x}$.

9. We use the substitution $w = 3 - t$, $dw = -dt$.

$$\int \sin(3-t)dt = -\int \sin(w)dw = -(-\cos(w)) + C = \cos(3-t) + C.$$

Check: $\frac{d}{dt}(\cos(3-t) + C) = -\sin(3-t)(-1) = \sin(3-t)$.

13. We use the substitution $w = 1 + 2x^3$, $dw = 6x^2\,dx$.

$$\int x^2(1+2x^3)^2\,dx = \int w^2(\tfrac{1}{6}\,dw) = \frac{1}{6}(\frac{w^3}{3}) + C = \frac{1}{18}(1+2x^3)^3 + C.$$

Check: $\frac{d}{dx}\left[\frac{1}{18}(1+2x^2)^3 + C\right] = \frac{1}{18}[3(1+2x^3)^2(6x^2)] = x^2(1+2x^3)^2.$

17. In this case, it seems easier not to substitute.

$$\int y^2(1+y)^2\,dy = \int y^2(y^2+2y+1)\,dy = \int (y^4 + 2y^3 + y^2)\,dy$$

$$= \frac{y^5}{5} + \frac{y^4}{2} + \frac{y^3}{3} + C.$$

Check: $\frac{d}{dy}\left(\frac{y^5}{5} + \frac{y^4}{2} + \frac{y^3}{3} + C\right) = y^4 + 2y^3 + y^2 = y^2(y+1)^2.$

21. We use the substitution $w = 4 - x$, $dw = -dx$.

$$\int \frac{1}{\sqrt{4-x}}\,dx = -\int \frac{1}{\sqrt{w}}\,dw = -2\sqrt{w} + C = -2\sqrt{4-x} + C.$$

Check: $\frac{d}{dx}(-2\sqrt{4-x} + C) = -2 \cdot \frac{1}{2} \cdot \frac{1}{\sqrt{4-x}} \cdot -1 = \frac{1}{\sqrt{4-x}}.$

25. We use the substitution $w = \cos\theta + 5$, $dw = -\sin\theta\, d\theta$.

$$\int \sin\theta(\cos\theta + 5)^7\, d\theta = -\int w^7\, dw = -\frac{1}{8}w^8 + C$$
$$= -\frac{1}{8}(\cos\theta + 5)^8 + C.$$

Check:

$$\frac{d}{d\theta}\left[-\frac{1}{8}(\cos\theta + 5)^8 + C\right] = -\frac{1}{8}\cdot 8(\cos\theta + 5)^7\cdot(-\sin\theta)$$
$$= \sin\theta(\cos\theta + 5)^7$$

29. We use the substitution $w = \sin 5\theta$, $dw = 5\cos 5\theta\, d\theta$.

$$\int \sin^6 5\theta \cos 5\theta\, d\theta = \frac{1}{5}\int w^6\, dw = \frac{1}{5}(\frac{w^7}{7}) + C = \frac{1}{35}\sin^7 5\theta + C.$$

Check: $\frac{d}{d\theta}(\frac{1}{35}\sin^7 5\theta + C) = \frac{1}{35}[7\sin^6 5\theta](5\cos 5\theta) = \sin^6 5\theta \cos 5\theta$.

Note that we could also use Problem 27 to solve this problem, substituting $w = 5\theta$ and $dw = 5\, d\theta$ to get:

$$\int \sin^6 5\theta \cos 5\theta\, d\theta = \frac{1}{5}\int \sin^6 w \cos w\, dw$$
$$= \frac{1}{5}(\frac{\sin^7 w}{7}) + C = \frac{1}{35}\sin^7 5\theta + C.$$

33. It seems easier not to substitute.

$$\int \frac{(t+1)^2}{t^2}\, dt = \int \frac{(t^2 + 2t + 1)}{t^2}\, dt$$
$$= \int \left(1 + \frac{2}{t} + \frac{1}{t^2}\right)\, dt = t + 2\ln|t| - \frac{1}{t} + C.$$

Check: $\frac{d}{dt}(t + 2\ln|t| - \frac{1}{t} + C) = 1 + \frac{2}{t} + \frac{1}{t^2} = \frac{(t+1)^2}{t^2}$.

37. We use the substitution $w = \sqrt{x}$, $dw = \frac{1}{2\sqrt{x}}\, dx$.

$$\int \frac{\cos\sqrt{x}}{\sqrt{x}}\, dx = \int \cos w(2\, dw) = 2\sin w + C = 2\sin\sqrt{x} + C.$$

Check: $\frac{d}{dx}(2\sin\sqrt{x} + C) = 2\cos\sqrt{x}\left(\frac{1}{2\sqrt{x}}\right) = \frac{\cos\sqrt{x}}{\sqrt{x}}$.

41. We use the substitution $w = x^2 + 2x + 19$, $dw = 2(x+1)dx$.

$$\int \frac{(x+1)dx}{x^2 + 2x + 19} = \frac{1}{2}\int \frac{dw}{w} = \frac{1}{2}\ln|w| + C = \frac{1}{2}\ln(x^2 + 2x + 19) + C.$$

(We can drop the absolute value signs, since $x^2 + 2x + 19 = (x+1)^2 + 18 > 0$ for all x.)

Check: $\frac{d}{dx}[\frac{1}{2}\ln(x^2 + 2x + 19)] = \frac{1}{2}\frac{1}{x^2 + 2x + 19}(2x + 2) = \frac{x+1}{x^2 + 2x + 19}$.

45. We use the substitution $w = 3t$, $dw = 3\, dt$.

$$\int \sinh 3t\, dt = \frac{1}{3}\int \sinh w\, dw = \frac{1}{3}\cosh w + C = \frac{1}{3}\cosh 3t + C.$$

Check: $\frac{d}{dt}\left(\frac{1}{3}\cosh 3t + C\right) = \frac{1}{3}(3\sinh 3t) = \sinh 3t$.

49. Use the substitution $w = \cosh x$ and $dw = \sinh x \, dx$ so

$$\int \cosh^2 x \sinh x \, dx = \int w^2 \, dw = \frac{1}{3}w^3 + C = \frac{1}{3}\cosh^3 x + C.$$

Check this answer by taking the derivative: $\dfrac{d}{dx}\left[\dfrac{1}{3}\cosh^3 x + C\right] = \cosh^2 x \sinh x.$

53. Make the substitution $w = x^2$, $dw = 2x \, dx$. We have

$$\int 2x\cos(x^2)\,dx = \int \cos w \, dw = \sin w + C = \sin x^2 + C.$$

57. Make the substitution $w = x^2 + 1$, $dw = 2x \, dx$. We have

$$\int \frac{x}{x^2+1}dx = \frac{1}{2}\int \frac{dw}{w} = \frac{1}{2}\ln|w| + C = \frac{1}{2}\ln(x^2+1) + C.$$

(Notice that since $x^2 + 1 \geq 0$, $|x^2 + 1| = x^2 + 1$.)

61. Since $d(-\cos\theta)/d\theta = \sin\theta$, we have

$$\int_0^{\pi/2} e^{-\cos\theta}\sin\theta \, d\theta = e^{-\cos\theta}\Big|_0^{\pi/2} = e^{-\cos(\pi/2)} - e^{-\cos(0)} = 1 - \frac{1}{e}.$$

65. We substitute $w = \sqrt{x}$. Then $dw = \frac{1}{2}x^{-1/2}dx$.

$$\int_{x=1}^{x=4}\frac{\cos\sqrt{x}}{\sqrt{x}}dx = \int_{w=1}^{w=2}\cos w(2\,dw)$$

$$= 2(\sin w)\Big|_1^2 = 2(\sin 2 - \sin 1).$$

69. $\displaystyle\int_1^3 \frac{1}{x}dx = \ln x\Big|_1^3 = \ln 3.$

73. Let $w = \sqrt{y+1}$, so $y = w^2 - 1$ and $dy = 2w\,dw$. Thus

$$\int y\sqrt{y+1}\,dy = \int (w^2-1)w2w\,dw = 2\int w^4 - w^2\,dw$$

$$= \frac{2}{5}w^5 - \frac{2}{3}w^3 + C = \frac{2}{5}(y+1)^{5/2} - \frac{2}{3}(y+1)^{3/2} + C.$$

77. Let $w = \sqrt{x-2}$, so $x = w^2 + 2$ and $dx = 2w\,dw$. Thus

$$\int x^2\sqrt{x-2}\,dx = \int (w^2+2)^2 w2w\,dw = 2\int w^6 + 4w^4 + 4w^2\,dw$$

$$= \frac{2}{7}w^7 + \frac{8}{5}w^5 + \frac{8}{3}w^3 + C$$

$$= \frac{2}{7}(x-2)^{7/2} + \frac{8}{5}(x-2)^{5/2} + \frac{8}{3}(x-2)^{3/2} + C.$$

Problems

81. Multiplying the numerator and denominator by e^t gives

$$\int \frac{1}{1+e^{-t}}dt = \int \frac{e^t}{e^t+1}dt.$$

To calculate this integral, we substitute $w = e^t + 1$, so $dw = e^t \, dt$:

$$\int \frac{e^t}{e^t+1}dt = \int \frac{1}{w}dw = \ln|w| + C = \ln(1+e^t) + C.$$

85. If we let $t = \pi - x$ in the first integral, we get $dt = -dx$ and $x = \pi - t$. Also, the limits $x = 0$ and $x = \pi$ become $t = \pi$ and $t = 0$. Thus

$$\int_0^\pi x \cos(\pi - x)\, dx = -\int_\pi^0 (\pi - t) \cos t\, dt = \int_0^\pi (\pi - t) \cos t\, dt.$$

89. Using substitution, we have

$$\int_1^3 f'(x) e^{f(x)}\, dx = e^{f(x)} \Big|_1^3 = e^{f(3)} - e^{f(1)} = e^{11} - e^7.$$

93. Using substitution, we can show

$$\int_0^1 f'(x)(f(x))^2\, dx = \frac{(f(x))^3}{3} \Big|_0^1 = \frac{(f(1))^3}{3} - \frac{(f(0))^3}{3} = \frac{7^3}{3} - \frac{5^3}{3} = \frac{218}{3}.$$

97. For the first integral, let $w = \sin x$, $dw = \cos x\, dx$. Then

$$\int e^{\sin x} \cos x\, dx = \int e^w\, dw.$$

For the second integral, let $w = \arcsin x$, $dw = \frac{1}{\sqrt{1-x^2}}dx$. Then

$$\int \frac{e^{\arcsin x}}{\sqrt{1-x^2}}\, dx = \int e^w\, dw.$$

101. By substitution with $w = g(x)$, we have

$$\int g'(x) e^{g(x)}\, dx = \int e^w\, dw = e^w + C = e^{g(x)} + C.$$

105. Letting $w = \sin t$, $dw = \cos t\, dt$, we have

$$\int \frac{\cos t}{\sin t}\, dt = \int \underbrace{(\sin t)}_{w}^{-1} \underbrace{\cos t\, dt}_{dw} = \int 1 \cdot w^{-1}\, dw,$$

so $w = \sin t$, $k = 1$, $n = -1$.

109. Since $w = \sin 2x$ we have $dw = 2 \cos 2x\, dx$, so $\cos 2x\, dx = 0.5\, dw$. Finding the limits of integration in terms of w gives:

$$w_0 = \sin\left(2 \cdot \frac{\pi}{12}\right) = \sin\left(\frac{\pi}{6}\right) = 0.5$$

$$w_1 = \sin\left(2 \cdot \frac{\pi}{4}\right) = \sin\left(\frac{\pi}{2}\right) = 1.$$

This gives

$$\int_{\pi/12}^{\pi/4} \sin^7(2x) \cos(2x)\, dx = \int_{0.5}^1 0.5 w^7\, dw,$$

so $k = 0.5$, $n = 7$, $w_0 = 0.5$, $w_1 = 1$.

113. We have

$$\int \frac{z^2\, dz}{e^{-z^3}} = \int e^{z^3} z^2\, dz.$$

Letting $w = z^3$, so $dw = 3z^2\, dz$, we have

$$\int \frac{z^2\, dz}{e^{-z^3}} = \int \frac{1}{3} e^w\, dw,$$

so $w = z^3$, $k = 1/3$.

117. (a) $2\sqrt{x} + C$

(b) $2\sqrt{x+1} + C$

(c) To get this last result, we make the substitution $w = \sqrt{x}$. Normally we would like to substitute $dw = \frac{1}{2\sqrt{x}}\,dx$, but in this case we cannot since there are no spare $\frac{1}{\sqrt{x}}$ terms around. Instead, we note $w^2 = x$, so $2w\,dw = dx$. Then

$$\int \frac{1}{\sqrt{x}+1}\,dx = \int \frac{2w}{w+1}\,dw$$
$$= 2\int \frac{(w+1)-1}{w+1}\,dw$$
$$= 2\int \left(1 - \frac{1}{w+1}\right)\,dw$$
$$= 2(w - \ln|w+1|) + C$$
$$= 2\sqrt{x} - 2\ln(\sqrt{x}+1) + C.$$

We also note that we can drop the absolute value signs, since $\sqrt{x}+1 \geq 0$ for all x.

121. The area under $f(x) = \sinh(x/2)$ between $x = 0$ and $x = 2$ is given by

$$A = \int_0^2 \sinh\left(\frac{x}{2}\right)\,dx = 2\cosh\left(\frac{x}{2}\right)\Big|_0^2 = 2\cosh 1 - 2.$$

125. See Figure 7.1. The period of $V = V_0 \sin(\omega t)$ is $2\pi/\omega$, so the area under the first arch is given by

$$\text{Area} = \int_0^{\pi/\omega} V_0 \sin(\omega t)\,dt$$
$$= -\frac{V_0}{\omega}\cos(\omega t)\Big|_0^{\pi/\omega}$$
$$= -\frac{V_0}{\omega}\cos(\pi) + \frac{V_0}{\omega}\cos(0)$$
$$= -\frac{V_0}{\omega}(-1) + \frac{V_0}{\omega}(1) = \frac{2V_0}{\omega}.$$

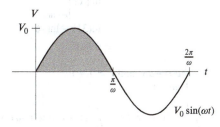

Figure 7.1

129. (a) If $w = t/2$, then $dw = (1/2)dt$. When $t = 0$, $w = 0$; when $t = 4$, $w = 2$. Thus,

$$\int_0^4 g(t/2)\,dt = \int_0^2 g(w)\,2dw = 2\int_0^2 g(w)\,dw = 2 \cdot 5 = 10.$$

(b) If $w = 2 - t$, then $dw = -dt$. When $t = 0$, $w = 2$; when $t = 2$, $w = 0$. Thus,

$$\int_0^2 g(2-t)\,dt = \int_2^0 g(w)\,(-dw) = +\int_0^2 g(w)\,dw = 5.$$

133. (a) We first try the substitution $w = \sin\theta$, $dw = \cos\theta\, d\theta$. Then
$$\int \sin\theta\cos\theta\, d\theta = \int w\, dw = \frac{w^2}{2} + C = \frac{\sin^2\theta}{2} + C.$$

(b) If we instead try the substitution $w = \cos\theta$, $dw = -\sin\theta\, d\theta$, we get
$$\int \sin\theta\cos\theta\, d\theta = -\int w\, dw = -\frac{w^2}{2} + C = -\frac{\cos^2\theta}{2} + C.$$

(c) Once we note that $\sin 2\theta = 2\sin\theta\cos\theta$, we can also say
$$\int \sin\theta\cos\theta\, d\theta = \frac{1}{2}\int \sin 2\theta\, d\theta.$$
Substituting $w = 2\theta$, $dw = 2\, d\theta$, the above equals
$$\frac{1}{4}\int \sin w\, dw = -\frac{\cos w}{4} + C = -\frac{\cos 2\theta}{4} + C.$$

(d) All these answers are correct. Although they have different forms, they differ from each other only in terms of a constant, and thus they are all acceptable antiderivatives. For example, $1 - \cos^2\theta = \sin^2\theta$, so $\frac{\sin^2\theta}{2} = -\frac{\cos^2\theta}{2} + \frac{1}{2}$. Thus the first two expressions differ only by a constant C.

Similarly, $\cos 2\theta = \cos^2\theta - \sin^2\theta = 2\cos^2\theta - 1$, so $-\frac{\cos 2\theta}{4} = -\frac{\cos^2\theta}{2} + \frac{1}{4}$, and thus the second and third expressions differ only by a constant. Of course, if the first two expressions and the last two expressions differ only in the constant C, then the first and last only differ in the constant as well.

137. Letting $w = \ln\left(x^2 + 1\right)$, we have $dw = \left(x^2 + 1\right)^{-1} \cdot 2x\, dx$, so
$$\frac{x}{x^2 + 1}\, dx = \frac{1}{2}\, dw.$$

This means we can write
$$\int_2^5 \frac{f\left(\ln\left(x^2 + 1\right)\right) x}{x^2 + 1}\, dx = \int_2^5 \overbrace{f\left(\ln\left(x^2 + 1\right)\right)}^{f(w)} \overbrace{\frac{x}{x^2 + 1}\, dx}^{(1/2)\, dw} = \int_{x=2}^{x=5} f(w) \cdot \frac{1}{2}\, dw = \int_{w=\ln 5}^{w=\ln 26} \frac{1}{2} \cdot f(w)\, dw.$$
so $k = 1/2$, $a = \ln 5$, $b = \ln 26$.

141. Letting $w = F(x)$, so $dw = f(x)\, dx$, we have:
$$\int_1^3 \frac{f(x)}{F(x)}\, dx = \int_{x=1}^{x=3} \frac{dw}{w}$$
$$= \ln|w| \Big|_{w=F(1)}^{w=F(3)}$$
$$= \ln|w| \Big|_{w=16}^{w=31}$$
$$= \ln(31) - \ln(16)$$
$$= 0.661.$$

145. We substitute $w = 1 - x$ into $I_{m,n}$. Then $dw = -dx$, and $x = 1 - w$.
When $x = 0$, $w = 1$, and when $x = 1$, $w = 0$, so
$$I_{m,n} = \int_0^1 x^m (1-x)^n dx = \int_1^0 (1-w)^m w^n (-dw)$$
$$= -\int_1^0 w^n (1-w)^m dw = \int_0^1 w^n (1-w)^m dw = I_{n,m}.$$

149. (a) A time period of Δt hours with flow rate of $f(t)$ cubic meters per hour has a flow of $f(t)\Delta t$ cubic meters. Summing the flows, we get total flow $\approx \Sigma f(t)\Delta t$, so
$$\text{Total flow} = \int_0^{72} f(t)\, dt \text{ cubic meters.}$$

(b) Since 1 day is 24 hours, $t = 24T$. The constant 24 has units hours per day, so $24T$ has units hours/day $\times$ day $=$ hours. Applying the substitution $t = 24T$ to the integral in part (a), we get
$$\text{Total flow} = \int_0^3 f(24T)\, 24\, dT \text{ cubic meters.}$$

153. (a) $E(t) = 1.4e^{0.07t}$

(b)

$$\text{Average Yearly Consumption} = \frac{\text{Total Consumption for the Century}}{100 \text{ years}}$$

$$= \frac{1}{100} \int_0^{100} 1.4e^{0.07t} \, dt$$

$$= (0.014) \left[\frac{1}{0.07} e^{0.07t} \Big|_0^{100} \right]$$

$$= (0.014) \left[\frac{1}{0.07} (e^7 - e^0) \right]$$

$$= 0.2(e^7 - 1) \approx 219 \text{ million megawatt-hours.}$$

(c) We are looking for t such that $E(t) \approx 219$:

$$1.4e^{0.07t} \approx 219$$
$$e^{0.07t} = 156.4.$$

Taking natural logs,

$$0.07t = \ln 156.4$$
$$t \approx \frac{5.05}{0.07} \approx 72.18.$$

Thus, consumption was closest to the average during 1972.

(d) Between the years 1900 and 2000 the graph of $E(t)$ looks like

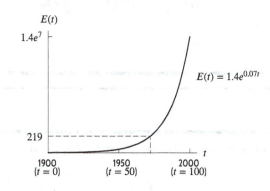

From the graph, we can see the t value such that $E(t) = 219$. It lies to the right of $t = 50$, and is thus in the second half of the century.

Strengthen Your Understanding

157. To do guess-and-check or substitution, we need an extra factor of $f'(x)$ in the integrand. For example, if $f(x) = 2x$, the left side is

$$\int (2x)^2 \, dx = \int 4x^2 \, dx = \frac{4x^3}{3} + C,$$

while the right side is $8x^3/3 + C$.

161. The integrand should have a factor that is a constant multiple of $3x^2 - 3$, for example $\int \sin(x^3 - 3x)(x^2 - 1) \, dx$.

Additional Problems (online only)

165. Since $v = \dfrac{dh}{dt}$, it follows that $h(t) = \displaystyle\int v(t) \, dt$ and $h(0) = h_0$. Since

$$v(t) = \frac{mg}{k} \left(1 - e^{-\frac{k}{m}t} \right) = \frac{mg}{k} - \frac{mg}{k} e^{-\frac{k}{m}t},$$

we have

$$h(t) = \int v(t)\,dt = \frac{mg}{k}\int dt - \frac{mg}{k}\int e^{-\frac{k}{m}t}\,dt.$$

The first integral is simply $\frac{mg}{k}t + C$. To evaluate the second integral, make the substitution $w = -\frac{k}{m}t$. Then

$$dw = -\frac{k}{m}\,dt,$$

so

$$\int e^{-\frac{k}{m}t}\,dt = \int e^w\left(-\frac{m}{k}\right)dw = -\frac{m}{k}e^w + C = -\frac{m}{k}e^{-\frac{k}{m}t} + C.$$

Thus

$$h(t) = \int v\,dt = \frac{mg}{k}t - \frac{mg}{k}\left(-\frac{m}{k}e^{-\frac{k}{m}t}\right) + C$$

$$= \frac{mg}{k}t + \frac{m^2 g}{k^2}e^{-\frac{k}{m}t} + C.$$

Since $h(0) = h_0$,

$$h_0 = \frac{mg}{k}\cdot 0 + \frac{m^2 g}{k^2}e^0 + C;$$

$$C = h_0 - \frac{m^2 g}{k^2}.$$

Thus

$$h(t) = \frac{mg}{k}t + \frac{m^2 g}{k^2}e^{-\frac{k}{m}t} - \frac{m^2 g}{k^2} + h_0$$

$$h(t) = \frac{mg}{k}t - \frac{m^2 g}{k^2}\left(1 - e^{-\frac{k}{m}t}\right) + h_0.$$

Solutions for Section 7.2

Exercises

1. (a) Since we can change x^2 into a multiple of x^3 by integrating, let $v' = x^2$ and $u = e^x$. Using $v = x^3/3$ and $u' = e^x$ we get

$$\int x^2 e^x dx = \int uv'dx = uv - \int u'v\,dx$$

$$= \frac{x^3 e^x}{3} - \frac{1}{3}\int x^3 e^x dx.$$

(b) Since we can change x^2 into a multiple of x by differentiating, let $u = x^2$ and $v' = e^x$. Using $u' = 2x$ and $v = e^x$ we have

$$\int x^2 e^x dx = \int uv'dx = uv - \int u'v\,dx$$

$$= x^2 e^x - 2\int xe^x dx.$$

5. Let $u = t$ and $v' = e^{5t}$, so $u' = 1$ and $v = \frac{1}{5}e^{5t}$.

Then $\int te^{5t}\,dt = \frac{1}{5}te^{5t} - \int \frac{1}{5}e^{5t}\,dt = \frac{1}{5}te^{5t} - \frac{1}{25}e^{5t} + C.$

9. Integration by parts with $u = \ln x$, $v' = x$ gives

$$\int x \ln x\,dx = \frac{x^2}{2}\ln x - \int \frac{1}{2}x\,dx = \frac{1}{2}x^2 \ln x - \frac{1}{4}x^2 + C.$$

13. Let $u = \sin\theta$ and $v' = \sin\theta$, so $u' = \cos\theta$ and $v = -\cos\theta$. Then

$$\int \sin^2\theta\, d\theta = -\sin\theta\cos\theta + \int \cos^2\theta\, d\theta$$

$$= -\sin\theta\cos\theta + \int (1 - \sin^2\theta)\, d\theta$$

$$= -\sin\theta\cos\theta + \int 1\, d\theta - \int \sin^2\theta\, d\theta.$$

By adding $\int \sin^2\theta\, d\theta$ to both sides of the above equation, we find that $2\int \sin^2\theta\, d\theta = -\sin\theta\cos\theta + \theta + C$, so $\int \sin^2\theta\, d\theta = -\frac{1}{2}\sin\theta\cos\theta + \frac{\theta}{2} + C'$.

17. Let $u = y$ and $v' = (y+3)^{1/2}$, so $u' = 1$ and $v = \frac{2}{3}(y+3)^{3/2}$:

$$\int y\sqrt{y+3}\, dy = \frac{2}{3}y(y+3)^{3/2} - \int \frac{2}{3}(y+3)^{3/2}\, dy = \frac{2}{3}y(y+3)^{3/2} - \frac{4}{15}(y+3)^{5/2} + C.$$

21. Let $u = \ln x$, $v' = x^{-2}$. Then $v = -x^{-1}$ and $u' = x^{-1}$. Integrating by parts, we get:

$$\int x^{-2}\ln x\, dx = -x^{-1}\ln x - \int (-x^{-1}) \cdot x^{-1}\, dx$$

$$= -x^{-1}\ln x - x^{-1} + C.$$

25. Let $u = \ln x$ and $v' = x^{1/2}$, so $u' = 1/x$ and $v = \frac{2}{3}x^{3/2}$. Then

$$\int \sqrt{x}\ln x\, dx = \frac{2}{3}x^{3/2}\ln x - \int \frac{2}{3}x^{3/2} \cdot \frac{1}{x}\, dx = \frac{2}{3}x^{3/2}\ln x - \frac{2}{3}\int x^{1/2}\, dx = \frac{2}{3}x^{3/2}\ln x - \frac{4}{9}x^{3/2} + C.$$

29. Let $u = \arctan 7z$ and $v' = 1$, so $u' = \dfrac{7}{1 + 49z^2}$ and $v = z$. Now $\displaystyle\int \frac{7z\, dz}{1 + 49z^2}$ can be evaluated by the substitution $w = 1 + 49z^2$, $dw = 98z\, dz$, so

$$\int \frac{7z\, dz}{1 + 49z^2} = 7\int \frac{\frac{1}{98}\, dw}{w} = \frac{1}{14}\int \frac{dw}{w} = \frac{1}{14}\ln|w| + C = \frac{1}{14}\ln(1 + 49z^2) + C$$

So

$$\int \arctan 7z\, dz = z\arctan 7z - \frac{1}{14}\ln(1 + 49z^2) + C.$$

33. Let $u = x, u' = 1$ and $v' = \sinh x, v = \cosh x$. Integrating by parts, we get

$$\int x\sinh x\, dx = x\cosh x - \int \cosh x\, dx$$

$$= x\cosh x - \sinh x + C.$$

37. If we let $w = 2x + 1$, then $dw = 2\, dx$, which gives $dx = dw/2$. Therefore,

$$\int (2x+1)^2 \ln(2x+1)\, dx = \int \frac{w^2}{2}\ln w\, dw.$$

Using integration by parts with $u = \ln w$ and $v' = w^2/2$ gives $u' = 1/w$ and $v = w^3/6$, so we have

$$\int (2x+1)^2 \ln(2x+1)\, dx = \int \frac{w^2}{2}\ln w\, dw$$

$$= \frac{w^3}{6} \cdot \ln w - \int \frac{w^3}{6} \cdot \frac{1}{w}\, dw$$

$$= \frac{1}{6}w^3 \ln w - \frac{1}{6}\int w^2 \, dw$$

$$= \frac{1}{6}w^3 \ln w - \frac{1}{18}w^3 + C$$

$$= \frac{1}{6}(2x+1)^3 \ln(2x+1) - \frac{1}{18}(2x+1)^3 + C$$

$$= \frac{1}{18}(2x+1)^3(3\ln(2x+1) - 1) + C.$$

41. We use integration by parts. Let $u = z$ and $v' = e^{-z}$, so $u' = 1$ and $v = -e^{-z}$. Then

$$\int_0^{10} ze^{-z} \, dz = -ze^{-z}\Big|_0^{10} + \int_0^{10} e^{-z} \, dz$$

$$= -10e^{-10} + (-e^{-z})\Big|_0^{10}$$

$$= -11e^{-10} + 1$$

$$\approx 0.9995.$$

45. We use integration by parts. Let $u = \arcsin z$ and $v' = 1$, so $u' = \dfrac{1}{\sqrt{1-z^2}}$ and $v = z$. Then

$$\int_0^1 \arcsin z \, dz = z \arcsin z\Big|_0^1 - \int_0^1 \frac{z}{\sqrt{1-z^2}} \, dz = \frac{\pi}{2} - \int_0^1 \frac{z}{\sqrt{1-z^2}} \, dz.$$

To find $\displaystyle\int_0^1 \frac{z}{\sqrt{1-z^2}} \, dz$, we substitute $w = 1 - z^2$, so $dw = -2z \, dz$.
Then

$$\int_{z=0}^{z=1} \frac{z}{\sqrt{1-z^2}} \, dz = -\frac{1}{2}\int_{w=1}^{w=0} w^{-\frac{1}{2}} \, dw = \frac{1}{2}\int_{w=0}^{w=1} w^{-\frac{1}{2}} \, dw = w^{\frac{1}{2}}\Big|_0^1 = 1.$$

Thus our final answer is $\frac{\pi}{2} - 1 \approx 0.571$.

Problems

49. Using a log property, we have

$$\int \ln\left((5-3x)^2\right) \, dx = \int 2\ln(5-3x) \, dx.$$

So we let

$$w = 5 - 3x$$
$$dw = -3dx$$
$$dx = -\frac{1}{3} \, dw,$$

which gives,

$$\int \ln(5-3x)^2 \, dx = \int 2\ln \overbrace{(5-3x)}^{w} \overbrace{dx}^{-\frac{1}{3} dw}$$

$$= \int -\frac{2}{3} \cdot \ln w \, dw.$$

Thus, $w = 5 - 3x, k = -2/3$.

53. Since $\int \arctan z \, dz = \int 1 \cdot \arctan z \, dz$, we take $u = \arctan z$, $v' = 1$, so $u' = 1/(1 + z^2)$ and $v = z$. Then

$$\int \arctan z \, dz = z \arctan z - \int \frac{z}{1 + z^2} \, dz = z \arctan z - \frac{1}{2} \ln(1 + z^2) + C.$$

Thus, we have

$$\int_0^2 \arctan z \, dz = \left(z \arctan z - \frac{1}{2} \ln(z^2 + 1) \right) \Big|_0^2 = 2 \arctan 2 - \frac{1}{2} \ln 5.$$

57. The graph of $f(x) = x \sin x$ is in Figure 7.2. The first positive zero is at $x = \pi$, so, using integration by parts,

$$\text{Area} = \int_0^\pi x \sin x \, dx$$

$$= -x \cos x \Big|_0^\pi + \int_0^\pi \cos x \, dx$$

$$= -x \cos x \Big|_0^\pi + \sin x \Big|_0^\pi$$

$$= -\pi \cos \pi - (-0 \cos 0) + \sin \pi - \sin 0 = \pi.$$

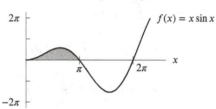

Figure 7.2

61. Let $u = e^\theta$ and $v' = \cos \theta$, so $u' = e^\theta$ and $v = \sin \theta$. Then $\int e^\theta \cos \theta \, d\theta = e^\theta \sin \theta - \int e^\theta \sin \theta \, d\theta$.

In Problem 60 we found that $\int e^x \sin x \, dx = \frac{1}{2} e^x (\sin x - \cos x) + C$.

$$\int e^\theta \cos \theta \, d\theta = e^\theta \sin \theta - \left[\frac{1}{2} e^\theta (\sin \theta - \cos \theta) \right] + C$$

$$= \frac{1}{2} e^\theta (\sin \theta + \cos \theta) + C.$$

65. Using integration by parts with $u(x) = x$ and $v'(x) = f''(x)$ gives $u'(x) = 1$ and $v(x) = f'(x)$, so

$$\int x f''(x) \, dx = x f'(x) - \int f'(x) \, dx = x f'(x) - f(x) + C.$$

69. We integrate by parts. Since we know what the answer is supposed to be, it's easier to choose u and v'. Let $u = \cos^{n-1} x$ and $v' = \cos x$, so $u' = (n-1) \cos^{n-2} x (-\sin x)$ and $v = \sin x$. Then

$$\int \cos^n x \, dx = \cos^{n-1} x \sin x + (n-1) \int \cos^{n-2} x \sin^2 x \, dx$$

$$= \cos^{n-1} x \sin x + (n-1) \int \cos^{n-2} x (1 - \cos^2 x) \, dx$$

$$= \cos^{n-1} x \sin x - (n-1) \int \cos^n x \, dx + (n-1) \int \cos^{n-2} x \, dx.$$

Thus, by adding $(n-1) \int \cos^n x \, dx$ to both sides of the equation, we find

$$n \int \cos^n x \, dx = \cos^{n-1} x \sin x + (n-1) \int \cos^{n-2} x \, dx,$$

$$\text{so} \int \cos^n dx = \frac{1}{n} \cos^{n-1} x \sin x + \frac{n-1}{n} \int \cos^{n-2} x \, dx.$$

73. We use integration by parts with $u = x^3$ and $v' = g'(x)$, so $u' = 3x^2$ and $v = g(x)$, giving:

$$\int x^3 g'(x)\,dx = uv - \int u'v\,dx$$

$$= x^3 g(x) - \int 3x^2 g(x)\,dx.$$

We are given that

$$\int x^3 g'(x)\,dx = f(x)g(x) - \int x^2 \cos x\,dx.$$

This means:

$$f(x)g(x) - \int x^2 \cos x\,dx = x^3 g(x) - \int 3x^2 g(x)\,dx$$

$$\text{so}\quad f(x) = x^3$$

$$g(x) = \frac{1}{3}\cos x.$$

77. (a) We have

$$F(a) = \int_0^a x^2 e^{-x}\,dx$$

$$= -x^2 e^{-x}\Big|_0^a + \int_0^a 2xe^{-x}\,dx$$

$$= (-x^2 e^{-x} - 2xe^{-x})\Big|_0^a + 2\int_0^a e^{-x}\,dx$$

$$= (-x^2 e^{-x} - 2xe^{-x} - 2e^{-x})\Big|_0^a$$

$$= -a^2 e^{-a} - 2ae^{-a} - 2e^{-a} + 2.$$

(b) $F(a)$ is increasing because $x^2 e^{-x}$ is positive, so as a increases, the area under the curve from 0 to a also increases and thus the integral increases.

(c) We have $F'(a) = a^2 e^{-a}$, so

$$F''(a) = 2ae^{-a} - a^2 e^{-a} = a(2 - a)e^{-a}.$$

We see that $F''(a) > 0$ for $0 < a < 2$, so F is concave up on this interval.

81. Since $\mathrm{Li}(x) = \displaystyle\int_2^x \frac{1}{\ln t}\,dt$, we see from the Construction Theorem that $\mathrm{Li}(x)$ is an antiderivative of $1/\ln(x)$, which means that $(\mathrm{Li}(x))' = 1/\ln(x)$. We need to perform integration by parts where $u = \mathrm{Li}(x), v = \ln x$. We have:

$$u' = \frac{1}{\ln x} \quad \text{and} \quad v' = \frac{1}{x},$$

which means

$$\int \underbrace{\mathrm{Li}(x)x^{-1}}_{uv'}\,dx = uv - \int u'v\,dx \qquad\qquad \text{integration by parts}$$

$$= \underbrace{\mathrm{Li}(x)\ln x}_{uv} - \int \underbrace{\frac{1}{\ln(x)}\cdot \ln x}_{u'v}\,dx$$

$$= \mathrm{Li}(x)\ln x - \int dx$$

$$= \mathrm{Li}(x)\ln x - x + C.$$

Strengthen Your Understanding

85. The integral $\int \theta^2 \sin\theta\,d\theta$ requires integration by parts twice. Many other answers are possible.

89. True. If we let $u = t^2$ and $v' = e^{3-t}$ and integrate by parts twice, we obtain

$$\int t^2 e^{3-t}\, dt = -t^2 e^{3-t} + 2(-t e^{3-t} - e^{3-t}) + C.$$

Additional Problems (online only)

93. (a) We know that $\dfrac{dE}{dt} = r$, so the total energy E used in the first T hours is given by $E = \displaystyle\int_0^T t e^{-at}\, dt$. We use integration by parts. Let $u = t$, $v' = e^{-at}$. Then $u' = 1$, $v = -\dfrac{1}{a}e^{-at}$.

$$\begin{aligned}
E &= \int_0^T t e^{-at}\, dt \\
&= -\frac{t}{a}e^{-at}\Big|_0^T - \int_0^T \left(-\frac{1}{a}e^{-at}\right) dt \\
&= -\frac{1}{a}T e^{-aT} + \frac{1}{a}\int_0^T e^{-at}\, dt \\
&= -\frac{1}{a}T e^{-aT} + \frac{1}{a^2}(1 - e^{-aT}).
\end{aligned}$$

(b)
$$\lim_{T\to\infty} E = -\frac{1}{a}\lim_{T\to\infty}\left(\frac{T}{e^{aT}}\right) + \frac{1}{a^2}\left(1 - \lim_{T\to\infty}\frac{1}{e^{aT}}\right).$$

Since $a > 0$, the second limit on the right hand side in the above expression is 0. In the first limit, although both the numerator and the denominator go to infinity, the denominator e^{aT} goes to infinity more quickly than T does. So in the end the denominator e^{aT} is much greater than the numerator T. Hence $\displaystyle\lim_{T\to\infty}\frac{T}{e^{aT}} = 0$. (You can check this by graphing $y = \dfrac{T}{e^{aT}}$ on a calculator or computer for some values of a.) Thus $\displaystyle\lim_{T\to\infty} E = \frac{1}{a^2}$.

Solutions for Section 7.3

Exercises

1. There is no formula.

5. There is no formula.

9. Formula VI-31, then Formula VI-29, both with $a = 3$.

13. Formula V-25 with $a = 3, b = 4, c = -2$.

17. The integrand, a polynomial, x^3, multiplied by $\sin 5x$, is in the form of III-15. There are only three successive derivatives of x^3 before 0 is reached (namely, $3x^2$, $6x$, and 6), so there will be four terms. The signs in the terms will be $-++-$, as given in III-15, so we get

$$\int x^3 \sin 5x\, dx = -\frac{1}{5}x^3 \cos 5x + \frac{1}{25}\cdot 3x^2 \sin 5x + \frac{1}{125}\cdot 6x \cos 5x - \frac{1}{625}\cdot 6 \sin 5x + C.$$

21. $-\dfrac{1}{4}\sin^3 x \cos x - \dfrac{3}{8}\sin x \cos x + \dfrac{3}{8}x + C.$
(Use IV-17.)

25. $\left(\dfrac{1}{3}x^4 - \dfrac{4}{9}x^3 + \dfrac{4}{9}x^2 - \dfrac{8}{27}x + \dfrac{8}{81}\right)e^{3x} + C.$
(Let $a = 3, p(x) = x^4$ in III-14.)

29. We first factor out the 16 and then use formula V-28 to get

$$\int \frac{dx}{\sqrt{25-16x^2}} = \int \frac{dx}{\sqrt{16(25/16-x^2)}} = \frac{1}{4}\int \frac{dx}{\sqrt{(5/4)^2-x^2}}$$

$$= \frac{1}{4}\arcsin\left(\frac{x}{5/4}\right) + C$$

$$= \frac{1}{4}\arcsin\left(\frac{4x}{5}\right) + C.$$

33. Let $m = 3$ in IV-21.

$$\int \frac{1}{\cos^3 x}\, dx = \frac{1}{2}\frac{\sin x}{\cos^2 x} + \frac{1}{2}\int \frac{1}{\cos x}\, dx$$

$$= \frac{1}{2}\frac{\sin x}{\cos^2 x} + \frac{1}{4}\ln\left|\frac{\sin x + 1}{\sin x - 1}\right| + C \text{ by IV-22.}$$

37.

$$\int y^2 \sin 2y\, dy = -\frac{1}{2}y^2 \cos 2y + \frac{1}{4}(2y)\sin 2y + \frac{1}{8}(2)\cos 2y + C$$

$$= -\frac{1}{2}y^2 \cos 2y + \frac{1}{2}y\sin 2y + \frac{1}{4}\cos 2y + C.$$

(Use $a = 2$, $p(y) = y^2$ in III-15.)

41. Substitute $w = 2\theta$, $dw = 2\, d\theta$. Then use IV-19, letting $m = 2$.

$$\int \frac{1}{\sin^2 2\theta}\, d\theta = \frac{1}{2}\int \frac{1}{\sin^2 w}\, dw = \frac{1}{2}(-\frac{\cos w}{\sin w}) + C = -\frac{1}{2\tan w} + C = -\frac{1}{2\tan 2\theta} + C.$$

45.

$$\int \frac{dz}{z(z-3)} = -\frac{1}{3}(\ln|z| - \ln|z-3|) + C.$$

(Let $a = 0$, $b = 3$ in V-26.)

49. We can use IV-17 in the table with $n = 3$ to obtain

$$\int \sin^3 x\, dx = -\frac{1}{3}\sin^2 x \cos x + \frac{2}{3}\int \sin x\, dx = -\frac{1}{3}\sin^2 x \cos x - \frac{2}{3}\cos x + C.$$

Alternatively, we can use the method of IV-23 in the table. The answer obtained by this method looks different than the previous one, but it is equivalent.

Using the Pythagorean Identity, we rewrite the integrand:

$$\sin^3 x = (\sin^2 x)\sin x = (1 - \cos^2 x)\sin x = \sin x - \cos^2 x \sin x.$$

Thus, we have

$$\int \sin^3 x\, dx = \int (\sin x - \cos^2 x \sin x)\, dx$$

$$= \int \sin x\, dx - \int \cos^2 x \sin x\, dx.$$

The first of these new integrals can be easily found. The second can be found using the substitution $w = \cos x$ so $dw = -\sin x\, dx$. The second integral becomes

$$\int \cos^2 x \sin x\, dx = -\int w^2 dw$$

$$= -\frac{1}{3}w^3 + C$$

$$= -\frac{1}{3}\cos^3 x + C,$$

so the final answer is

$$\int \sin^3 x \, dx = \int \sin x \, dx - \int \cos^2 x \sin x \, dx$$
$$= -\cos x + (1/3)\cos^3 x + C.$$

53.

$$\int \sin^3 3\theta \cos^2 3\theta \, d\theta = \int (\sin 3\theta)(\cos^2 3\theta)(1 - \cos^2 3\theta) \, d\theta$$
$$= \int \sin 3\theta(\cos^2 3\theta - \cos^4 3\theta) \, d\theta.$$

Using rule IV-23, we let $w = \cos 3\theta$, $dw = -3 \sin 3\theta \, d\theta$.

$$\int \sin 3\theta(\cos^2 3\theta - \cos^4 3\theta) \, d\theta = -\frac{1}{3} \int (w^2 - w^4) \, dw$$
$$= -\frac{1}{3}(\frac{w^3}{3} - \frac{w^5}{5}) + C$$
$$= -\frac{1}{9}(\cos^3 3\theta) + \frac{1}{15}(\cos^5 3\theta) + C.$$

57. Using III-13:

$$\int_1^2 (x - 2x^3) \ln x \, dx = \int_1^2 x \ln x \, dx - 2 \int_1^2 x^3 \ln x \, dx$$
$$= (\frac{1}{2}x^2 \ln x - \frac{1}{4}x^2)\Big|_1^2 - (\frac{1}{2}x^4 \ln x - \frac{1}{8}x^4)\Big|_1^2$$
$$= 2 \ln 2 - \frac{3}{4} - (8 \ln 2 - \frac{15}{8})$$
$$= \frac{9}{8} - 6 \ln 2 = -3.034.$$

61. Let $w = x^2$, $dw = 2x \, dx$. When $x = 0$, $w = 0$, and when $x = \frac{1}{\sqrt{2}}$, $w = \frac{1}{2}$. Then

$$\int_0^{\frac{1}{\sqrt{2}}} \frac{x \, dx}{\sqrt{1 - x^4}} = \int_0^{\frac{1}{2}} \frac{\frac{1}{2} dw}{\sqrt{1 - w^2}} = \frac{1}{2} \arcsin w \Big|_0^{\frac{1}{2}} = \frac{1}{2}(\arcsin \frac{1}{2} - \arcsin 0) = \frac{\pi}{12}.$$

Problems

65. Letting $w = 2x + 1$, $dw = 2 \, dx$, $dx = 0.5 \, dw$, we find that $k = 0.5$, $w = 2x + 1$, $n = 3$:

$$\int (2x + 1)^3 \ln(2x + 1) \, dx = \int w^3 \ln(w)0.5 \, dw = \int 0.5w^3 \ln w \, dw.$$

69. We can use form (iii) by writing this as:

$$\int \frac{dx}{(x^2 - 5x + 6)^3(x^2 - 4x + 4)^2(x^2 - 6x + 9)^2} = \int \frac{dx}{(x^2 - 5x + 6)^3 \, ((x - 2)(x - 2))^2 \, ((x - 3)(x - 3))^2} \qquad \text{factor}$$
$$= \int \frac{dx}{(x^2 - 5x + 6)^3(x - 2)^4(x - 3)^4} \qquad \text{regroup}$$
$$= \int \frac{dx}{(x^2 - 5x + 6)^3 \, ((x - 2)(x - 3))^4} \qquad \text{regroup}$$
$$= \int \frac{dx}{(x^2 - 5x + 6)^3 \, (x^2 - 5x + 6)^4} \qquad \text{multiply out}$$
$$= \int \frac{dx}{(x^2 - 5x + 6)^7} \qquad \text{simplify,}$$

where $a = 1, b = -5, c = 6, n = 7$.

73. Using II-10 in the integral table, if $m \neq \pm n$, then

$$\int_{-\pi}^{\pi} \sin m\theta \sin n\theta \, d\theta = \frac{1}{n^2 - m^2}[m \cos m\theta \sin n\theta - n \sin m\theta \cos n\theta]\Big|_{-\pi}^{\pi}$$

$$= \frac{1}{n^2 - m^2}[(m \cos m\pi \sin n\pi - n \sin m\pi \cos n\pi) - (m \cos(-m\pi) \sin(-n\pi) - n \sin(-m\pi) \cos(-n\pi))]$$

But $\sin k\pi = 0$ for all integers k, so each term reduces to 0, making the whole integral reduce to 0.

Strengthen Your Understanding

77. If $a = 3$, then $x^2 + 4x + 3 = (x + 1)(x + 3)$, so by Formula V-26, the answer involves ln not arctan. In general, the antiderivative involves an arctan only if the quadratic has no real roots.

81. To evaluate $\int 1/\sqrt{2x - x^2} \, dx$, we must first complete the square for $2x - x^2$ to get $1 - (x - 1)^2$. Then we can substitute $w = x - 1$ and use Formula V-28 in the table.

85. False. Factoring gives

$$\int \frac{dx}{x^2 + 4x - 5} = \int \frac{dx}{(x + 5)(x - 1)} = \frac{1}{6} \int \left(\frac{1}{x - 1} - \frac{1}{x + 5}\right) dx = \frac{1}{6}(\ln|x - 1| - \ln|x + 5|) + C.$$

Solutions for Section 7.4

Exercises

1. Since $6x + x^2 = x(6 + x)$, we take

$$\frac{x + 1}{6x + x^2} = \frac{A}{x} + \frac{B}{6 + x}.$$

So,

$$x + 1 = A(6 + x) + Bx$$
$$x + 1 = (A + B)x + 6A,$$

giving

$$A + B = 1$$
$$6A = 1.$$

Thus $A = 1/6$, and $B = 5/6$ so

$$\frac{x + 1}{6x + x^2} = \frac{1/6}{x} + \frac{5/6}{6 + x}.$$

5. Since $y^3 - 4y = y(y - 2)(y + 2)$, we take

$$\frac{8}{y^3 - 4y} = \frac{A}{y} + \frac{B}{y - 2} + \frac{C}{y + 2}.$$

So,

$$8 = A(y - 2)(y + 2) + By(y + 2) + Cy(y - 2)$$
$$8 = (A + B + C)y^2 + (2B - 2C)y - 4A,$$

giving

$$A + B + C = 0$$
$$2B - 2C = 0$$
$$-4A = 8.$$

Thus $A = -2$, $B = C = 1$ so

$$\frac{8}{y^3 - 4y} = \frac{-2}{y} + \frac{1}{y-2} + \frac{1}{y+2}.$$

9. Using the result of Problem 2, we have

$$\int \frac{20}{25 - x^2}\,dx = \int \frac{2}{5-x}\,dx + \int \frac{2}{5+x}\,dx = -2\ln|5 - x| + 2\ln|5 + x| + C.$$

13. Using the result of Exercise 6, we have

$$\int \frac{2(1+s)}{s(s^2 + 3s + 2)}\,ds = \int \left(\frac{1}{s} - \frac{1}{s+2}\right)ds = \ln|s| - \ln|s + 2| + C.$$

17. We let

$$\frac{10x + 2}{x^3 - 5x^2 + x - 5} = \frac{10x + 2}{(x-5)(x^2+1)} = \frac{A}{x-5} + \frac{Bx + C}{x^2+1}$$

giving

$$10x + 2 = A(x^2 + 1) + (Bx + C)(x - 5)$$
$$10x + 2 = (A + B)x^2 + (C - 5B)x + A - 5C$$

so

$$A + B = 0$$
$$C - 5B = 10$$
$$A - 5C = 2.$$

Thus, $A = 2$, $B = -2$, $C = 0$, so

$$\int \frac{10x + 2}{x^3 - 5x^2 + x - 5}\,dx = \int \frac{2}{x-5}\,dx - \int \frac{2x}{x^2+1}\,dx = 2\ln|x - 5| - \ln\left|x^2 + 1\right| + K.$$

21. Since $x = \sin t + 2$, we have

$$4x - 3 - x^2 = 4(\sin t + 2) - 3 - (\sin t + 2)^2 = 1 - \sin^2 t = \cos^2 t$$

and $dx = \cos t\,dt$, so substitution gives

$$\int \frac{1}{\sqrt{4x - 3 - x^2}} = \int \frac{1}{\sqrt{\cos^2 t}}\cos t\,dt = \int dt = t + C = \arcsin(x - 2) + C.$$

Problems

25. We have

$$w = (3x + 2)(x - 1)$$
$$= 3x^2 - x - 2$$
$$\text{so} \quad dw = (6x - 1)\,dx$$
$$\text{which means} \quad (12x - 2)\,dx = 2\,dw,$$
$$\text{giving} \quad \int \frac{12x - 2}{(3x + 2)(x - 1)}\,dx = 2\int \frac{dw}{w},$$

so $w = (3x + 2)(x - 1) = 3x^2 - x - 2$ and $k = 2$.

29. (a) We have

$$\frac{3x+6}{x^2+3x} = \frac{2(x+3)}{x(x+3)} + \frac{x}{x(x+3)} = \frac{2}{x} + \frac{1}{x+3}.$$

Thus

$$\int \frac{3x+6}{x^2+3x}\,dx = \int \left(\frac{2}{x} + \frac{1}{x+3}\right)\,dx = 2\ln|x| + \ln|x+3| + C.$$

(b) Let $a = 0, b = -3, c = 3$, and $d = 6$ in V-27.

$$\int \frac{3x+6}{x^2+3x}\,dx = \int \frac{3x+6}{x(x+3)}\,dx$$
$$= \frac{1}{3}(6\ln|x| + 3\ln|x+3|) + C = 2\ln|x| + \ln|x+3| + C.$$

33. Since $x^2 + 2x + 2 = (x+1)^2 + 1$, we have

$$\int \frac{x+1}{x^2+2x+2}\,dx = \int \frac{x+1}{(x+1)^2+1}\,dx.$$

Substitute $w = (x+1)^2$, so $dw = 2(x+1)\,dx$.

This integral can also be calculated without completing the square, by substituting $w = x^2 + 2x + 2$, so $dw = 2(x+1)\,dx$.

37. Since $\theta^2 - 4\theta = (\theta - 2)^2 - 4$, we have

$$\int (2-\theta)\cos(\theta^2 - 4\theta)\,d\theta = \int -(\theta - 2)\cos((\theta - 2)^2 - 4)\,d\theta.$$

Substitute $w = (\theta - 2)^2 - 4$, so $dw = 2(\theta - 2)\,d\theta$.

This integral can also be computed without completing the square, by substituting $w = \theta^2 - 4\theta$, so $dw = (2\theta - 4)\,d\theta$.

41. The denominator $x^2 - 3x + 2$ can be factored as $(x-1)(x-2)$. Splitting the integrand into partial fractions with denominators $(x-1)$ and $(x-2)$, we have

$$\frac{x}{x^2-3x+2} = \frac{x}{(x-1)(x-2)} = \frac{A}{x-1} + \frac{B}{x-2}.$$

Multiplying by $(x-1)(x-2)$ gives the identity

$$x = A(x-2) + B(x-1)$$

so

$$x = (A+B)x - 2A - B.$$

Since this equation holds for all x, the constant terms on both sides must be equal. Similarly, the coefficient of x on both sides must be equal. So

$$-2A - B = 0$$
$$A + B = 1.$$

Solving these equations gives $A = -1, B = 2$ and the integral becomes

$$\int \frac{x}{x^2-3x+2}\,dx = -\int \frac{1}{x-1}\,dx + 2\int \frac{1}{x-2}\,dx = -\ln|x-1| + 2\ln|x-2| + C.$$

45. Using partial fractions, we have:

$$\frac{3x+1}{x^2-3x+2} = \frac{3x+1}{(x-1)(x-2)} = \frac{A}{x-1} + \frac{B}{x-2}.$$

Multiplying by $(x-1)$ and $(x-2)$, this becomes

$$3x+1 = A(x-2) + B(x-1)$$
$$= (A+B)x - 2A - B$$

which produces the system of equations

$$\begin{cases} A + B = 3 \\ -2A - B = 1. \end{cases}$$

Solving this system yields $A = -4$ and $B = 7$. So,

$$\int \frac{3x+1}{x^2 - 3x + 2} dx = \int \left(-\frac{4}{x-1} + \frac{7}{x-2} \right) dx$$

$$= -4 \int \frac{dx}{x-1} + 7 \int \frac{dx}{x-2}$$

$$= -4 \ln|x-1| + 7 \ln|x-2| + C.$$

49. Let $y = 5 \tan \theta$ so $dy = (5/\cos^2 \theta) d\theta$. Since $1 + \tan^2 \theta = 1/\cos^2 \theta$, we have

$$\int \frac{y^2}{25 + y^2} dy = \int \frac{25 \tan^2 \theta}{25(1 + \tan^2 \theta)} \cdot \frac{5}{\cos^2 \theta} d\theta = 5 \int \tan^2 \theta \, d\theta.$$

Using $1 + \tan^2 \theta = 1/\cos^2 \theta$ again gives

$$\int \frac{y^2}{25 + y^2} dy = 5 \int \tan^2 \theta \, d\theta = 5 \int \left(\frac{1}{\cos^2 \theta} - 1 \right) d\theta = 5 \tan \theta - 5\theta + C.$$

In addition, since $\theta = \arctan(y/5)$, we get

$$\int \frac{y^2}{25 + y^2} dy = y - 5 \arctan\left(\frac{y}{5} \right) + C.$$

53. Using the substitution $w = e^x$, we get $dw = e^x dx$, so we have

$$\int \frac{e^x}{(e^x - 1)(e^x + 2)} dx = \int \frac{dw}{(w-1)(w+2)}.$$

But

$$\frac{1}{(w-1)(w+2)} = \frac{1}{3} \left(\frac{1}{w-1} - \frac{1}{w+2} \right),$$

so

$$\int \frac{e^x}{(e^x - 1)(e^x + 2)} dx = \int \frac{1}{3} \left(\frac{1}{w-1} - \frac{1}{w+2} \right) dw$$

$$= \frac{1}{3} (\ln|w-1| - \ln|w+2|) + C$$

$$= \frac{1}{3} (\ln|e^x - 1| - \ln|e^x + 2|) + C.$$

57. Let $3x = 5 \sin \theta$. Then $dx = (5/3) \cos \theta \, d\theta$ and $\sqrt{25 - 9x^2} = \sqrt{25 - 25 \sin^2 \theta} = 5 \cos \theta$, so

$$\int \frac{\sqrt{25 - 9x^2}}{x} dx = \int \frac{5 \cos \theta}{(5/3) \sin \theta} \left(\frac{5}{3} \cos \theta \, d\theta \right) = 5 \int \frac{\cos^2 \theta}{\sin \theta} d\theta$$

$$= 5 \int \frac{1 - \sin^2 \theta}{\sin \theta} d\theta = 5 \int \left(\frac{1}{\sin \theta} - \sin \theta \right) d\theta.$$

From formula IV-21, we get

$$\int \left(\frac{1}{\sin \theta} - \sin \theta \right) d\theta = \frac{1}{2} \ln \left| \frac{\cos \theta - 1}{\cos \theta + 1} \right| + \cos \theta + C.$$

We have $\sin \theta = 3x/5$ and $\theta = \arcsin(3x/5)$. From the triangle in Figure 7.3, which shows $\sin \theta = 3x/5$, we see that $\cos \theta = \sqrt{25 - 9x^2}/5$. The figure is for $0 < \theta < \pi/2$. (Negative angles (and 0, and $\pm \pi/2$) can be addressed using the coordinate plane definition: $\sin \theta = $ y-coordinate/Distance from origin.). Therefore

$$\int \frac{\sqrt{25 - 9x^2}}{x} dx = 5 \left(\frac{1}{2} \ln \left| \frac{\cos \theta - 1}{\cos \theta + 1} \right| + \cos \theta \right) + C = \frac{5}{2} \ln \left| \frac{5 - \sqrt{25 - 9x^2}}{5 + \sqrt{25 - 9x^2}} \right| + \sqrt{25 - 9x^2} + C$$

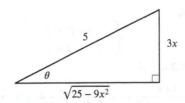

Figure 7.3: In this triangle,
$\sin\theta = 3x/5$

61. Let $2x = 5\tan\theta$ so $dx = (5/2)(1/\cos^2\theta)d\theta$. Since

$$(25 + 4x^2)^{3/2} = (25 + 25\tan^2\theta)^{3/2} = (25/\cos^2\theta)^{3/2} = 125/\cos^3\theta,$$

we have

$$\int \frac{1}{(25+4x^2)^{3/2}}\,dx = \int \frac{\cos^3\theta}{125}\left(\frac{5}{2}\frac{1}{\cos^2\theta}\,d\theta\right) = \frac{1}{50}\int \cos\theta\,d\theta = \frac{1}{50}\sin\theta + C.$$

We need to know $\sin\theta$ in terms of x. From the triangle in Figure 7.4, which shows $2x/5 = \tan\theta$, we see that $\sin\theta = 2x/\sqrt{25+4x^2}$. The figure is for $0 < \theta < \pi/2$. (Negative angles and 0 can be addressed using the coordinate plane definition: $\tan\theta = y\text{-coordinate}/x\text{-coordinate}$.) Thus

$$\int \frac{1}{(25+4x^2)^{3/2}}\,dx = \frac{1}{50}\sin\theta + C = \frac{1}{50}\frac{2x}{\sqrt{25+4x^2}} + C = \frac{1}{25}\frac{x}{\sqrt{25+4x^2}} + C.$$

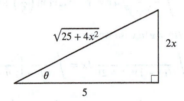

Figure 7.4: In this triangle,
$\tan\theta = 2x/5$

65. Let $x = 2\sin\theta$. Then $dx = 2\cos\theta\,d\theta$ and $\sqrt{4-x^2} = \sqrt{4 - 4\sin^2\theta} = 2\cos\theta$, so

$$\int \frac{x^3}{\sqrt{4-x^2}}\,dx = \int \frac{2^3\sin^3\theta}{2\cos\theta}\cdot 2\cos\theta\,d\theta = 8\int \sin^3\theta\,d\theta$$

$$= 8\int (1 - \cos^2\theta)\sin\theta\,d\theta = -8\cos\theta + \frac{8}{3}\cos^3\theta + C.$$

To convert back to x, we use the fact that $\sin\theta = x/2$ and $\theta = \arcsin(x/2)$. From the triangle in Figure 7.5, which shows $\sin\theta = x/2$, we see that $\cos\theta = \sqrt{4-x^2}/2$. The figure is for $0 < \theta < \pi/2$. (Negative angles (and 0 and $\pm\pi/2$) can be addressed using the coordinate plane definition: $\sin\theta = y\text{-coordinate}/\text{Distance from origin}$.) Therefore

$$\int \frac{x^3}{\sqrt{4-x^2}}\,dx = -8\cos\theta + \frac{8}{3}\cos^3\theta + C = -8\frac{\sqrt{4-x^2}}{2} + \frac{8}{3}\left(\frac{\sqrt{4-x^2}}{2}\right)^3 + C = -4\sqrt{4-x^2} + \frac{1}{3}(4-x^2)^{3/2} + C.$$

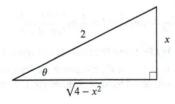

Figure 7.5: In this triangle, $x = 2\sin\theta$

69. We have

$$\text{Area} = \int_0^{\sqrt{2}} \frac{x^3}{\sqrt{4 - x^2}}\, dx.$$

Let $x = 2 \sin \theta$ so $dx = 2 \cos \theta\, d\theta$ and $\sqrt{4 - x^2} = \sqrt{4 - 4 \sin^2 \theta} = 2 \cos \theta$. When $x = 0, \theta = 0$ and when $x = \sqrt{2}, \theta = \pi/4$.

$$\int_0^{\sqrt{2}} \frac{x^3}{\sqrt{4 - x^2}}\, dx = \int_0^{\pi/4} \frac{(2 \sin \theta)^3}{\sqrt{4 - (2 \sin \theta)^2}} 2 \cos \theta\, d\theta$$

$$= 8 \int_0^{\pi/4} \sin^3 \theta\, d\theta = 8 \int_0^{\pi/4} (\sin \theta - \sin \theta \cos^2 \theta)\, d\theta$$

$$= 8 \left(-\cos \theta + \frac{\cos^3 \theta}{3} \right) \Bigg|_0^{\pi/4} = 8 \left(\frac{2}{3} - \frac{5}{6\sqrt{2}} \right).$$

73. Using partial fractions, we write

$$\frac{2x}{x^2 - 1} = \frac{A}{x + 1} + \frac{B}{x - 1}$$
$$2x = A(x - 1) + B(x + 1) = (A + B)x - A + B.$$

So, $A + B = 2$ and $-A + B = 0$, giving $A = B = 1$. Thus

$$\int \frac{2x}{x^2 - 1}\, dx = \int \left(\frac{1}{x + 1} + \frac{1}{x - 1} \right) dx = \ln|x + 1| + \ln|x - 1| + C.$$

Using the substitution $w = x^2 - 1$, we get $dw = 2x\, dx$, so we have

$$\int \frac{2x}{x^2 - 1}\, dx = \int \frac{dw}{w} = \ln|w| + C = \ln\left|x^2 - 1\right| + C.$$

The properties of logarithms show that the two results are the same:

$$\ln|x + 1| + \ln|x - 1| = \ln|(x + 1)(x - 1)| = \ln\left|x^2 - 1\right|.$$

77. (a) If $a \neq b$, we have

$$\int \frac{x}{(x - a)(x - b)}\, dx = \int \frac{1}{a - b} \left(\frac{a}{x - a} - \frac{b}{x - b} \right) dx = \frac{1}{a - b}(a \ln|x - a| - b \ln|x - b|) + C.$$

(b) If $a = b$, we have

$$\int \frac{x}{(x - a)^2}\, dx = \int \left(\frac{1}{x - a} + \frac{a}{(x - a)^2} \right) dx = \ln|x - a| - \frac{a}{x - a} + C.$$

Strengthen Your Understanding

81. We use $x = 2 \tan \theta$.

85. An example is the integral

$$\int \frac{dx}{\sqrt{9 - 4x^2}}.$$

If we make the substitution $x = \frac{3}{2} \sin \theta$, $dx = \frac{3}{2} \cos \theta\, d\theta$, then we obtain

$$\int \frac{dx}{\sqrt{9 - 4x^2}} = \int \frac{\frac{3}{2} \cos \theta}{\sqrt{9 - 9 \sin^2 \theta}}\, d\theta = \int \frac{\frac{3}{2} \cos \theta}{\sqrt{9 \cos^2 \theta}}\, d\theta = \int \frac{\frac{3}{2} \cos \theta}{3 \cos \theta}\, d\theta$$

$$= \int \frac{1}{2}\, d\theta = \frac{1}{2}\theta + C = \frac{1}{2} \cdot \arcsin \left(\frac{2}{3}x \right) + C$$

89. (c) and (b). $\dfrac{x^2}{1 - x^2} = -1 + \dfrac{1}{1 - x^2} = 1 + \dfrac{1}{(1 - x)(1 + x)}$.

Solutions for Section 7.5

Exercises

1. (a) The approximation LEFT(2) uses two rectangles, with the height of each rectangle determined by the left-hand end-point. See Figure 7.6. We see that this approximation is an underestimate.

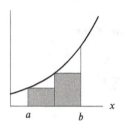

Figure 7.6

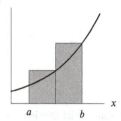

Figure 7.7

(b) The approximation RIGHT(2) uses two rectangles, with the height of each rectangle determined by the right-hand endpoint. See Figure 7.7. We see that this approximation is an overestimate.

(c) The approximation TRAP(2) uses two trapezoids, with the height of each trapezoid given by the secant line connecting the two endpoints. See Figure 7.8. We see that this approximation is an overestimate.

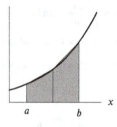

Figure 7.8

(d) The approximation MID(2) uses two rectangles, with the height of each rectangle determined by the height at the mid-point. Alternately, we can view MID(2) as a trapezoid rule where the height is given by the tangent line at the midpoint. Both interpretations are shown in Figure 7.9. We see from the tangent line interpretation that this approximation is an underestimate

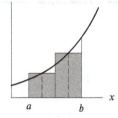

Figure 7.9

5. (a) The approximation LEFT(2) uses two rectangles, with the height of each rectangle determined by the left-hand end-point. See Figure 7.10. We see that this approximation is an underestimate (that is, it is more negative).

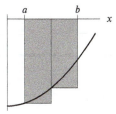

Figure 7.10

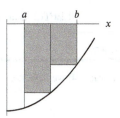

Figure 7.11

(b) The approximation RIGHT(2) uses two rectangles, with the height of each rectangle determined by the right-hand endpoint. See Figure 7.11. We see that this approximation is an overestimate (that is, it is less negative).

(c) The approximation TRAP(2) uses two trapezoids, with the height of each trapezoid given by the secant line connecting the two endpoints. See Figure 7.12. We see that this approximation is an overestimate (that is, it is less negative).

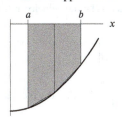

Figure 7.12

(d) The approximation MID(2) uses two rectangles, with the height of each rectangle determined by the height at the midpoint. Alternately, we can view MID(2) as a trapezoid rule where the height is given by the tangent line at the midpoint. Both interpretations are shown in Figure 7.13. We see from the tangent line interpretation that this approximation is an underestimate (that is, it is more negative).

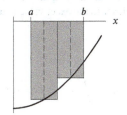

Figure 7.13

9. (a) We have

$$\begin{aligned}
\text{LEFT}(2) &= 2 \cdot f(0) + 2 \cdot f(2) \\
&= 2 \cdot 1 + 2 \cdot 5 \\
&= 12 \\
\text{RIGHT}(2) &= 2 \cdot f(2) + 2 \cdot f(4) \\
&= 2 \cdot 5 + 2 \cdot 17 \\
&= 44
\end{aligned}$$

(b)

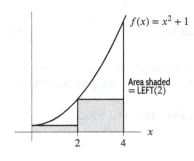

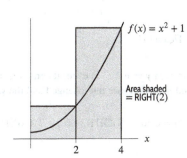

LEFT(2) is an underestimate, while RIGHT(2) is an overestimate.

Problems

13. For $n = 4$, we have $\Delta x = (2 - 0)/4 = 0.5$, so

$$MID(4) = f(0.25)0.5 + f(0.75)(0.5) + f(1.25)(0.5) + f(1.75)(0.5)$$
$$= 0.5(5.8 + 9.3 + 10.8 + 10.3)$$
$$= 18.1.$$

17. Judging from the table, $g(t)$ is an increasing function whose graph is concave down. This means both a left-hand sum and the trapezoid rule underestimate the value of this integral. So I and IV. A right-hand sum and the midpoint rule overestimate it.

21. Let $s(t)$ be the distance traveled at time t and $v(t)$ be the velocity at time t. Then the distance traveled during the interval $0 \leq t \leq 6$ is

$$s(6) - s(0) = s(t)\Big|_0^6$$
$$= \int_0^6 s'(t)\,dt \quad \text{(by the Fundamental Theorem)}$$
$$= \int_0^6 v(t)\,dt.$$

We estimate the distance by estimating this integral.

From the table, we find: LEFT(6) = 31, RIGHT(6) = 39, TRAP(6) = 35.

25. (a) Since the function is decreasing, LEFT is an overestimate and RIGHT is an underestimate. See Figure 7.14. Since the graph is concave down, secant lines lie below the graph so TRAP is an underestimate and tangent lines lie above the graph so MID is an overestimate. In Figure 7.15 we see that MID and TRAP are closer to the exact value than LEFT and RIGHT. Thus we have:

$$\text{RIGHT} < \text{TRAP} < \text{Exact} < \text{MID} < \text{LEFT}.$$

Thus we have

$$\text{RIGHT}(1) = 0.368 \quad \text{LEFT}(1) = 1 \quad \text{TRAP}(1) = 0.684 \quad \text{MID}(1) = 0.882 \quad \text{and} \quad \int_0^1 e^{-x^2/2}dx = 0.856.$$

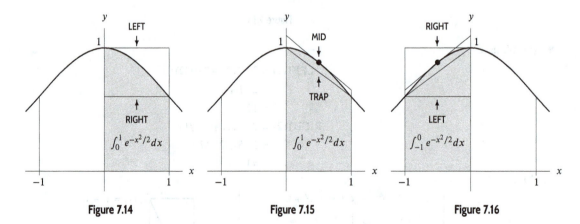

Figure 7.14   **Figure 7.15**   **Figure 7.16**

(b) Since the function $y = e^{-x^2/2}$ is even, its graph is symmetric about the y-axis. See Figure 7.16. Thus the values of LEFT(1)) and RIGHT(1) are interchanged and the values of the other three quantities remain the same:

$$\text{LEFT}(1) = 0.368 \quad \text{RIGHT}(1) = 1 \quad \text{TRAP}(1) = 0.684 \quad \text{MID}(1) = 0.882 \quad \text{and} \quad \int_{-1}^0 e^{-x^2/2}dx = 0.856.$$

29. $f(x)$ is concave down, so MID gives an overestimate and TRAP gives an underestimate.

33. (a) Since $f(x)$ is closer to horizontal (that is, $|f'| < |g'|$), LEFT and RIGHT will be more accurate with $f(x)$.

(b) Since $g(x)$ has more curvature, MID and TRAP will be more accurate with $f(x)$.

37. (a) We have $\int_0^1 7x^6 \, dx = x^7 \Big|_0^1 = 1$.

(b) Using a computer or calculator to find the approximations, and then calculating Error = Exact − Approximation, we have the values in Table 7.1:

Table 7.1

	LEFT(5)	RIGHT(5)	TRAP(5)	MID(5)	SIMP(5)
VALUE	0.438144	1.838144	1.138144	0.931623	1.0004633
ERROR	0.561856	−0.838144	−0.138144	0.068377	−0.0004633

(c) Using a computer or calculator gives the values in Table 7.2:

Table 7.2

	LEFT(10)	RIGHT(10)	TRAP(10)	MID(10)	SIMP(10)
VALUE	0.6848835	1.3848835	1.0348835	0.9826019	1.000029115
ERROR	0.3151165	−0.3848835	−0.0348835	0.0173981	−0.000029115

(d) The ratios are found by dividing corresponding errors. For LEFT, we have

$$\text{Ratio} = \frac{\text{LEFT}(5)}{\text{LEFT}(10)} = \frac{0.561856}{0.3151165} = 1.78.$$

Similar calculations gives

$$\text{LEFT} = 1.78, \quad \text{RIGHT} = 2.18, \quad \text{TRAP} = 3.96, \quad \text{MID} = 3.93, \quad \text{SIMP} = 15.91.$$

The values are as expected: the error in the LEFT and RIGHT approximations improve by about the same factor of 2; the error in the TRAP and MID approximations improve by about the square of this factor, 4, and the SIMP approximation improves by about the fourth power of this factor, 16.

41. From Figure 7.17 it appears that the errors in the trapezoid and midpoint rules depend on how much the curve is bent up or down. In other words, the concavity, and hence the magnitude of the second derivative, f'', has an effect on the errors of these two rules.

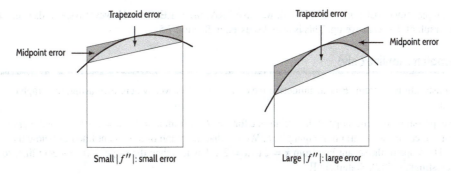

Figure 7.17: The error in the trapezoid and midpoint rules depends on how bent the curve is

45.

$$\text{TRAP}(n) = \frac{\text{LEFT}(n) + \text{RIGHT}(n)}{2}$$

$$= \frac{\text{LEFT}(n) + \text{LEFT}(n) + f(b)\Delta x - f(a)\Delta x}{2}$$

$$= \text{LEFT}(n) + \frac{1}{2}(f(b) - f(a))\Delta x$$

Strengthen Your Understanding

49. The midpoint rule is exact if the integrand is a linear function.

53. Since we want RIGHT(10) to be an underestimate of $\int_0^1 f(x)dx$, we use a function $f(x)$ that is decreasing on $[0, 1]$. Since we want MID(10) to be an overestimate of the integral, we make $f(x)$ concave down on $[0, 1]$. The function $f(x) = 1 - x^2$ has both of these features. Since $f(x) = 1 - x^2$ is both decreasing and concave down on $[0, 1]$, we have

$$\text{RIGHT}(10) < \int_0^1 f(x)dx < \text{MID}(10).$$

57. False. The subdivision size $\Delta x = (1/10)(6 - 2) = 4/10$.

61. True. We have

$$\text{LEFT}(n) - \text{RIGHT}(n) = (f(x_0) + f(x_1) + \cdots + f(x_{n-1}))\Delta x - (f(x_1) + f(x_2) + \cdots + f(x_n))\Delta x.$$

On the right side of the equation, all terms cancel except the first and last, so:

$$\text{LEFT}(n) - \text{RIGHT}(n) = (f(x_0) - f(x_n))\Delta x = (f(2) - f(6))\Delta x.$$

This is also discussed in Section 5.1.

65. False. This is true if f is an increasing function or if f is a decreasing function, but it is not true in general. For example, suppose that $f(2) = f(6)$. Then LEFT(n) = RIGHT(n) for all n, which means that if $\int_2^6 f(x)\,dx$ lies between LEFT(n) and RIGHT(n), then it must equal LEFT(n), which is not always the case.

For example, if $f(x) = (x - 4)^2$ and $n = 1$, then $f(2) = f(6) = 4$, so

$$\text{LEFT}(1) = \text{RIGHT}(1) = 4 \cdot (6 - 2) = 16.$$

However

$$\int_2^6 (x - 4)^2\,dx = \left.\frac{(x - 4)^3}{3}\right|_2^6 = \frac{2^3}{3} - \left(-\frac{2^3}{3}\right) = \frac{16}{3}.$$

In this example, since LEFT(n) = RIGHT(n), we have TRAP(n) = LEFT(n). However trapezoids overestimate the area, since the graph of f is concave up. This is also discussed in Section 7.5.

Additional Problems (online only)

69. By the Construction theorem, F is an antiderivative of f. Since f is everywhere increasing, the graph of F is everywhere concave up.

(a) The expression gives the slope of the secant line for $y = f(x)$ from $x = 2$ to $x = 2 + h$. Since the graph of f is concave down, we can draw the graph in Figure 7.18. We see that the graph of the secant line lies below the tangent line.

The slope of the secant line from $x = 2$ to $x = 2 + h$ is less than the slope of the tangent line, so this expression underestimates $f'(2)$. Statement II.

(b) The expression gives the slope of the secant line for $y = F(x)$ from $x = 2$ to $x = 2 + h$. Since the graph of F is concave up, we can draw a graph similar to that in part (a). We see that the graph of the secant line lies above the tangent line

The slope of the secant line from $x = 2$ to $x = 2 + h$ is greater than the slope of the tangent line, so this expression overestimates $F'(2)$. Since $F'(2) = f(2)$, this means the expression overestimates $f(2)$. Statement III.

(c) This is a right-hand sum with $\Delta x = 0.2$ approximating $\int_1^2 f(x)\,dx$, which we know by definition equals $F(2)$.

Since the graph of f is increasing, the right-hand sum gives an overestimate of the integral. So this expression overestimates $F(2)$. Statement V.

(d) This is a midpoint sum with $\Delta x = 0.2$ approximating $\int_1^2 f(x)\,dx$, which we know by definition equals $F(2)$.

Since the graph of f is concave down, the midpoint rule gives an overestimate of the integral. So this expression overestimates $F(2)$. Statement V.

(e) If we rewrite the expression, we see that:

$$F'(1) + f'(1) = f(1) + f'(1) \cdot 1 \quad \text{since } F'(1) = f(1)$$
$$= f(1) + f'(1)\Delta x \quad \text{where } \Delta x = 1.$$

This gives the tangent line approximation for $f(2)$.

We can make a graph as in Figure 7.19. Since the graph of f is concave-down, it lies below its tangent-line, so this overestimates $f(2)$. Statement III.

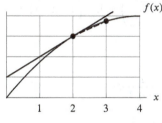

Figure 7.18

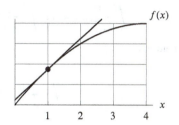

Figure 7.19

Solutions for Section 7.6

Exercises

1. (a) See Figure 7.20. The area extends out infinitely far along the positive x-axis.

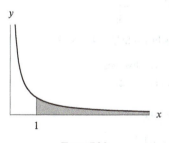

Figure 7.20

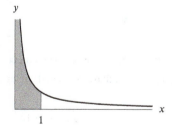

Figure 7.21

(b) See Figure 7.21. The area extends up infinitely far along the positive y-axis.

5. We have

$$\int_1^\infty \frac{1}{5x+2}\,dx = \lim_{b\to\infty} \int_1^b \frac{1}{5x+2}\,dx = \lim_{b\to\infty} \left(\frac{1}{5}\ln(5x+2) \right)\Big|_1^b = \lim_{b\to\infty} \left(\frac{1}{5}\ln(5b+2) - \frac{1}{5}\ln(7) \right).$$

As $b \leftarrow \infty$, we know that $\ln(5b+2) \to \infty$, and so this integral diverges.

9. We have
$$\int_0^\infty xe^{-x^2}\,dx = \lim_{b\to\infty}\int_0^b xe^{-x^2}\,dx = \lim_{b\to\infty}\left(\frac{-1}{2}e^{-x^2}\right)\Big|_0^b = \lim_{b\to\infty}\left(\frac{-1}{2}e^{-b^2} - \frac{-1}{2}\right) = 0 + \frac{1}{2} = \frac{1}{2}.$$
This integral converges to $1/2$.

13.
$$\int_{-\infty}^0 \frac{e^x}{1+e^x}\,dx = \lim_{b\to-\infty}\int_b^0 \frac{e^x}{1+e^x}\,dx$$
$$= \lim_{b\to-\infty} \ln|1+e^x|\Big|_b^0$$
$$= \lim_{b\to-\infty} [\ln|1+e^0| - \ln|1+e^b|]$$
$$= \ln(1+1) - \ln(1+0) = \ln 2.$$

17. This integral is improper because $1/\sqrt{x}$ is undefined at $x = 0$. Then
$$\int_0^4 \frac{1}{\sqrt{x}}\,dx = \lim_{b\to0^+}\int_b^4 \frac{1}{\sqrt{x}}\,dx = \lim_{b\to0^+}\left(2\sqrt{x}\,\Big|_b^4\right) = \lim_{b\to0^+}\left(4 - 2\sqrt{b}\right) = 4.$$

The integral converges.

21.
$$\int_1^\infty \frac{1}{x^2+1}\,dx = \lim_{b\to\infty}\int_1^b \frac{1}{x^2+1}\,dx$$
$$= \lim_{b\to\infty}\arctan(x)\Big|_1^b$$
$$= \lim_{b\to\infty}[\arctan(b) - \arctan(1)]$$
$$= \pi/2 - \pi/4 = \pi/4.$$

25. We use V-26 with $a = 4$ and $b = -4$:
$$\int_0^4 \frac{-1}{u^2-16}\,du = \lim_{b\to4^-}\int_0^b \frac{-1}{u^2-16}\,du$$
$$= \lim_{b\to4^-}\int_0^b \frac{-1}{(u-4)(u+4)}\,du$$
$$= \lim_{b\to4^-}\frac{-(\ln|u-4| - \ln|u+4|)}{8}\Big|_0^b$$
$$= \lim_{b\to4^-} -\frac{1}{8}\left(\ln|b-4| + \ln 4 - \ln|b+4| - \ln 4\right).$$

As $b \to 4^-$, $\ln|b-4| \to -\infty$, so the limit does not exist and the integral diverges.

29. This is a proper integral; use V-26 in the integral table with $a = 4$ and $b = -4$.
$$\int_{16}^{20} \frac{1}{y^2-16}\,dy = \int_{16}^{20} \frac{1}{(y-4)(y+4)}\,dy$$
$$= \frac{\ln|y-4| - \ln|y+4|}{8}\Big|_{16}^{20}$$
$$= \frac{\ln 16 - \ln 24 - (\ln 12 - \ln 20)}{8}$$
$$= \frac{\ln 320 - \ln 288}{8} = \frac{1}{8}\ln(10/9) = 0.01317.$$

33. $\displaystyle\int_4^\infty \frac{dx}{(x-1)^2} = \lim_{b\to\infty}\int_4^b \frac{dx}{(x-1)^2} = \lim_{b\to\infty} -\frac{1}{(x-1)}\Big|_4^b = \lim_{b\to\infty}\left[-\frac{1}{b-1} + \frac{1}{3}\right] = \frac{1}{3}.$

37. The integral can be evaluated by parts. Let $u = t$ and $v' = e^{-2t}$, so $u' = 1$ and $v = -\frac{1}{2}e^{-2t}$. Then:

$$\int_0^\infty te^{-2t}\,dt = \lim_{b\to\infty}\int_0^b te^{-2t}\,dt = \lim_{b\to\infty}\left(-\frac{t}{2}e^{-2t}\Big|_0^b + \frac{1}{2}\int_0^b e^{-2t}\,dt\right)$$

$$= \lim_{b\to\infty}\left(-\frac{t}{2}e^{-2t} - \frac{1}{4}e^{-2t}\right)\Big|_0^b$$

$$= \lim_{b\to\infty}\left(-\frac{b}{2}e^{-2b} - \frac{1}{4}e^{-2b} + \frac{1}{4}\right).$$

As $b \to \infty$, since e^{-2b} dominates b, we see that $be^{-2b} \to 0$. Since also $e^{-2b} \to 0$ as $b \to \infty$, the integral converges:

$$\int_0^\infty te^{-2t}\,dt = \lim_{b\to\infty}\left(-\frac{b}{2}e^{-2b} - \frac{1}{4}e^{-2b} + \frac{1}{4}\right) = \frac{1}{4}.$$

Problems

41. (a) There is no simple antiderivative for this integrand, so we use numerical methods. We find

$$P(1) = \frac{1}{\sqrt{\pi}}\int_0^1 e^{-t^2}\,dt = 0.421.$$

(b) To calculate this improper integral, use numerical methods. If you cannot input infinity into your calculator, increase the upper limit until the value of the integral settles down. We find

$$P(\infty) = \frac{1}{\sqrt{\pi}}\int_0^\infty e^{-t^2}\,dt = 0.500.$$

45. We have:

$$f(3) = \int_1^\infty t^{-3}\,dt = \lim_{b\to\infty}\int_1^b t^{-3}\,dt = \lim_{b\to\infty}-\frac{1}{2}t^{-2}\Big|_1^b = \lim_{b\to\infty}-\frac{1}{2}\left(b^{-2} - 1\right) = \frac{1}{2}.$$

(Since $\lim_{b\to\infty} b^{-2} = 0$.)

49. The energy required is

$$E = \int_1^\infty \frac{kq_1q_2}{r^2}\,dr = kq_1q_2 \lim_{b\to\infty}-\frac{1}{r}\Big|_1^b$$

$$= (9\times 10^9)(1)(1)(1) = 9\times 10^9 \text{ joules}$$

53. Make the substitution $w = 2x$, so $dw = 2\,dx$ and $x = w/2$. When $x = 0$, $w = 0$ and when $x \to \infty$, we have $w \to \infty$. We have

$$\int_0^\infty \frac{x^4 e^{2x}}{(e^{2x} - 1)^2}\,dx = \int_0^\infty \frac{(w/2)^4 e^w}{(e^w - 1)^2}\,dw/2$$

$$= \frac{1}{32}\int_0^\infty \frac{w^4 e^w}{(e^w - 1)^2}\,dw$$

$$= \frac{1}{32}\frac{4\pi^4}{15} = \frac{\pi^4}{120}.$$

Strengthen Your Understanding

57. If $f(x) = 1/x$, then $\int_1^\infty f(x)\,dx$ diverges, but $\lim_{x\to\infty} f(x) = 0$.

61. diverges.

True. Suppose that f has period p. Then $\int_0^p f(x)\,dx$, $\int_p^{2p} f(x)\,dx$, $\int_{2p}^{3p} f(x)\,dx$,... are all equal. If we let $k = \int_0^p f(x)\,dx$, then $\int_0^{np} f(x)\,dx = nk$, for any positive integer n. Since $f(x)$ is positive, so is k. Thus as n approaches ∞, the value of $\int_0^{np} f(x)\,dx = nk$ approaches ∞. That means that $\lim_{b\to\infty} \int_0^b f(x)\,dx$ is not finite; that is, the integral diverges.

65. False. For example, let $f(x) = x$ and $g(x) = -x$. Then $f(x) + g(x) = 0$, so $\int_0^\infty (f(x) + g(x))\,dx$ converges, even though $\int_0^\infty f(x)\,dx$ and $\int_0^\infty g(x)\,dx$ diverge.

69. False. We have

$$\int_0^b (a + f(x))\,dx = \int_0^b a\,dx + \int_0^b f(x)\,dx.$$

Since $\int_0^\infty f(x)\,dx$ converges, the second integral on the right side of the equation has a finite limit as b approaches infinity. But the first integral on the right side has an infinite limit as b approaches infinity, since $a \neq 0$. Thus the right side all together has an infinite limit, which means that $\int_0^\infty (a + f(x))\,dx$ diverges.

Solutions for Section 7.7

Exercises

1. For large x, the integrand behaves like $1/x^2$ because

$$\frac{x^2}{x^4 + 1} \approx \frac{x^2}{x^4} = \frac{1}{x^2}.$$

Since $\displaystyle\int_1^\infty \frac{dx}{x^2}$ converges, we expect our integral to converge. More precisely, since $x^4 + 1 > x^4$, we have

$$\frac{x^2}{x^4 + 1} < \frac{x^2}{x^4} = \frac{1}{x^2}.$$

Since $\displaystyle\int_1^\infty \frac{dx}{x^2}$ is convergent, the comparison test tells us that $\displaystyle\int_1^\infty \frac{x^2}{x^4 + 1}\,dx$ converges also.

5. The integrand is continuous for all $x \geq 1$, so whether the integral converges or diverges depends only on the behavior of the function as $x \to \infty$. As $x \to \infty$, polynomials behave like the highest powered term. Thus, as $x \to \infty$, the integrand $\dfrac{x}{x^2 + 2x + 4}$ behaves like $\dfrac{x}{x^2}$ or $\dfrac{1}{x}$. Since $\displaystyle\int_1^\infty \frac{1}{x}\,dx$ diverges, we predict that the given integral will diverge.

9. The integrand is continuous for all $x \geq 1$, so whether the integral converges or diverges depends only on the behavior of the function as $x \to \infty$. As $x \to \infty$, polynomials behave like the highest powered term. Thus, as $x \to \infty$, the integrand $\dfrac{x^2 + 4}{x^4 + 3x^2 + 11}$ behaves like $\dfrac{x^2}{x^4}$ or $\dfrac{1}{x^2}$. Since $\displaystyle\int_1^\infty \frac{1}{x^2}\,dx$ converges, we predict that the given integral will converge.

13. Since

$$0 \leq \frac{\sin^2 x}{\sqrt{x}} \leq \frac{1}{\sqrt{x}} \qquad \text{and} \qquad \int_0^1 \frac{1}{\sqrt{x}}\,dx \text{ converges,}$$

we conclude that $\displaystyle\int_0^1 \frac{\sin^2 x}{\sqrt{x}}\,dx$ converges.

17. The integrand is unbounded as $t \to 5$. We substitute $w = t - 5$, so $dw = dt$. When $t = 5$, $w = 0$ and when $t = 8$, $w = 3$.

$$\int_5^8 \frac{6}{\sqrt{t - 5}}\,dt = \int_0^3 \frac{6}{\sqrt{w}}\,dw.$$

Since

$$\int_0^3 \frac{6}{\sqrt{w}}\,dw = \lim_{a\to 0^+} 6 \int_a^3 \frac{1}{\sqrt{w}}\,dw = 6 \lim_{a\to 0^+} 2w^{1/2} \Big|_a^3 = 12 \lim_{a\to 0^+} (\sqrt{3} - \sqrt{a}) = 12\sqrt{3},$$

our integral converges.

21. Since $\dfrac{1}{u+u^2} < \dfrac{1}{u^2}$ for $u \geq 1$, and since $\displaystyle\int_1^\infty \dfrac{du}{u^2}$ converges, $\displaystyle\int_1^\infty \dfrac{du}{u+u^2}$ converges.

25. Since $\dfrac{1}{1+e^y} \leq \dfrac{1}{e^y} = e^{-y}$ and $\displaystyle\int_0^\infty e^{-y}\,dy$ converges, the integral $\displaystyle\int_0^\infty \dfrac{dy}{1+e^y}$ converges.

29. Since $\dfrac{3+\sin\alpha}{\alpha} \geq \dfrac{2}{\alpha}$ for $\alpha \geq 4$, and since $\displaystyle\int_4^\infty \dfrac{2}{\alpha}\,d\alpha$ diverges, then $\displaystyle\int_4^\infty \dfrac{3+\sin\alpha}{\alpha}\,d\alpha$ diverges.

Problems

33. First let's calculate the indefinite integral $\displaystyle\int \dfrac{dx}{x(\ln x)^p}$. Let $\ln x = w$, then $\dfrac{dx}{x} = dw$. So

$$\int \frac{dx}{x(\ln x)^p} = \int \frac{dw}{w^p}$$
$$= \begin{cases} \ln|w| + C, & \text{if } p = 1 \\ \frac{1}{1-p}w^{1-p} + C, & \text{if } p \neq 1 \end{cases}$$
$$= \begin{cases} \ln|\ln x| + C, & \text{if } p = 1 \\ \frac{1}{1-p}(\ln x)^{1-p} + C, & \text{if } p \neq 1. \end{cases}$$

Notice that $\lim\limits_{x\to\infty} \ln x = +\infty$.

(a) $p = 1$:

$$\int_2^\infty \frac{dx}{x\ln x} = \lim_{b\to\infty}\left(\ln|\ln b| - \ln|\ln 2|\right) = +\infty.$$

(b) $p < 1$:

$$\int_2^\infty \frac{dx}{x(\ln x)^p} = \frac{1}{1-p}\left(\lim_{b\to\infty}(\ln b)^{1-p} - (\ln 2)^{1-p}\right) = +\infty.$$

(c) $p > 1$:

$$\int_2^\infty \frac{dx}{x(\ln x)^p} = \frac{1}{1-p}\left(\lim_{b\to\infty}(\ln b)^{1-p} - (\ln 2)^{1-p}\right)$$
$$= \frac{1}{1-p}\left(\lim_{b\to\infty}\frac{1}{(\ln b)^{p-1}} - (\ln 2)^{1-p}\right)$$
$$= -\frac{1}{1-p}(\ln 2)^{1-p}.$$

Thus, $\displaystyle\int_2^\infty \dfrac{dx}{x(\ln x)^p}$ is convergent for $p > 1$, divergent for $p \leq 1$.

Strengthen Your Understanding

37. We cannot compare the integrals, because the first integrand is sometimes less than the second and sometimes greater than the second, depending on the sign of $\sin x$.

41. We know that the integral $\int_1^\infty 3/(2x^2)\,dx$ converges because

$$\int_1^\infty \frac{3}{2x^2}\,dx = \lim_{b\to\infty}\int_1^b \frac{3}{2x^2}\,dx = \lim_{b\to\infty}\left(-\frac{3}{2b} + \frac{3}{2}\right) = \frac{3}{2}.$$

So we know that if $f(x)$ is positive and $f(x) \leq 3/(2x^2)$ for all $x \geq 1$, then $\int_1^\infty f(x)\,dx$ converges. So the function $f(x) = 3/(2x^2 + 1)$ is a good example.

45. True. Since $a < b$, each term in the Riemann sum approximating the integral of f is less than the corresponding term in the Riemann sum approximating the integral of g.

The condition $a < b$ is necessary to make all the Δxs positive. If they are negative, the inequality reverses.

Solutions for Chapter 7 Review

Exercises

1. $\frac{1}{3}(t+1)^3$

5. Using the power rule gives $\frac{3}{2}w^2 + 7w + C$.

9. Let $5z = w$, then $5dz = dw$, which means $dz = \frac{1}{5}dw$, so

$$\int e^{5z} \, dz = \int e^w \cdot \frac{1}{5}dw = \frac{1}{5}\int e^w \, dw = \frac{1}{5}e^w + C = \frac{1}{5}e^{5z} + C,$$

where C is a constant.

13. The power rule gives $\frac{2}{5}x^{5/2} + \frac{3}{5}x^{5/3} + C$

17. Dividing by x^2 gives

$$\int \left(\frac{x^3 + x + 1}{x^2}\right) dx = \int \left(x + \frac{1}{x} + \frac{1}{x^2}\right) dx = \frac{1}{2}x^2 + \ln|x| - \frac{1}{x} + C.$$

21. Integration by parts twice gives

$$\int x^2 e^{2x} \, dx = \frac{x^2 e^{2x}}{2} - \int 2x e^{2x} \, dx = \frac{x^2}{2}e^{2x} - \frac{x}{2}e^{2x} + \frac{1}{4}e^{2x} + C$$
$$= (\frac{1}{2}x^2 - \frac{1}{2}x + \frac{1}{4})e^{2x} + C.$$

Or use the integral table, III-14 with $p(x) = x^2$ and $a = 1$.

25. We integrate by parts, using $u = (\ln x)^2$ and $v' = 1$. Then $u' = 2\frac{\ln x}{x}$ and $v = x$, so

$$\int (\ln x)^2 \, dx = x(\ln x)^2 - 2\int \ln x \, dx.$$

But, integrating by parts or using the integral table, $\int \ln x \, dx = x \ln x - x + C$. Therefore,

$$\int (\ln x)^2 \, dx = x(\ln x)^2 - 2x \ln x + 2x + C.$$

Check:

$$\frac{d}{dx}\left[x(\ln x)^2 - 2x \ln x + 2x + C\right] = (\ln x)^2 + x\frac{2\ln x}{x} - 2\ln x - 2x\frac{1}{x} + 2 = (\ln x)^2.$$

29. Expanding the numerator and dividing, we have

$$\int \frac{(u+1)^3}{u^2} \, du = \int \frac{(u^3 + 3u^2 + 3u + 1)}{u^2} \, du = \int \left(u + 3 + \frac{3}{u} + \frac{1}{u^2}\right) du$$
$$= \frac{u^2}{2} + 3u + 3\ln|u| - \frac{1}{u} + C.$$

Check:

$$\frac{d}{du}\left(\frac{u^2}{2} + 3u + 3\ln|u| - \frac{1}{u} + C\right) = u + 3 + 3/u + 1/u^2 = \frac{(u+1)^3}{u^2}.$$

33. Multiplying out and integrating term by term:

$$\int t^{10}(t - 10) \, dt = \int (t^{11} - 10t^{10}) \, dt = \int t^{11} dt - 10\int t^{10} \, dt = \frac{1}{12}t^{12} - \frac{10}{11}t^{11} + C.$$

37. Integrating term by term:

$$\int \left(x^2 + 2x + \frac{1}{x} \right) dx = \frac{1}{3}x^3 + x^2 + \ln|x| + C,$$

where C is a constant.

41. If $u = \sin(5\theta)$, $du = \cos(5\theta) \cdot 5 \, d\theta$, so

$$\int \sin(5\theta) \cos(5\theta) d\theta = \frac{1}{5} \int \sin(5\theta) \cdot 5 \cos(5\theta) d\theta = \frac{1}{5} \int u \, du$$

$$= \frac{1}{5} \left(\frac{u^2}{2} \right) + C = \frac{1}{10} \sin^2(5\theta) + C$$

or

$$\int \sin(5\theta) \cos(5\theta) d\theta = \frac{1}{2} \int 2 \sin(5\theta) \cos(5\theta) d\theta = \frac{1}{2} \int \sin(10\theta) d\theta \quad \text{(using } \sin(2x) = 2 \sin x \cos x)$$

$$= \frac{-1}{20} \cos(10\theta) + C.$$

45. Let $w = \cos 2\theta$. Then $dw = -2 \sin 2\theta \, d\theta$, hence

$$\int \cos^3 2\theta \sin 2\theta \, d\theta = -\frac{1}{2} \int w^3 \, dw = -\frac{w^4}{8} + C = -\frac{\cos^4 2\theta}{8} + C.$$

Check:

$$\frac{d}{d\theta} \left(-\frac{\cos^4 2\theta}{8} \right) = -\frac{(4 \cos^3 2\theta)(-\sin 2\theta)(2)}{8} = \cos^3 2\theta \sin 2\theta.$$

49. Let $\sin \theta = w$, then $\cos \theta \, d\theta = dw$, so

$$\int \cos \theta \sqrt{1 + \sin \theta} \, d\theta = \int \sqrt{1 + w} \, dw$$

$$= \frac{(1 + w)^{3/2}}{3/2} + C = \frac{2}{3}(1 + \sin \theta)^{3/2} + C,$$

where C is a constant.

53. Let $w = 3z + 5$ and $dw = 3 \, dz$. Then

$$\int (3z + 5)^3 \, dz = \frac{1}{3} \int w^3 dw = \frac{1}{12} w^4 + C = \frac{1}{12}(3z + 5)^4 + C.$$

57. Let $w = \ln x$, then $dw = (1/x)dx$ so that

$$\int \frac{1}{x} \sin(\ln x) \, dx = \int \sin w \, dw = -\cos w + C = -\cos(\ln x) + C.$$

61. Let $w = \ln x$. Then $dw = (1/x)dx$ which gives

$$\int \frac{dx}{x \ln x} = \int \frac{dw}{w} = \ln|w| + C = \ln|\ln x| + C.$$

65. Using integration by parts, let $r = u$ and $dt = e^{ku}du$, so $dr = du$ and $t = (1/k)e^{ku}$. Thus

$$\int u e^{ku} \, du = \frac{u}{k} e^{ku} - \frac{1}{k} \int e^{ku} \, du = \frac{u}{k} e^{ku} - \frac{1}{k^2} e^{ku} + C.$$

69. Integrate by parts, $r = \ln u$ and $dt = u^2 \, du$, so $dr = (1/u) \, du$ and $t = (1/3)u^3$. We have

$$\int u^2 \ln u \, du = \frac{1}{3}u^3 \ln u - \frac{1}{3} \int u^2 \, du = \frac{1}{3}u^3 \ln u - \frac{1}{9}u^3 + C.$$

73. Integration by parts will be used twice here. First let $u = y^2$ and $dv = \sin(cy)\,dy$, then $du = 2y\,dy$ and $v = -(1/c)\cos(cy)$. Thus

$$\int y^2 \sin(cy)\,dy = -\frac{y^2}{c}\cos(cy) + \frac{2}{c}\int y\cos(cy)\,dy.$$

Now use integration by parts to evaluate the integral in the right hand expression. Here let $u = y$ and $dv = \cos(cy)dy$ which gives $du = dy$ and $v = (1/c)\sin(cy)$. Then we have

$$\int y^2 \sin(cy)\,dy = -\frac{y^2}{c}\cos(cy) + \frac{2}{c}\left(\frac{y}{c}\sin(cy) - \frac{1}{c}\int \sin(cy)\,dy\right)$$

$$= -\frac{y^2}{c}\cos(cy) + \frac{2y}{c^2}\sin(cy) + \frac{2}{c^3}\cos(cy) + C.$$

77. Factor $\sqrt{3}$ out of the integrand and use VI-30 of the integral table with $u = 2x$ and $du = 2dx$ to get

$$\int \sqrt{3 + 12x^2}\,dx = \int \sqrt{3}\sqrt{1 + 4x^2}\,dx$$

$$= \frac{\sqrt{3}}{2}\int \sqrt{1 + u^2}\,du$$

$$= \frac{\sqrt{3}}{4}\left(u\sqrt{1+u^2} + \int \frac{1}{\sqrt{1+u^2}}\,du\right).$$

Then from VI-29, simplify the integral on the right to get

$$\int \sqrt{3 + 12x^2}\,dx = \frac{\sqrt{3}}{4}\left(u\sqrt{1+u^2} + \ln|u + \sqrt{1+u^2}|\right) + C$$

$$= \frac{\sqrt{3}}{4}\left(2x\sqrt{1 + (2x)^2} + \ln|2x + \sqrt{1 + (2x)^2}|\right) + C.$$

81. We can factor the denominator into $ax(x + \frac{b}{a})$, so

$$\int \frac{dx}{ax^2 + bx} = \frac{1}{a}\int \frac{1}{x(x + \frac{b}{a})}$$

Now we can use V-26 (with $A = 0$ and $B = -\frac{b}{a}$ to give

$$\frac{1}{a}\int \frac{1}{x(x + \frac{b}{a})} = \frac{1}{a}\cdot\frac{a}{b}\left(\ln|x| - \ln\left|x + \frac{b}{a}\right|\right) + C = \frac{1}{b}\left(\ln|x| - \ln\left|x + \frac{b}{a}\right|\right) + C.$$

85. If $u = 1 - x$, $du = -1\,dx$, so

$$\int 10^{1-x}dx = -1\int 10^{1-x}(-1\,dx) = -1\int 10^u\,du = -1\frac{10^u}{\ln 10} + C = -\frac{1}{\ln 10}10^{1-x} + C.$$

89. By long division, $\dfrac{z^3}{z - 5} = z^2 + 5z + 25 + \dfrac{125}{z - 5}$, so

$$\int \frac{z^3}{z - 5}dz = \int \left(z^2 + 5z + 25 + \frac{125}{z - 5}\right)dz = \frac{z^3}{3} + \frac{5z^2}{2} + 25z + 125\int \frac{1}{z - 5}dz$$

$$= \frac{z^3}{3} + \frac{5}{2}z^2 + 25z + 125\ln|z - 5| + C.$$

93. Dividing and integrating term by term gives

$$\int \frac{x + 1}{\sqrt{x}}dx = \int \left(\frac{x}{\sqrt{x}} + \frac{1}{\sqrt{x}}\right)dx = \int (x^{1/2} + x^{-1/2})dx = \frac{x^{3/2}}{\frac{3}{2}} + \frac{x^{1/2}}{\frac{1}{2}} + C = \frac{2}{3}x^{3/2} + 2\sqrt{x} + C.$$

97. If $u = z^2 - 5$, $du = 2z\,dz$, then

$$\int \frac{z}{(z^2-5)^3}\,dz = \int (z^2-5)^{-3} z\,dz = \frac{1}{2}\int (z^2-5)^{-3} 2z\,dz = \frac{1}{2}\int u^{-3}\,du = \frac{1}{2}\left(\frac{u^{-2}}{-2}\right) + C$$

$$= \frac{1}{-4(z^2-5)^2} + C.$$

101. We use the substitution $w = x^2 + x$, $dw = (2x+1)\,dx$.

$$\int (2x+1)e^{x^2} e^x\,dx = \int (2x+1)e^{x^2+x}\,dx = \int e^w\,dw$$

$$= e^w + C = e^{x^2+x} + C.$$

Check: $\dfrac{d}{dx}(e^{x^2+x} + C) = e^{x^2+x} \cdot (2x+1) = (2x+1)e^{x^2} e^x.$

105. Let $x = 2\theta$, then $dx = 2d\theta$. Thus

$$\int \sin^2(2\theta)\cos^3(2\theta)\,d\theta = \frac{1}{2}\int \sin^2 x \cos^3 x\,dx.$$

We let $w = \sin x$ and $dw = \cos x\,dx$. Then

$$\frac{1}{2}\int \sin^2 x \cos^3 x\,dx = \frac{1}{2}\int \sin^2 x \cos^2 x \cos x\,dx$$

$$= \frac{1}{2}\int \sin^2 x(1 - \sin^2 x)\cos x\,dx$$

$$= \frac{1}{2}\int w^2(1 - w^2)\,dw = \frac{1}{2}\int (w^2 - w^4)\,dw$$

$$= \frac{1}{2}\left(\frac{w^3}{3} - \frac{w^5}{5}\right) + C = \frac{1}{6}\sin^3 x - \frac{1}{10}\sin^5 x + C$$

$$= \frac{1}{6}\sin^3(2\theta) - \frac{1}{10}\sin^5(2\theta) + C.$$

109. Use the substitution $w = \sinh x$ and $dw = \cosh x\,dx$ so

$$\int \sinh^2 x \cosh x\,dx = \int w^2\,dw = \frac{w^3}{3} + C = \frac{1}{3}\sinh^3 x + C.$$

Check this answer by taking the derivative: $\dfrac{d}{dx}\left[\dfrac{1}{3}\sinh^3 x + C\right] = \sinh^2 x \cosh x.$

113. We substitute $w = \cos\theta + 5$, $dw = -\sin\theta\,d\theta$. Then

$$\int_{\theta=0}^{\theta=\pi} \sin\theta\,d\theta(\cos\theta+5)^7\,d\theta = -\int_{w=6}^{w=4} w^7\,dw = \int_{w=4}^{w=6} w^7\,dw = \frac{w^8}{8}\bigg|_4^6 = 201{,}760.$$

117. In Problem 25, we found that

$$\int (\ln x)^2\,dx = x(\ln x)^2 - 2x\ln x + 2x + C.$$

Thus

$$\int_1^e (\ln x)^2\,dx = [x(\ln x)^2 - 2x\ln x + 2x]\bigg|_1^e = e - 2 \approx 0.71828.$$

This matches the approximation given by Simpson's rule with 10 intervals.

121. We substitute $w = \sqrt[3]{x} = x^{\frac{1}{3}}$. Then $dw = \frac{1}{3}x^{-\frac{2}{3}}dx = \dfrac{1}{3\sqrt[3]{x^2}}\,dx.$

$$\int_1^8 \frac{e^{\sqrt[3]{x}}}{\sqrt[3]{x^2}}dx = \int_{x=1}^{x=8} e^w(3\,dw) = 3e^w\Big|_{x=1}^{x=8} = 3e^{\sqrt[3]{x}}\Big|_1^8 = 3(e^2 - e).$$

125. Since $\dfrac{1}{x^2 - 1} = \dfrac{1}{(x-1)(x+1)}$, let's imagine that our fraction is the result of adding together two terms, one with a denominator of $x - 1$, the other with a denominator of $x + 1$:

$$\frac{1}{(x-1)(x+1)} = \frac{A}{x-1} + \frac{B}{x+1}.$$

To find A and B, we multiply by the least common multiple of both sides to clear the fractions. This yields

$$1 = A(x+1) + B(x-1)$$
$$= (A+B)x + (A-B).$$

Since the two sides are equal for all values of x in the domain, and there is no x term on the left-hand side, $A + B = 0$. Similarly, since A and $-B$ are constant terms on the right-hand side, and 1 is the constant term on the left-hand side, $A - B = 1$. Therefore, we have the system of equations

$$A + B = 0$$
$$A - B = 1.$$

Solving this gives us $A = 1/2$ and $B = -1/2$, so

$$\frac{1}{x^2 - 1} = \frac{1}{2(x-1)} - \frac{1}{2(x+1)}.$$

Now, we find the integral

$$\int \frac{1}{x^2 - 1}dx = \int \left(\frac{1}{2(x-1)} - \frac{1}{2(x+1)} \right) dx$$
$$= \frac{1}{2}\ln|x-1| - \frac{1}{2}\ln|x+1| + C.$$

129. Let $x = 5\sin t$. Then $dx = 5\cos t\,dt$, so substitution gives

$$\int \frac{1}{\sqrt{25 - x^2}} = \int \frac{5\cos t}{\sqrt{25 - 25\sin^2 t}}dt = \int dt = t + C = \arcsin\left(\frac{x}{5}\right) + C.$$

133. The denominator can be factored to give $x(x-1)(x+1)$. Splitting the integrand into partial fractions with denominators x, $x - 1$, and $x + 1$, we have

$$\frac{3x + 1}{x(x-1)(x+1)} = \frac{A}{x-1} + \frac{B}{x+1} + \frac{C}{x}.$$

Multiplying by $x(x-1)(x+1)$ gives the identity

$$3x + 1 = Ax(x+1) + Bx(x-1) + C(x-1)(x+1)$$

so

$$3x + 1 = (A + B + C)x^2 + (A - B)x - C.$$

Since this equation holds for all x, the constant terms on both sides must be equal. Similarly, the coefficient of x and x^2 on both sides must be equal. So

$$-C = 1$$
$$A - B = 3$$
$$A + B + C = 0.$$

Solving these equations gives $A = 2$, $B = -1$ and $C = -1$. The integral becomes

$$\int \frac{3x + 1}{x(x+1)(x-1)}dx = \int \frac{2}{x-1}dx - \int \frac{1}{x+1}dx - \int \frac{1}{x}dx$$
$$= 2\ln|x-1| - \ln|x+1| - \ln|x| + K.$$

We use K as the constant of integration, since we already used C in the problem.

137. We use the trigonometric substitution $bx = a \sin \theta$. Then $dx = \frac{a}{b} \cos \theta \, d\theta$, and we have

$$\int \frac{1}{\sqrt{a^2 - (bx)^2}} \, dx = \int \frac{1}{\sqrt{a^2 - (a \sin \theta)^2}} \cdot \frac{a}{b} \cos \theta \, d\theta = \int \frac{1}{a\sqrt{1 - \sin^2 \theta}} \cdot \frac{a}{b} \cos \theta \, d\theta$$

$$= \frac{1}{b} \int \frac{\cos \theta}{\sqrt{\cos^2 \theta}} \, d\theta = \frac{1}{b} \int 1 \, d\theta = \frac{1}{b} \theta + C = \frac{1}{b} \arcsin \left(\frac{bx}{a} \right) + C.$$

141. This integral is improper because $5/x^2$ is undefined at $x = 0$. Then

$$\int_0^3 \frac{5}{x^2} \, dx = \lim_{b \to 0^+} \int_b^3 \frac{5}{x^2} \, dx = \lim_{b \to 0^+} \left(-5x^{-1} \Big|_b^3 \right) = \lim_{b \to 0^+} \left(\frac{-5}{3} + \frac{5}{b} \right).$$

As $b \to 0^+$, this goes to infinity and the integral diverges.

145. $\int \frac{dx}{x \ln x} = \ln |\ln x| + C.$ (Substitute $w = \ln x$, $dw = \frac{1}{x} dx$).
Thus

$$\int_{10}^\infty \frac{dx}{x \ln x} = \lim_{b \to \infty} \int_{10}^b \frac{dx}{x \ln x} = \lim_{b \to \infty} \ln |\ln x| \Big|_{10}^b = \lim_{b \to \infty} \ln(\ln b) - \ln(\ln 10).$$

As $b \to \infty$, $\ln(\ln b) \to \infty$, so this diverges.

149. It is easy to see that this integral converges:

$$\frac{1}{4 + z^2} < \frac{1}{z^2}, \quad \text{and so} \quad \int_2^\infty \frac{1}{4 + z^2} \, dz < \int_2^\infty \frac{1}{z^2} \, dz = \frac{1}{2}.$$

We can also find its exact value.

$$\int_2^\infty \frac{1}{4 + z^2} \, dz = \lim_{b \to \infty} \int_2^b \frac{1}{4 + z^2} \, dz$$

$$= \lim_{b \to \infty} \left(\frac{1}{2} \arctan \frac{z}{2} \Big|_2^b \right)$$

$$= \lim_{b \to \infty} \left(\frac{1}{2} \arctan \frac{b}{2} - \frac{1}{2} \arctan 1 \right)$$

$$= \frac{1}{2} \frac{\pi}{2} - \frac{1}{2} \frac{\pi}{4} = \frac{\pi}{8}.$$

Note that $\frac{\pi}{8} < \frac{1}{2}$.

153. Let $\phi = 2\theta$. Then $d\phi = 2 \, d\theta$, and

$$\int_0^{\pi/4} \tan 2\theta \, d\theta = \int_0^{\pi/2} \frac{1}{2} \tan \phi \, d\phi = \int_0^{\pi/2} \frac{1}{2} \frac{\sin \phi}{\cos \phi} \, d\phi$$

$$= \lim_{b \to (\pi/2)^-} \int_0^b \frac{1}{2} \frac{\sin \phi}{\cos \phi} \, d\phi = \lim_{b \to (\pi/2)^-} -\frac{1}{2} \ln |\cos \phi| \Big|_0^b.$$

As $b \to \pi/2$, $\cos \phi \to 0$, so $\ln |\cos \phi| \to -\infty$. Thus the integral diverges.

One could also see this by noting that $\cos x \approx \pi/2 - x$ and $\sin x \approx 1$ for x close to $\pi/2$: therefore, $\tan x \approx 1/(\frac{\pi}{2} - x)$, the integral of which diverges.

157. Since $0 \le \sin x < 1$ for $0 \le x \le 1$, we have

$$(\sin x)^{\frac{3}{2}} < (\sin x)$$

$$\text{so} \quad \frac{1}{(\sin x)^{\frac{3}{2}}} > \frac{1}{(\sin x)}$$

$$\text{or} \quad (\sin x)^{-\frac{3}{2}} > (\sin x)^{-1}$$

Thus $\int_0^1 (\sin x)^{-1} dx = \lim_{a \to 0} \ln \left| \frac{1}{\sin x} - \frac{1}{\tan x} \right| \Big|_a^1$, which is infinite.

Hence, $\int_0^1 (\sin x)^{-\frac{3}{2}} dx$ is infinite.

Problems

161. As is evident from Figure 7.22 showing the graphs of $y = \sin x$ and $y = \cos x$, the crossings occur at $x = \frac{\pi}{4}, \frac{5\pi}{4}, \frac{9\pi}{4}, \ldots$, and the regions bounded by any two consecutive crossings have the same area. So picking two consecutive crossings, we get an area of

$$\text{Area} = \int_{\frac{\pi}{4}}^{\frac{5\pi}{4}} (\sin x - \cos x)\, dx$$

$$= 2\sqrt{2}.$$

(Note that we integrated $\sin x - \cos x$ here because for $\frac{\pi}{4} \leq x \leq \frac{5\pi}{4}$, $\sin x \geq \cos x$.)

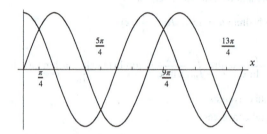

Figure 7.22

165. Using partial fractions we have

$$\frac{1}{(2x - 3)(3x - 2)} = \frac{A}{2x - 3} + \frac{B}{3x - 2}$$

$$A(3x - 2) + B(2x - 3) = 1$$

$$3Ax - 2A + 2Bx - 3B = 1$$

$$(3A + 2B)x - 2A - 3B = 1.$$

We see that:

$$3A + 2B = 0$$

$$\text{so} \quad A = -\frac{2}{3} \cdot B$$

$$\text{and} \quad -2A - 3B = 1$$

$$-2\left(-\frac{2}{3} \cdot B\right) - 3B = 1 \qquad \text{since } A = -\frac{2}{3}B$$

$$\frac{4}{3} \cdot B - 3B = 1$$

$$-\frac{5}{3}B = 1$$

$$B = -\frac{3}{5} = -0.6$$

$$A = -\frac{2}{3}\left(-\frac{3}{5}\right) \qquad \text{since } A = -\frac{2}{3}B$$

$$= \frac{2}{5} = 0.4.$$

Thus, $\displaystyle\int \frac{dx}{(2x - 3)(3x - 2)} = \int \left(\frac{0.4}{2x - 3} + \frac{-0.6}{3x - 2}\right) dx.$

169. After the substitution $w = x^2$, the second integral becomes

$$\frac{1}{2} \int \frac{dw}{\sqrt{1 - w^2}}.$$

173. If we let $w = 2x$ in the first integral, we get $dw = 2dx$. Also, the limits $w = 0$ and $w = 2$ become $x = 0$ and $x = 1$. Thus

$$\int_0^2 e^{-w^2}\, dw = \int_0^1 e^{-4x^2}\, 2\, dx.$$

177. (a) Since $h(z)$ is even, we know that $\int_0^1 h(z)\, dz = \int_{-1}^0 h(z)\, dz$. Since $\int_{-1}^1 h(z)\, dz = \int_{-1}^0 h(z)\, dz + \int_0^1 h(z)\, dz$, we see that $\int_{-1}^1 h(z)\, dz = 2\int_0^1 h(z)\, dz = 7$. Thus $\int_0^1 h(z)\, dz = 3.5$

   (b) If $w = z + 3$, then $dw = dz$. When $z = -4$, $w = -1$; when $z = -2$, $w = 1$. Thus,

$$\int_{-4}^{-2} 5h(z+3)\, dt = 5\int_{-1}^1 h(w)\,(dw) = 5 \cdot 7 = 35.$$

181. This calculation cannot be correct because the integrand is positive everywhere, yet the value given for the integral is negative.

   The calculation is incorrect because the integral is improper but has not been treated as such. The integral is improper because the integrand $1/x^2$ is undefined at $x = 0$. To determine whether the integral converges we split the integral into two improper integrals:

$$\int_{-2}^2 \frac{1}{x^2}\, dx = \int_{-2}^0 \frac{1}{x^2}\, dx + \int_0^2 \frac{1}{x^2}\, dx.$$

To decide whether the second integral converges, we compute

$$\int_0^2 \frac{1}{x^2}\, dx = \lim_{a\to 0^+} \int_a^2 \frac{1}{x^2}\, dx = \lim_{a\to 0^+}\left(-\frac{1}{2} + \frac{1}{a}\right).$$

The limit does not exist, and $\int_0^2 (1/x^2)\, dx$ diverges, so the original the integral $\int_{-2}^2 1/x^2\, dx$ diverges.

185. (a) For the left-hand rule, error is approximately proportional to $\frac{1}{n}$. If we let n_p be the number of subdivisions needed for accuracy to p places, then there is a constant k such that

$$5 \times 10^{-5} = \frac{1}{2} \times 10^{-4} \approx \frac{k}{n_4}$$

$$5 \times 10^{-9} = \frac{1}{2} \times 10^{-8} \approx \frac{k}{n_8}$$

$$5 \times 10^{-13} = \frac{1}{2} \times 10^{-12} \approx \frac{k}{n_{12}}$$

$$5 \times 10^{-21} = \frac{1}{2} \times 10^{-20} \approx \frac{k}{n_{20}}$$

Thus the ratios $n_4 : n_8 : n_{12} : n_{20} \approx 1 : 10^4 : 10^8 : 10^{16}$, and assuming the computer time necessary is proportional to n_p, the computer times are approximately

4 places:	2 seconds	
8 places:	2×10^4 seconds	$\approx$ 6 hours
12 places:	2×10^8 seconds	$\approx$ 6 years
20 places:	2×10^{16} seconds	$\approx$ 600 million years

   (b) For the trapezoidal rule, error is approximately proportional to $\frac{1}{n^2}$. If we let N_p be the number of subdivisions needed for accuracy to p places, then there is a constant C such that

$$5 \times 10^{-5} = \frac{1}{2} \times 10^{-4} \approx \frac{C}{N_4{}^2}$$

$$5 \times 10^{-9} = \frac{1}{2} \times 10^{-8} \approx \frac{C}{N_8{}^2}$$

$$5 \times 10^{-13} = \frac{1}{2} \times 10^{-12} \approx \frac{C}{N_{12}{}^2}$$

$$5 \times 10^{-21} = \frac{1}{2} \times 10^{-20} \approx \frac{C}{N_{20}{}^2}$$

Thus the ratios $N_4{}^2 : N_8{}^2 : N_{12}{}^2 : N_{20}{}^2 \approx 1 : 10^4 : 10^8 : 10^{16}$, and the ratios $N_4 : N_8 : N_{12} : N_{20} \approx 1 : 10^2 : 10^4 : 10^8$. So the computer times are approximately

4 places:	2 seconds	
8 places:	2×10^2 seconds	$\approx$ 3 minutes
12 places:	2×10^4 seconds	$\approx$ 6 hours
20 places:	2×10^8 seconds	$\approx$ 6 years

189. (a) Since the rate is given by $r(t) = 2te^{-2t}$ ml/sec, by the Fundamental Theorem of Calculus, the total quantity is given by the definite integral:

$$\text{Total quantity} \approx \int_0^\infty 2te^{-2t}\, dt = 2 \lim_{b \to \infty} \int_0^b te^{-2t}\, dt.$$

Integration by parts with $u = t$, $v' = e^{-2t}$ gives

$$\text{Total quantity} \approx 2 \lim_{b \to \infty} \left(-\frac{t}{2}e^{-2t} - \frac{1}{4}e^{-2t} \right)\Big|_0^b$$

$$= 2 \lim_{b \to \infty} \left(\frac{1}{4} - \left(\frac{b}{2} + \frac{1}{4} \right) e^{-2b} \right) = 2 \cdot \frac{1}{4} = 0.5 \text{ ml}.$$

(b) At the end of 5 seconds,

$$\text{Quantity received} = \int_0^5 2te^{-2t}\, dt \approx 0.49975 \text{ ml}.$$

Since $0.49975/0.5 = 0.9995 = 99.95\%$, the patient has received 99.95% of the dose in the first 5 seconds.

CAS Challenge Problems

193. (a) A CAS gives

$$\int \frac{\ln x}{x}\, dx = \frac{(\ln x)^2}{2}$$

$$\int \frac{(\ln x)^2}{x}\, dx = \frac{(\ln x)^3}{3}$$

$$\int \frac{(\ln x)^3}{x}\, dx = \frac{(\ln x)^4}{4}$$

(b) Looking at the answers to part (a),

$$\int \frac{(\ln x)^n}{x}\, dx = \frac{(\ln x)^{n+1}}{n+1} + C.$$

(c) Let $w = \ln x$. Then $dw = (1/x)dx$, and

$$\int \frac{(\ln x)^n}{x}\, dx = \int w^n\, dw = \frac{w^{n+1}}{n+1} + C = \frac{(\ln x)^{n+1}}{n} + C.$$

197. (a) A possible answer from the CAS is

$$\int \frac{x^4}{(1+x^2)^2}\, dx = x + \frac{x}{2\left(1+x^2\right)} - \frac{3}{2}\arctan(x).$$

Different systems may give the answer in different form.

(b) Differentiating gives

$$\frac{d}{dx}\left(x + \frac{x}{2\left(1+x^2\right)} - \frac{3}{2}\arctan(x) \right) = 1 - \frac{x^2}{\left(1+x^2\right)^2} - \frac{1}{1+x^2}.$$

(c) Putting the result of part (b) over a common denominator, we get

$$1 - \frac{x^2}{\left(1 + x^2\right)^2} - \frac{1}{1 + x^2} = \frac{\left(1 + x^2\right)^2 - x^2 - (1 + x^2)}{(1 + x^2)^2}$$

$$= \frac{1 + 2x^2 + x^4 - x^2 - 1 - x^2}{(1 + x^2)^2} = \frac{x^4}{(1 + x^2)^2}.$$

CHAPTER EIGHT

Solutions for Section 8.1

Exercises

1. (a) The strip stretches between $y = 0$ and $y = 2x$, so

$$\text{Area of region} \approx \sum (2x)\Delta x.$$

(b) We have

$$\text{Area} = \int_0^3 (2x)\,dx = x^2 \Big|_0^3 = 9.$$

5. Each strip is a rectangle of length 3 and width Δx, so

$$\text{Area of strip} = 3\Delta x, \quad \text{so}$$
$$\text{Area of region} = \int_0^5 3\,dx = 3x \Big|_0^5 = 15.$$

Check: This area can also be computed using Length $\times$ Width $= 5 \cdot 3 = 15$.

9. The strip has width Δy, so the variable of integration is y. The length of the strip is x. Since $x^2 + y^2 = 10$ and the region is in the first quadrant, solving for x gives $x = \sqrt{10 - y^2}$. Thus

$$\text{Area of strip} \approx x\Delta y = \sqrt{10 - y^2}\,dy.$$

The region stretches from $y = 0$ to $y = \sqrt{10}$, so

$$\text{Area of region} = \int_0^{\sqrt{10}} \sqrt{10 - y^2}\,dy.$$

Evaluating using VI-30 from the Table of Integrals, we have

$$\text{Area} = \frac{1}{2}\left(y\sqrt{10 - y^2} + 10\arcsin\left(\frac{y}{\sqrt{10}}\right)\right)\Bigg|_0^{\sqrt{10}} = 5(\arcsin 1 - \arcsin 0) = \frac{5}{2}\pi.$$

Check: This area can also be computed using the formula $\frac{1}{4}\pi r^2 = \frac{1}{4}\pi(\sqrt{10})^2 = \frac{5}{2}\pi$.

13. (a) Region I is bounded by the y-axis, or $x = 0$, so it is C. Region II is bounded by the line $y = 2$, so it is B. Region III is bounded by $x = 2$ so it is A. Region IV is bounded by the x-axis, or $y = 0$, so it is D.
    (b) The line $y + x = 2$ can be written as $y = 2 - x$, and this line and the line $y = x$ intersect at $x = 1, y = 1$.
    For Region II:
$$\text{Area} = \int_0^1 (2 - (2 - x))\,dx + \int_1^2 (2 - x)\,dx = \int_0^1 x\,dx + \int_1^2 (2 - x)\,dx.$$
    For Region III:
$$\text{Area} = \int_1^2 (x - (2 - x))\,dx = \int_1^2 (2x - 2)\,dx.$$

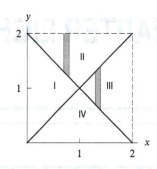

Figure 8.1

17. Each slice is a rectangular slab of length 10 m and width that decreases with height. See Figure 8.2. At height y, the length x is given by the Pythagorean Theorem

$$y^2 + x^2 = 7^2.$$

Solving gives $x = \sqrt{7^2 - y^2}$ m. Thus the width of the slab is $2x = 2\sqrt{7^2 - y^2}$ and

$$\text{Volume of slab} = \text{Length} \cdot \text{Width} \cdot \text{Height} = 10 \cdot 2\sqrt{7^2 - y^2} \cdot \Delta y = 20\sqrt{7^2 - y^2}\,\Delta y \text{ m}^3.$$

Summing over all slabs, we have

$$\text{Total volume} \approx \sum 20\sqrt{7^2 - y^2}\,\Delta y \text{ m}^3.$$

Taking a limit as $\Delta y \to 0$, we get

$$\text{Total volume} = \lim_{\Delta y \to 0} \sum 20\sqrt{7^2 - y^2}\,\Delta y = \int_0^7 20\sqrt{7^2 - y^2}\,dy \text{ m}^3.$$

To evaluate, we use the table of integrals or the fact that $\displaystyle\int_0^7 \sqrt{7^2 - y^2}\,dy$ represents the area of a quarter circle of radius 7, so

$$\text{Total volume} = \int_0^7 20\sqrt{7^2 - y^2}\,dy = 20 \cdot \frac{1}{4}\pi 7^2 = 245\pi \text{ m}^3.$$

Check: the volume of a half cylinder can also be calculated using the formula $V = \frac{1}{2}\pi r^2 h = \frac{1}{2}\pi 7^2 \cdot 10 = 245\pi \text{ m}^3$.

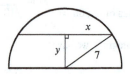

Figure 8.2

21. Each slice is a triangle with width 4 cm, height 3 cm, and thickness Δx, so

$$\text{Volume of triangular slice} = \text{Area of triangle} \cdot \Delta x = \frac{1}{2}4 \cdot 3 \cdot \Delta x = 6\Delta x \text{ cm}^3.$$

Summing over all the triangular slices, we have

$$\text{Total volume} \approx \sum 6\Delta x \text{ cm}^3.$$

Taking a limit as $\Delta x \to 0$, we get

$$\text{Total volume} = \lim_{\Delta x \to 0} \sum 6\Delta x = \int_0^{15} 6\,dx \text{ cm}^3.$$

Evaluating gives

$$\text{Total volume} = 6x\Big|_0^{15} = 90 \text{ cm}^3.$$

Problems

25. Triangle of base and height 7 and 5. See Figure 8.3. (Either 7 or 5 can be the base. A non-right triangle is also possible.)

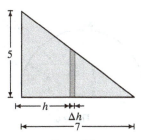

Figure 8.3

29. (a) The area is shown in Figure 8.4(a) with two vertical slices shown. Because some of the slices are bounded on the top by $y = x^2$ while others are bounded on the top by $y = 6 - x$, we use two separate integrals to calculate this area using vertical slices. The slices between $x = 0$ and $x = 2$ are bounded on the top by $y = x^2$ and are bounded on the bottom by the x-axis, which is $y = 0$. This part of the area is given by

$$\text{Left part of the area} = \int_0^2 x^2\, dx = \frac{1}{3}x^3 \Big|_0^2 = \frac{1}{3}\cdot 2^3 - 0 = 2.667.$$

The slices between $x = 2$ and $x = 6$ are bounded on the top by $y = 6 - x$ and are bounded on the bottom by the x-axis, which is $y = 0$. This part of the area is given by

$$\text{Right part of the area} = \int_2^6 (6 - x)\, dx = 6x - \frac{1}{2}x^2 \Big|_2^6$$
$$= 6\cdot 6 - \frac{1}{2}\cdot 6^2 - \left(6\cdot 2 - \frac{1}{2}\cdot 2^2\right) = 8.$$

The total area is given by

$$\text{Total area} = \int_0^2 x^2\, dx + \int_2^6 (6 - x)\, dx = 2.667 + 8 = 10.667.$$

(a)
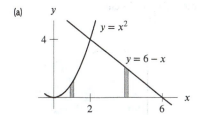

(b)
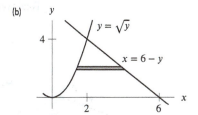

Figure 8.4

(b) The area is shown in Figure 8.4(b) with one horizontal slice shown. The slices are all bounded on the right by $y = 6-x$, which is $x = 6 - y$ distance from the y-axis, and are bounded on the left by $y = x^2$, which is $x = \sqrt{y}$ distance from the y-axis. The area ranges from $y = 0$ to $y = 4$ so the area is given by

$$A = \int_0^4 ((6 - y) - \sqrt{y})\, dy = 6y - \frac{1}{2}y^2 - \frac{2}{3}\cdot y^{3/2} \Big|_0^4$$
$$= 6\cdot 4 - \frac{1}{2}\cdot 4^2 - \frac{2}{3}\cdot 4^{3/2} - 0$$
$$= 10.667.$$

Notice that the two ways of computing the area give the same answer, as we expect.

33. The integral $\int_0^8 \pi \left(64 - h^2\right) dh$ represents the volume for a hemisphere with radius 8, see Figure 8.5, so the integral

$2\int_0^8 \pi \left(64 - h^2\right) dh$ represents the volume for a sphere with radius 8.

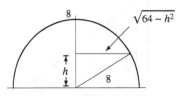

Figure 8.5

37. (a) The thickness of each slice is the height of the hemisphere divided by the number of slices, so $\Delta h = 4/20 = 0.2$.

(b) From Figure 8.6, using similar triangles, we see that the radius of the sphere at height h is $r = \sqrt{16 - h^2}$, so the volume, ΔV, of a cylinder approximating the volume of this slice is

$$\Delta V = \pi r^2 \Delta h = \pi \left(\sqrt{16 - h^2}\right)^2 \cdot 0.2 = 0.2\pi \left(16 - h^2\right).$$

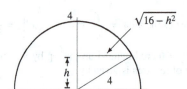

Figure 8.6

41. (a) A vertical slice has a triangular shape and thickness Δx. See Figure 8.7.

$$\text{Volume of slice} = \text{Area of triangle} \cdot \Delta x = \frac{1}{2} \text{ Base } \cdot \text{ Height } \cdot \Delta x = \frac{1}{2} \cdot 2 \cdot 3\Delta x = 3\Delta x \text{ cm}^3.$$

Thus,

$$\text{Total volume} = \lim_{\Delta x \to 0} \sum 3\Delta x = \int_0^4 3\, dx = 3x \bigg|_0^4 = 12 \text{ cm}^3.$$

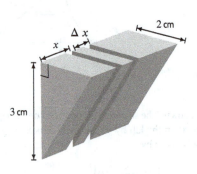
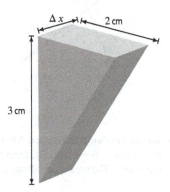

Figure 8.7

(b) A horizontal slice has a rectangular shape and thickness Δh. See Figure 8.8. Using similar triangles, we see that

$$\frac{w}{2} = \frac{3 - h}{3},$$

so

$$w = \frac{2}{3}(3 - h) = 2 - \frac{2}{3}h.$$

Thus

$$\text{Volume of slice} \approx 4w\Delta h = 4\left(2 - \frac{2}{3}h\right)\Delta h = \left(8 - \frac{8}{3}h\right)\Delta h.$$

So,

$$\text{Total volume} = \lim_{\Delta h \to 0} \sum \left(8 - \frac{8}{3}h\right)\Delta h = \int_0^3 \left(8 - \frac{8}{3}h\right) dh = \left(8h - \frac{4h^2}{3}\right)\Big|_0^3 = 12 \text{ cm}^3.$$

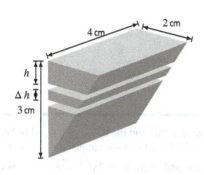 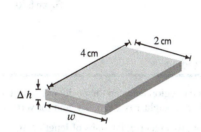 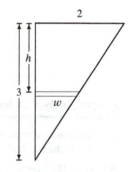

Figure 8.8

---

45. We slice the tank horizontally. There is an outside radius r_{out} and an inside radius r_{in} and, at height h,

$$\text{Volume of a slice} \approx \pi(r_{\text{out}})^2\Delta h - \pi(r_{\text{in}})^2\Delta h.$$

See Figure 8.9. We see that $r_{\text{out}} = 3$ for every slice. We use similar triangles to find r_{in} in terms of the height h:

$$\frac{r_{\text{in}}}{h} = \frac{3}{6} \quad \text{so} \quad r_{\text{in}} = \frac{1}{2}h.$$

At height h,

$$\text{Volume of slice} \approx \pi(3)^2\Delta h - \pi\left(\frac{1}{2}h\right)^2\Delta h.$$

To find the total volume, we integrate this quantity from $h = 0$ to $h = 6$.

$$V = \int_0^6 \left(\pi(3)^2 - \pi\left(\frac{1}{2}h\right)^2\right) dh = \pi \int_0^6 \left(9 - \frac{1}{4}h^2\right) dh = \pi\left(9h - \frac{h^3}{12}\right)\Big|_0^6 = 36\pi = 113.097 \text{ m}^3.$$

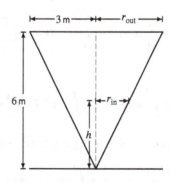

Figure 8.9

49. From Figure 8.10, using similar triangles, we see that the radius of the sphere at height h is $r = \sqrt{16 - h^2}$, so the volume, ΔV, of a cylinder approximating the volume of this slice is (IV) since

$$\Delta V = \pi r^2 \Delta h = \pi \left(\sqrt{16 - h^2} \right)^2 \Delta h = \pi \left(16 - h^2 \right) \Delta h.$$

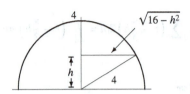

Figure 8.10

Strengthen Your Understanding

53. One possible answer is the region between the positive x-axis, the positive y-axis and the line $y = 1 - x$. For horizontal slices, the width of a slice at height y is $x = 1 - y$, and for vertical slices the height of a slice at position x is $y = 1 - x$.

57. True. Horizontal slicing gives rectangular slabs of length l, thickness Δy, and width $w = 2\sqrt{r^2 - y^2}$. So the volume of one slab is $2l\sqrt{r^2 - y^2}\Delta y$, and the integral is $\int_{-r}^{r} 2l\sqrt{r^2 - y^2}\,dy$.

Solutions for Section 8.2

Exercises

1. (a) The volume of a disk is given by

$$\Delta V \approx \pi(2x)^2 \Delta x = 4\pi x^2 \,\Delta x,$$

so

$$\text{Volume} = \int_0^3 4\pi x^2 \,dx.$$

(b) We have

$$\text{Volume} = 4\pi \frac{x^3}{3}\bigg|_0^3 = 36\pi.$$

5. The volume is given by

$$V = \int_0^1 \pi y^2 dx = \int_0^1 \pi x^4 dx = \pi \frac{x^5}{5}\bigg|_0^1 = \frac{\pi}{5}.$$

9. The volume is given by

$$V = \int_{-1}^1 \pi y^2\,dx = \int_{-1}^1 \pi(e^x)^2\,dx = \int_{-1}^1 \pi e^{2x}\,dx = \frac{\pi}{2}e^{2x}\bigg|_{-1}^1 = \frac{\pi}{2}(e^2 - e^{-2}).$$

13. Since the graph of $y = x^2$ is below the graph of $y = x$ for $0 \le x \le 1$, the volume is given by

$$V = \int_0^1 \pi x^2\,dx - \int_0^1 \pi(x^2)^2\,dx = \pi \int_0^1 (x^2 - x^4)\,dx = \pi \left(\frac{x^3}{3} - \frac{x^5}{5} \right)\bigg|_0^1 = \frac{2\pi}{15}.$$

17. For the region being rotated, see Figure 8.11. We slice the volume horizontally, which divides the solid into thin washers with

$$\text{Volume of slice } = \pi((r_{\text{out}})^2 - (r_{\text{in}})^2)\Delta y$$

where r_{out} is the outer radius and r_{in} the inner radius.

The outer radius is the horizontal distance from the y-axis to the curve $x = 2$, so $r_{\text{out}} = 2$. The inner radius is the horizontal distance from the y-axis to the curve $y = \ln x$ or $x = e^y$, so $r_{\text{in}} = e^y$.

Since the region lies between $y = 0$ and $y = \ln 2$, the volume V is given by

$$V = \int_0^{\ln 2} \pi \left(2^2 - (e^y)^2 \right) dy$$

$$= \int_0^{\ln 2} \left(4\pi - e^{2y}\pi \right) dy$$

$$= \left(4\pi y - \frac{e^{2y}}{2}\pi \right) \Big|_0^{\ln 2}$$

$$= 4\pi \ln 2 - \frac{e^{2\ln 2}}{2}\pi - \left(0 - \frac{e^0}{2}\pi \right)$$

$$= 4\pi \ln 2 - \frac{4}{2}\pi + \frac{\pi}{2} = 4\pi \ln 2 - \frac{3}{2}\pi.$$

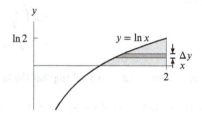

Figure 8.11

21. Since $f'(x) = 1/(x + 1)$, we evaluate the integral numerically to get

$$\text{Arc length } = \int_0^2 \sqrt{1 + \left(\frac{1}{x+1} \right)^2}\, dx = 2.302.$$

25. Since the curve is not retraced for $1 \le t \le 2$, its length is

$$\int_1^2 \sqrt{(x'(t))^2 + (y'(t))^2}\, dt = \int_1^2 \sqrt{5^2 + 4^2}\, dt = \int_1^2 \sqrt{41}\, dt = \sqrt{41}.$$

Notice that the curve is a straight line from the point $(8, 5)$ to $(13, 9)$. The arc length integral then becomes the formula for the distance between two points.

Problems

29. The two functions intersect at $(0, 0)$ and $(5, 25)$. We slice the volume with planes perpendicular to the x-axis. This divides the solid into thin washers with volume

$$\text{Volume of slice } = \pi((r_{out})^2 - (r_{in})^2)\Delta x.$$

The outer radius is the vertical distance from the x-axis to the curve $y = 5x$, so $r_{out} = 5x$. Similarly, the inner radius is the vertical distance from the x-axis to the curve $y = x^2$, so $r_{in} = x^2$. Integrating from $x = 0$ to $x = 5$ we have

$$V = \int_0^5 \pi((5x)^2 - (x^2)^2)\, dx.$$

33. The two functions intersect at $(0,0)$ and $(8,2)$. We slice the volume with planes perpendicular to the line $x = 9$. This divides the solid into thin washers with volume

$$\text{Volume of slice} = \pi r_{out}^2 \Delta y - \pi r_{in}^2 \Delta y.$$

The outer radius is the horizontal distance from the line $x = 9$ to the curve $x = y^3$, so $r_{out} = 9 - y^3$. Similarly, the inner radius is the horizontal distance from the line $x = 9$ to the curve $x = 4y$, so $r_{in} = 9 - 4y$. Integrating from $y = 0$ to $y = 2$ we have

$$V = \int_0^2 [\pi(9 - y^3)^2 - \pi(9 - 4y)^2] \, dy.$$

37. One arch of the sine curve lies between $x = 0$ and $x = \pi$. Since $d(\sin x)/dx = \cos x$, evaluating the integral numerically gives

$$\text{Arc length} = \int_0^\pi \sqrt{1 + \cos^2 x} \, dx = 3.820.$$

41.

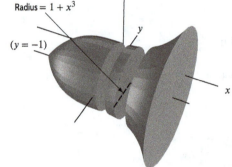

Radius $= 1 + x^3$

$(y = -1)$

We slice the region perpendicular to the x–axis. The Riemann sum we get is $\sum \pi(x^3+1)^2 \Delta x$. So the volume V is the integral

$$V = \int_{-1}^1 \pi(x^3 + 1)^2 \, dx$$

$$= \pi \int_{-1}^1 (x^6 + 2x^3 + 1) \, dx$$

$$= \pi \left(\frac{x^7}{7} + \frac{x^4}{2} + x \right) \Big|_{-1}^1$$

$$= (16/7)\pi \approx 7.18.$$

45. Slice the object into rings vertically, as is Figure 8.12. A typical ring has thickness Δx and outer radius $y = 1$ and inner radius $y = x^2$.

$$\text{Volume of slice} \approx \pi 1^2 \Delta x - \pi y^2 \Delta x = \pi(1 - x^4) \, \Delta x.$$

$$\text{Volume of solid} = \lim_{\Delta x \to 0} \sum \pi(1 - x^4) \, \Delta x = \int_0^1 \pi(1 - x^4) \, dx = \pi \left(x - \frac{x^5}{5} \right) \Big|_0^1 = \frac{4}{5}\pi.$$

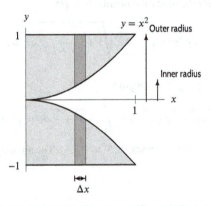

Figure 8.12: Cross-section of solid

49. An equilateral triangle of side s has height $\sqrt{3}s/2$ and

$$\text{Area} = \frac{1}{2} \cdot s \cdot \frac{\sqrt{3}s}{2} = \frac{\sqrt{3}}{4}s^2.$$

Slicing perpendicularly to the y-axis gives equilateral triangles whose thickness is Δy and whose side is $x = \sqrt{y}$. See Figure 8.13. Thus

$$\text{Volume of triangular slice} \approx \frac{\sqrt{3}}{4}(\sqrt{y})^2 \Delta y = \frac{\sqrt{3}}{4} y \, \Delta y.$$

$$\text{Volume of solid } = \int_0^1 \frac{\sqrt{3}}{4} y\, dy = \frac{\sqrt{3}}{4} \frac{y^2}{2}\bigg|_0^1 = \frac{\sqrt{3}}{8}.$$

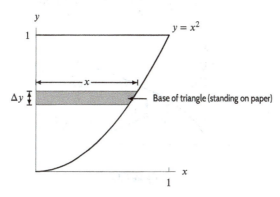

Figure 8.13: Base of solid

53. We now slice perpendicular to the x-axis. As stated in the problem, the cross-sections obtained thereby will be squares, with base length e^x. The volume of one square slice is $(e^x)^2\, dx$. (See Figure 8.14.) Adding the volumes of the slices yields

$$\text{Volume } = \int_{x=0}^{x=1} y^2\, dx = \int_0^1 e^{2x}\, dx = \frac{e^{2x}}{2}\bigg|_0^1 = \frac{e^2 - 1}{2} = 3.195.$$

Figure 8.14

57. For the region being rotated, see Figure 8.15. We slice the volume horizontally, which divides the solid into thin washers with

$$\text{Volume of slice } = \pi((r_{\text{out}})^2 - (r_{\text{in}})^2)\Delta y$$

where r_{out} is the outer radius and r_{in} the inner radius.

The outer radius is the horizontal distance from the y-axis to the curve $y = 2 - \sqrt{x}$ or $x = (2 - y)^2$, so $r_{\text{out}} = (2 - y)^2$. The inner radius is the horizontal distance from the y-axis to the curve $y = \sqrt{x}$ or $x = y^2$, so $r_{\text{in}} = y^2$.

Since the region lies between $y = 0$ and $y = 1$, where the curves intersect, the volume V is given by

$$V = \int_0^1 \pi\left(((2-y)^2)^2 - (y^2)^2\right) dy$$
$$= \int_0^1 \left((2-y)^4 \pi - y^4 \pi\right) dy$$

$$= \left(-\frac{(2-y)^5}{5}\pi - \frac{y^5}{5}\pi \right) \Bigg|_0^1$$

$$= \left(-\frac{1}{5}\pi - \frac{1}{5}\pi \right) - \left(-\frac{2^5}{5}\pi \right) = 6\pi.$$

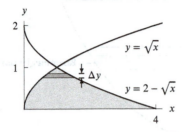

Figure 8.15

61. Solving for y gives $y^{2/3} = a^{2/3} - x^{2/3}$ so

$$y = \pm(a^{2/3} - x^{2/3})^{3/2}.$$

Positive y-values give the upper half of the curve; negative y-values give the lower half.

Slicing the volume perpendicular to the x-axis gives disks of radius y; see Figure 8.16. We have

$$\text{Volume of slice } = \pi y^2 \Delta x = \pi \left((a^{2/3} - x^{2/3})^{3/2} \right)^2 \Delta x = \pi \left(a^{2/3} - x^{2/3} \right)^3 \Delta x$$

$$= \pi(a^2 - 3a^{4/3}x^{2/3} + 3a^{2/3}x^{4/3} - x^2)\Delta x.$$

Adding the volumes of slices from $x = 0$ to $x = a$, multiplying by 2 to include the volume to the left of the y-axis, and taking the limit gives

$$\text{Total volume } = 2\pi \int_0^a y^2 \, dx = 2\pi \int_0^a \left(a^2 - 3a^{4/3}x^{2/3} + 3a^{2/3}x^{4/3} - x^2 \right) dx$$

$$= 2\pi \left(a^2 x - 3a^{4/3}\frac{x^{5/3}}{5/3} + 3a^{2/3}\frac{x^{7/3}}{7/3} - \frac{x^3}{3} \right) \Bigg|_0^a$$

$$= 2\pi \left(a^3 - \frac{9}{5}a^3 + \frac{9}{7}a^3 - \frac{1}{3}a^3 \right) = \frac{32\pi}{105}a^3.$$

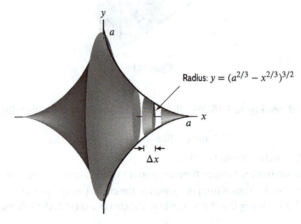

Figure 8.16

65. We can find the volume of the tree by slicing it into a series of thin horizontal cylinders of height dh and circumference C. The volume of each cylindrical disk will then be

$$V = \pi r^2 \, dh = \pi \left(\frac{C}{2\pi}\right)^2 dh = \frac{C^2 \, dh}{4\pi}.$$

Summing all such cylinders, we have the total volume of the tree as

$$\text{Total volume} = \frac{1}{4\pi} \int_0^{120} C^2 \, dh.$$

We can estimate this volume using a trapezoidal approximation to the integral with $\Delta h = 20$:

$$\text{LEFT estimate} = \frac{1}{4\pi}[20(31^2 + 28^2 + 21^2 + 17^2 + 12^2 + 8^2)] = \frac{1}{4\pi}(53660).$$

$$\text{RIGHT estimate} = \frac{1}{4\pi}[20(28^2 + 21^2 + 17^2 + 12^2 + 8^2 + 2^2)] = \frac{1}{4\pi}(34520).$$

$$\text{TRAP} = \frac{1}{4\pi}(44090) \approx 3509 \text{ cubic inches.}$$

69. (a) Each slice is a semicircular disk of thickness Δx as shown in Figure 8.17. In Figure 8.18, we see that the radius of the disk is $r = \sqrt{a^2 - x^2}$. Thus,

$$\text{Volume of slice} = \frac{1}{2}\pi r^2 \cdot \Delta x = \frac{\pi}{2}\left(\sqrt{a^2 - x^2}\right)^2 \cdot \Delta x = \frac{\pi}{2}(a^2 - x^2)\,\Delta x.$$

(b) Summing over the slices and taking a limit as $\Delta x \to 0$, we have

$$\text{Volume of hemisphere} = \lim_{\Delta x \to 0} \sum \frac{\pi}{2}(a^2 - x^2)\Delta x = \int_{-a}^{a} \frac{\pi}{2}(a^2 - x^2)\,dx.$$

(c) Evaluating the integral gives

$$\text{Volume of hemisphere} = \int_{-a}^{a} \frac{\pi}{2}(a^2 - x^2)\,dx = \frac{\pi}{2}\left(a^2 x - \frac{x^3}{3}\right)\Bigg|_{-a}^{a} = \frac{2\pi}{3}a^3.$$

This is half the volume of a sphere, as expected.

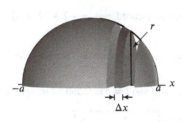

Figure 8.17

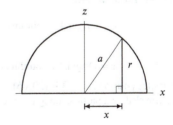

Figure 8.18

73. The volume of a perpendicular slice at x is:

$$\text{Volume of slice} \approx \pi y^2 \Delta x = \pi \left(kx(x - 2)\right)^2 \Delta x.$$

The total volume is found by adding these slices from $x = 0$ to $x = 2$:

$$\text{Total volume} \approx \sum \pi \left(kx(x - 2)\right)^2 \Delta x$$

Taking the limit gives the integral:

$$\text{Total volume} = \int_0^2 \pi \left(kx(x - 2)\right)^2 dx = \pi k^2 \left(\frac{1}{5}x^5 - x^4 + \frac{4}{3}x^3\right)\Bigg|_0^2 = \frac{16\pi}{15}k^2.$$

We know the volume is $192\pi/5$, so:

$$\frac{16\pi}{15}k^2 = \frac{192\pi}{5}$$

$$k^2 = \frac{192\pi}{5} \cdot \frac{15}{16\pi} = 36$$

$$k = 6. \qquad\qquad \text{since } k > 0.$$

77. The arc length of $g(x)$ from $x = 4$ to $x = 44$ is $\int_4^{44} \sqrt{1 + (g'(x))^2}\, dx$. We calculate the derivative of $g(x)$ using the chain rule:

$$g'(x) = 4f'(0.25x + 1) \cdot 0.25 = f'(0.25x + 1).$$

Therefore,

$$
\begin{aligned}
\text{Arc length} &= \int_4^{44} \sqrt{1 + (g'(x))^2}\, dx \\
&= \int_4^{44} \sqrt{1 + (f'(0.25x + 1))^2}\, dx \\
&= \int_{x=4}^{x=44} \sqrt{1 + (f'(w))^2}\, 4\, dw \qquad \text{let } w = 0.25x + 1, dw = 0.25\, dx \\
&= 4 \int_{w=2}^{w=12} \sqrt{1 + (f'(w))^2}\, dw \\
&= 4\, (\text{Arc length of } f(w) \text{ from } 2 \text{ to } 12) \\
&= 4 \cdot 20 = 80.
\end{aligned}
$$

81. (a) To find the change in position in the x-coordinate from $t = 0$ to $t = 3$, we compute the definite integral of dx/dt, the instantaneous velocity in the x-direction:

$$\text{Change in } x\text{-coordinate} = \int_0^3 3t^2\, dt = t^3 \Big|_0^3 = 27.$$

Similarly,

$$\text{Change in } y\text{-coordinate} = \int_0^3 12t\, dt = 6t^2 \Big|_0^3 = 54.$$

(b) From part (a), we know the change in x from $t = 0$ to $t = 3$ is 27 and the change in y from $t = 0$ to $t = 3$ is 54. Therefore, at time $t = 3$, we have

$$x = \text{Original } x\text{-coordinate} + \text{Change in } x\text{-coordinate} = -7 + 27 = 20.$$

and

$$y = \text{Original } y\text{-coordinate} + \text{Change in } y\text{-coordinate} = 11 + 54 = 65.$$

Thus, the particles position at $t = 3$ is $(20, 65)$.

(c) The distance traveled by a particle along a curve between $t = a$ and $t = b$ is found by integrating its speed:

$$\text{Distance traveled} = \int_a^b \sqrt{\left(\frac{dx}{dt}\right)^2 + \left(\frac{dy}{dt}\right)^2}\, dt.$$

In this case, we have $a = 0$ and $b = 3$ so

$$
\begin{aligned}
\text{Distance traveled} &= \int_0^3 \sqrt{\left(3t^2\right)^2 + (12t)^2}\, dt \\
&= \int_0^3 \sqrt{9t^4 + 144t^2}\, dt \\
&= \int_0^3 \sqrt{9t^2\left(t^2 + 16\right)}\, dt \qquad \text{factoring out } 9t^2 \\
&= \int_0^3 3t\sqrt{t^2 + 16}\, dt \\
&= \left(t^2 + 16\right)^{3/2} \Big|_0^3 \qquad \text{using substitution with } w = t^2 + 16 \\
&= (25)^{3/2} - (16)^{3/2} = 61.
\end{aligned}
$$

85. The problem appears complicated, because we are now working in three dimensions. However, if we take one dimension at a time, we will see that the solution is not too difficult. For example, let's just work at a constant depth, say 0. We apply the trapezoid rule to find the approximate area along the length of the boat. For example, by the trapezoid rule the approximate area at depth 0 from the front of the boat to 10 feet toward the back is $\frac{(2+8)\cdot 10}{2} = 50$. Overall, at depth 0 we have that the area for each length span is as follows:

Table 8.1

length span:	0–10	10–20	20–30	30–40	40–50	50–60
depth 0	50	105	145	165	165	130

We can fill in the whole chart the same way:

Table 8.2

	length span:	0–10	10–20	20–30	30–40	40–50	50–60
	0	50	105	145	165	165	130
	2	25	60	90	105	105	90
depth	4	15	35	50	65	65	50
	6	5	15	25	35	35	25
	8	0	5	10	10	10	10

Now, to find the volume, we just apply the trapezoid rule to the depths and areas. For example, according to the trapezoid rule the approximate volume as the depth goes from 0 to 2 and the length goes from 0 to 10 is $\frac{(50+25)\cdot 2}{2} = 75$. Again, we fill in a chart:

Table 8.3

	length span:	0–10	10–20	20–30	30–40	40–50	50–60
	0–2	75	165	235	270	270	220
depth	2–4	40	95	140	170	170	140
span	4–6	20	50	75	100	100	75
	6–8	5	20	35	45	45	35

Adding all this up, we find the volume is approximately 2595 cubic feet.

You might wonder what would have happened if we had done our trapezoids along the depth axis first instead of along the length axis. If you try this, you'd find that you come up with the same answers in the volume chart! For the trapezoid rule, it does not matter which axis you choose first.

89. The graph of f is concave down where $f''(x) < 0$:

$$f(x) = x^4 - 8x^3 + 18x^2 + 3x + 7$$
$$f'(x) = 4x^3 - 24x^2 + 36x + 3$$
$$f''(x) = 12x^2 - 48x + 36$$
$$= 12(x - 1)(x - 3).$$

Hence $f''(x) < 0$ for $1 < x < 3$. The arc length of this portion of the graph of f is given by

$$\text{Arc length} = \int_1^3 \sqrt{1 + (f'(x))^2}\, dx$$
$$= \int_1^3 \sqrt{1 + \left(4x^3 - 24x^2 + 36x + 3\right)^2}\, dx.$$

Strengthen Your Understanding

93. The arc length is given by $\int_0^{\pi/4} \sqrt{1 + (f'(x))^2}\, dx$ with $f(x) = \sin x$. Thus, the correct formula for the arc length is $\int_0^{\pi/4} \sqrt{1 + \cos^2 x}\, dx$.

97. One example is the region bounded by $y = 2x$ and the x-axis for $0 \le x \le 1$. When the region is rotated about the x-axis, we get a cone of radius 2 and height 1. When it is rotated about the y-axis, we get a cone of radius 1 and height 2. Since

$$\text{Volume around } x\text{-axis} = \frac{1}{3}\pi \cdot 2^2 \cdot 1 = \frac{4\pi}{3}$$

$$\text{Volume around } y\text{-axis} = \frac{1}{3}\pi \cdot 1^2 \cdot 2 = \frac{2\pi}{3},$$

the volume is greater around the x-axis than around the y-axis.

101. False. Suppose that the graph of f starts at the point $(0, 100)$ and then goes down to $(1, 0)$ and from there on goes along the x-axis. For example, if $f(x) = 100(x-1)^2$ on the interval $[0, 1]$ and $f(x) = 0$ on the interval $[1, 10]$, then f is differentiable on the interval $[0, 10]$. The arc length of the graph of f on the interval $[0, 1]$ is at least 100, while the arc length on the interval $[1, 10]$ is 9.

Additional Problems (online only)

105. (a) If $f(x) = \int_0^x \sqrt{g'(t)^2 - 1} \, dt$, then, by the Fundamental Theorem of Calculus, $f'(x) = \sqrt{g'(x)^2 - 1}$. So the arc length of f from 0 to x is

$$\int_0^x \sqrt{1 + (f'(t))^2} \, dt = \int_0^x \sqrt{1 + (\sqrt{g'(t)^2 - 1})^2} \, dt$$

$$= \int_0^x \sqrt{1 + g'(t)^2 - 1} \, dt$$

$$= \int_0^x g'(t) \, dt = g(x) - g(0) = g(x).$$

(b) If g is the arc length of any function f, then by the Fundamental Theorem of Calculus, $g'(x) = \sqrt{1 + f'(x)^2} \ge 1$. So if $g'(x) < 1$, g cannot be the arc length of a function.

(c) We find a function f whose arc length from 0 to x is $g(x) = 2x$. Using part (a), we see that

$$f(x) = \int_0^x \sqrt{(g'(t))^2 - 1} \, dt = \int_0^x \sqrt{2^2 - 1} \, dt = \sqrt{3}x.$$

This is the equation of a line. Does it make sense to you that the arc length of a line segment depends linearly on its right endpoint?

Solutions for Section 8.3

Exercises

1. With $r = 1$ and $\theta = 2\pi/3$, we find $x = r\cos\theta = 1 \cdot \cos(2\pi/3) = -1/2$ and $y = r\sin\theta = 1 \cdot \sin(2\pi/3) = \sqrt{3}/2$. The rectangular coordinates are $(-1/2, \sqrt{3}/2)$.

5. With $x = 1$ and $y = 1$, find r from $r = \sqrt{x^2 + y^2} = \sqrt{1^2 + 1^2} = \sqrt{2}$. Find θ from $\tan\theta = y/x = 1/1 = 1$. Thus, $\theta = \tan^{-1}(1) = \pi/4$. Since $(1, 1)$ is in the first quadrant this is a correct θ. The polar coordinates are $(\sqrt{2}, \pi/4)$.

9. (a) Table 8.4 contains values of $r = 1 - \sin\theta$, both exact and rounded to one decimal.

Table 8.4

θ	0	$\pi/3$	$\pi/2$	$2\pi/3$	π	$4\pi/3$	$3\pi/2$	$5\pi/3$	2π	$7\pi/3$	$5\pi/2$	$8\pi/3$
r	1	$1 - \sqrt{3}/2$	0	$1 - \sqrt{3}/2$	1	$1 + \sqrt{3}/2$	2	$1 + \sqrt{3}/2$	1	$1 - \sqrt{3}/2$	0	$1 - \sqrt{3}/2$
r	1	0.134	0	0.134	1	1.866	2	1.866	1	0.134	0	0.134

(b) See Figure 8.19.

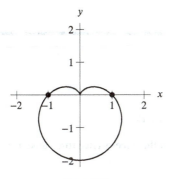

Figure 8.19

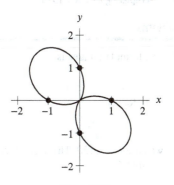

Figure 8.20

(c) The circle has equation $r = 1/2$. The cardioid is $r = 1 - \sin\theta$. Solving these two simultaneously gives

$$1/2 = 1 - \sin\theta,$$

or

$$\sin\theta = 1/2.$$

Thus, $\theta = \pi/6$ or $5\pi/6$. This gives the points $(x, y) = ((1/2)\cos\pi/6, (1/2)\sin\pi/6) = (\sqrt{3}/4, 1/4)$ and $(x, y) = ((1/2)\cos 5\pi/6, (1/2)\sin 5\pi/6) = (-\sqrt{3}/4, 1/4)$ as the location of intersection.

(d) The curve $r = 1 - \sin 2\theta$, pictured in Figure 8.20, has two regions instead of the one region that $r = 1 - \sin\theta$ has. This is because $1 - \sin 2\theta$ will be 0 twice for every 2π cycle in θ, as opposed to once for every 2π cycle in θ for $1 - \sin\theta$.

13. See Figures 8.21 and 8.22. The first curve will be similar to the second curve, except the cardioid (heart) will be rotated clockwise by 90° ($\pi/2$ radians). This makes sense because of the identity $\sin\theta = \cos(\theta - \pi/2)$.

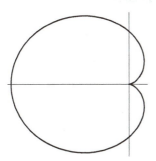

Figure 8.21: $r = 1 - \cos\theta$

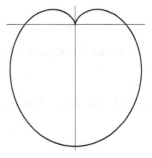

Figure 8.22: $r = 1 - \sin\theta$

17. The region is given by $\sqrt{8} \le r \le \sqrt{18}$ and $\pi/4 \le \theta \le \pi/2$.

21. Expressing x and y in terms of θ, we have

$$x = e^\theta \cos\theta \quad \text{and} \quad y = e^\theta \sin\theta.$$

The slope is given by

$$\frac{dy}{dx} = \frac{e^\theta \sin\theta + e^\theta \cos\theta}{e^\theta \cos\theta - e^\theta \sin\theta} = \frac{\sin\theta + \cos\theta}{\cos\theta - \sin\theta}.$$

At $\theta = \pi/2$, we have $\sin\theta = 1$ and $\cos\theta = 0$, so

$$\left.\frac{dy}{dx}\right|_{\theta=\pi/2} = \frac{1+0}{0-1} = -1.$$

Problems

25. The formula for area is

$$A = \frac{1}{2} \int_\alpha^\beta r^2 \, d\theta$$

Therefore, since

$$A = \frac{1}{2} \int_0^{\pi/3} \sin^2(3\theta) \, d\theta = \frac{1}{2} \int_0^{\pi/3} (\sin 3\theta)^2 \, d\theta,$$

we have $r = \sin 3\theta$. The integral represents the shaded area inside one petal of the three-petaled rose curve, $r = \sin 3\theta$, in Figure 8.23.

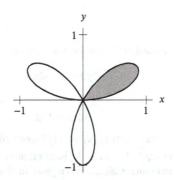

Figure 8.23: Graph of $r = \sin 3\theta$

29. (a) See Figure 8.24. In polar coordinates, the line $x = 1$ is $r \cos \theta = 1$, so its equation is

$$r = \frac{1}{\cos \theta}.$$

The circle of radius 2 centered at the origin has equation

$$r = 2.$$

(b) The line and circle intersect where

$$\frac{1}{\cos \theta} = 2$$

$$\cos \theta = \frac{1}{2}$$

$$\theta = -\frac{\pi}{3}, \frac{\pi}{3}.$$

Thus,

$$\text{Area} = \frac{1}{2} \int_{-\pi/3}^{\pi/3} \left(2^2 - \left(\frac{1}{\cos \theta} \right)^2 \right) d\theta.$$

(c) Evaluating gives

$$\text{Area} = \frac{1}{2} \int_{-\pi/3}^{\pi/3} \left(4 - \frac{1}{\cos^2 \theta} \right) d\theta = \frac{1}{2} (4\theta - \tan \theta) \Big|_{-\pi/3}^{\pi/3} = \frac{4\pi}{3} - \sqrt{3}.$$

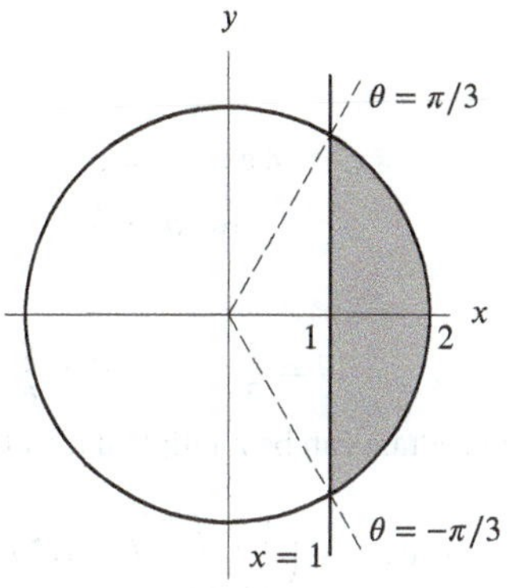

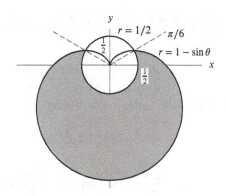

Figure 8.24

33. The two curves intersect where

$$1 - \sin\theta = \frac{1}{2}$$

$$\sin\theta = \frac{1}{2}$$

$$\theta = \frac{\pi}{6}, \frac{5\pi}{6}.$$

See Figure 8.25. We find the area of the right half and multiply that answer by 2 to get the entire area. The integrals can be computed numerically with a calculator or, as we show, using integration by parts or formula IV-17 in the integral tables.

$$\text{Area of right half} = \frac{1}{2}\int_{-\pi/2}^{\pi/6} \left((1 - \sin\theta)^2 - \left(\frac{1}{2}\right)^2\right) d\theta$$

$$= \frac{1}{2}\int_{-\pi/2}^{\pi/6} \left(1 - 2\sin\theta + \sin^2\theta - \frac{1}{4}\right) d\theta$$

$$= \frac{1}{2}\int_{-\pi/2}^{\pi/6} \left(\frac{3}{4} - 2\sin\theta + \sin^2\theta\right) d\theta$$

$$= \frac{1}{2}\left(\frac{3}{4}\theta + 2\cos\theta - \frac{1}{2}\sin\theta\cos\theta + \frac{1}{2}\theta\right)\Big|_{-\pi/2}^{\pi/6}$$

$$= \frac{1}{2}\left(\frac{5\pi}{6} + \frac{7\sqrt{3}}{8}\right).$$

Thus,

$$\text{Total area} = \frac{5\pi}{6} + \frac{7\sqrt{3}}{8}.$$

Figure 8.25

37. (a) See Figure 8.26.

(b) The curves intersect when $r^2 = 2$

$$4\cos 2\theta = 2$$
$$\cos 2\theta = \frac{1}{2}.$$

In the first quadrant:

$$2\theta = \frac{\pi}{3} \quad \text{so} \quad \theta = \frac{\pi}{6}.$$

Using symmetry, the area in the first quadrant can be multiplied by 4 to find the area of the total bounded region.

$$\begin{aligned}
\text{Area} &= 4\left(\frac{1}{2}\right)\int_0^{\pi/6}(4\cos 2\theta - 2)\,d\theta \\
&= 2\left(\frac{4\sin 2\theta}{2} - 2\theta\right)\Big|_0^{\pi/6} \\
&= 4\sin\frac{\pi}{3} - \frac{2}{3}\pi \\
&= 4\frac{\sqrt{3}}{2} - \frac{2}{3}\pi \\
&= 2\sqrt{3} - \frac{2}{3}\pi = 1.370.
\end{aligned}$$

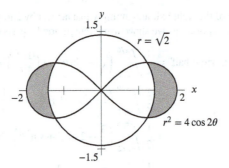

Figure 8.26

41. (a) Expressing x and y parametrically in terms of θ, we have

$$x = r\cos\theta = \frac{\cos\theta}{\theta} \quad \text{and} \quad y = r\sin\theta = \frac{\sin\theta}{\theta}.$$

The slope of the tangent line is given by

$$\frac{dy}{dx} = \frac{dy/d\theta}{dx/d\theta} = \left(\frac{\theta\cos\theta - \sin\theta}{\theta^2}\right)\Big/\left(\frac{-\theta\sin\theta - \cos\theta}{\theta^2}\right) = \frac{\sin\theta - \theta\cos\theta}{\cos\theta + \theta\sin\theta}.$$

At $\theta = \pi/2$, we have

$$\frac{dy}{dx}\Big|_{\theta=\pi/2} = \frac{1 - (\pi/2)0}{0 + (\pi/2)1} = \frac{2}{\pi}.$$

At $\theta = \pi/2$, we have $x = 0$, $y = 2/\pi$, so the equation of the tangent line is

$$y = \frac{2}{\pi}x + \frac{2}{\pi}.$$

(b) As $\theta \to 0$,

$$x = \frac{\cos\theta}{\theta} \to \infty \quad \text{and} \quad y = \frac{\sin\theta}{\theta} \to 1.$$

Thus, $y = 1$ is a horizontal asymptote. See Figure 8.27.

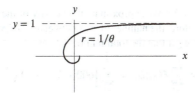

Figure 8.27

45. Parameterized by θ, the curve $r = f(\theta)$ is given by $x = f(\theta)\cos\theta$ and $y = f(\theta)\sin\theta$. Then

$$\text{Arc length} = \int_\alpha^\beta \sqrt{\left(\frac{dx}{d\theta}\right)^2 + \left(\frac{dy}{d\theta}\right)^2}\, d\theta$$

$$= \int_\alpha^\beta \sqrt{(f'(\theta)\cos\theta - f(\theta)\sin\theta)^2 + (f'(\theta)\sin\theta + f(\theta)\cos\theta)^2}\, d\theta$$

$$= \int_\alpha^\beta \sqrt{(f'(\theta))^2\cos^2\theta - 2f'(\theta)f(\theta)\cos\theta\sin\theta + (f(\theta))^2\sin^2\theta}$$

$$\overline{+(f'(\theta))^2\sin^2\theta + 2f'(\theta)f(\theta)\sin\theta\cos\theta + (f(\theta))^2\cos^2\theta}\, d\theta$$

$$= \int_\alpha^\beta \sqrt{(f'(\theta))^2(\cos^2\theta + \sin^2\theta) + (f(\theta))^2(\sin^2\theta + \cos^2\theta)}\, d\theta$$

$$= \int_\alpha^\beta \sqrt{(f'(\theta))^2 + (f(\theta))^2}\, d\theta.$$

Strengthen Your Understanding

49. The points on the polar curve with $\pi/2 < \theta < \pi$ are in quadrant IV, because $r = \sin(2\theta) < 0$.

53. The circle of radius k centered at the origin has equation $r = k$. For example, $r = 100$.

Solutions for Section 8.4

Exercises

1. Since density is e^{-x} gm/cm,

$$\text{Mass} = \int_0^{10} e^{-x}\, dx = -e^{-x}\Big|_0^{10} = 1 - e^{-10} \text{ gm.}$$

5. (a) Figure 8.28 shows a graph of the density function.

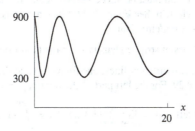

Figure 8.28

(b) Suppose we choose an x, $0 \leq x \leq 20$. We approximate the density of the number of the cars between x and $x + \Delta x$ miles as $\delta(x)$ cars per mile. Therefore, the number of cars between x and $x + \Delta x$ is approximately $\delta(x)\Delta x$. If we slice the 20 mile strip into N slices, we get that the total number of cars is

$$C \approx \sum_{i=1}^{N} \delta(x_i)\Delta x = \sum_{i=1}^{N} \left[600 + 300 \sin(4\sqrt{x_i + 0.15}) \right] \Delta x,$$

where $\Delta x = 20/N$. (This is a right-hand approximation; the corresponding left-hand approximation is $\sum_{i=0}^{N-1} \delta(x_i)\Delta x$.)

(c) As $N \to \infty$, the Riemann sum above approaches the integral

$$C = \int_0^{20} (600 + 300 \sin 4\sqrt{x + 0.15})\, dx.$$

If we calculate the integral numerically, we find $C \approx 11513$. We can also find the integral exactly as follows:

$$C = \int_0^{20} (600 + 300 \sin 4\sqrt{x + 0.15})\, dx$$

$$= \int_0^{20} 600\, dx + \int_0^{20} 300 \sin 4\sqrt{x + 0.15}\, dx$$

$$= 12000 + 300 \int_0^{20} \sin 4\sqrt{x + 0.15}\, dx.$$

Let $w = \sqrt{x + 0.15}$, so $x = w^2 - 0.15$ and $dx = 2w\, dw$. Then

$$\int_{x=0}^{x=20} \sin 4\sqrt{x + 0.15}\, dx = 2 \int_{w=\sqrt{0.15}}^{w=\sqrt{20.15}} w \sin 4w\, dw, \text{ (using integral table III-15)}$$

$$= 2\left[-\frac{1}{4}w \cos 4w + \frac{1}{16} \sin 4w \right]\Bigg|_{\sqrt{0.15}}^{\sqrt{20.15}}$$

$$\approx -1.624.$$

Using this, we have $C \approx 12000 + 300(-1.624) \approx 11513$, which matches our numerical approximation.

9. We slice the block horizontally. A slice has area $10 \cdot 3 = 30$ and thickness Δz. On such a slice, the density is approximately constant. Thus

$$\text{Mass of slice} \approx \text{Density} \cdot \text{Volume} \approx (2 - z) \cdot 30\Delta z,$$

so we have

$$\text{Mass of block} \approx \sum (2 - z)30\Delta z.$$

In the limit as $\Delta z \to 0$, the sum becomes an integral and the approximation becomes exact. Thus

$$\text{Mass of block} = \int_0^1 (2 - z)30\, dz = 30\left(2z - \frac{z^2}{2} \right)\Bigg|_0^1 = 30\left(2 - \frac{1}{2} \right) = 45.$$

Problems

13. Since the density varies with x, the region must be sliced perpendicular to the x-axis. This has the effect of making the density approximately constant on each strip. See Figure 8.29. Since a strip is of height y, its area is approximately $y\Delta x$. The density on the strip is $\delta(x) = 1 + x$ gm/cm^2. Thus

$$\text{Mass of strip} \approx \text{Density} \cdot \text{Area} \approx (1 + x)y\Delta x \text{ gm}.$$

Because the tops of the strips end on two different lines, one for $x \geq 0$ and the other for $x < 0$, the mass is calculated as the sum of two integrals. See Figure 8.29. For the left part of the region, $y = x + 1$, so

$$\text{Mass of left part} = \lim_{\Delta x \to 0} \sum (1 + x)y\Delta x = \int_{-1}^{0} (1 + x)(x + 1)\, dx$$

$$= \int_{-1}^{0} (1 + x)^2\, dx = \frac{(x + 1)^3}{3}\Bigg|_{-1}^{0} = \frac{1}{3} \text{ gm}.$$

From Figure 8.29, we see that for the right part of the region, $y = -x + 1$, so

$$\text{Mass of right part} = \lim_{\Delta x \to 0} \sum (1+x)y\Delta x = \int_0^1 (1+x)(-x+1)\,dx$$

$$= \int_0^1 (1-x^2)\,dx = x - \frac{x^3}{3}\Big|_0^1 = \frac{2}{3} \text{ gm.}$$

$$\text{Total mass} = \frac{1}{3} + \frac{2}{3} = 1 \text{ gm.}$$

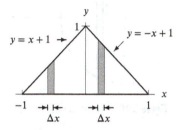

Figure 8.29

17. (a) Partition $[0, 10{,}000]$ into N subintervals of width Δr. The area in the ith subinterval is $\approx 2\pi r_i \Delta r$. So the total mass in the slick $= M \approx \sum_{i=1}^N 2\pi r_i \left(\frac{50}{1+r_i}\right)\Delta r$.

(b) $M = \int_0^{10{,}000} 100\pi \frac{r}{1+r}\,dr$. We may rewrite $\frac{r}{1+r}$ as $\frac{1+r}{1+r} - \frac{1}{1+r} = 1 - \frac{1}{1+r}$, so that

$$M = \int_0^{10{,}000} 100\pi(1 - \frac{1}{1+r})\,dr = 100\pi\left(r - \ln|1+r|\Big|_0^{10{,}000}\right)$$

$$= 100\pi(10{,}000 - \ln(10{,}001)) \approx 3.14 \times 10^6 \text{ kg.}$$

(c) We wish to find an R such that

$$\int_0^R 100\pi \frac{r}{1+r}\,dr = \frac{1}{2}\int_0^{10{,}000} 100\pi \frac{r}{1+r}\,dr \approx 1.57 \times 10^6.$$

So $100\pi(R - \ln|R+1|) \approx 1.57 \times 10^6$; $R - \ln|R+1| \approx 5000$. By trial and error, we find $R \approx 5009$ meters.

21. We have

$$\text{Total mass of the rod} = \int_0^3 (1+x^2)\,dx = \left[x + \frac{x^3}{3}\right]\Big|_0^3 = 12 \text{ grams.}$$

In addition,

$$\text{Moment} = \int_0^3 x(1+x^2)\,dx = \left[\frac{x^2}{2} + \frac{x^4}{4}\right]\Big|_0^3 = \frac{99}{4} \text{ gram-meters.}$$

Thus, the center of mass is at the position $\bar{x} = \frac{99/4}{12} = 2.06$ meters.

25. Since the region is symmetric about the x-axis, $\bar{y} = 0$.

To find $\bar{x}$, we first find the density. The area of the disk is $\pi/2$ m^2, so it has density $3/(\pi/2) = 6/\pi$ kg/m^2. We find the mass of the small strip of width Δx in Figure 8.30. The height of the strip is $\sqrt{1-x^2}$, so

$$\text{Area of the small strip} \approx A_x(x)\Delta x = 2 \cdot \sqrt{1-x^2}\Delta x \text{ m}^2.$$

When multiplied by the density $6/\pi$, we get

$$\text{Mass of the strip} \approx \frac{12}{\pi} \cdot \sqrt{1-x^2}\Delta x \text{ kg.}$$

We then sum the product of these masses with x, and take the limit as $\Delta x \to 0$ to get

$$\text{Moment} = \int_0^1 \frac{12}{\pi} x \sqrt{1 - x^2} \, dx = -\frac{4}{\pi}(1 - x^2)^{3/2} \Big|_0^1 = \frac{4}{\pi} \text{ meter.}$$

Finally, we divide by the total mass 3 kg to get the result $\bar{x} = 4/(3\pi)$ meters.

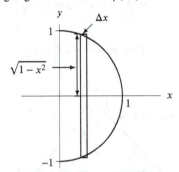

Figure 8.30: Area of a small strip

29. Stand the cone with the base horizontal, with center at the origin. Symmetry gives us that $\bar{x} = \bar{y} = 0$. Since the cone is fatter near its base we expect the center of mass to be nearer to the base.

  Slice the cone into disks parallel to the xy-plane.

  As we saw in Example 3 on page 404, a disk of thickness Δz at height z above the base has

$$\text{Volume of disk} = A_z(z)\Delta z \approx \pi(5 - z)^2 \Delta z \text{ cm}^3.$$

  Thus, since the density is δ,

$$\bar{z} = \frac{\int z\delta A_z(z)\, dz}{\text{Mass}} = \frac{\int_0^5 z \cdot \delta\pi(5 - z)^2\, dz}{\text{Mass}} \text{ cm.}$$

  To evaluate the integral in the numerator, we factor out the constant density δ and π to get

$$\int_0^5 z \cdot \delta\pi(5 - z)^2\, dz = \delta\pi \int_0^5 z(25 - 10z + z^2)\, dz = \delta\pi \left(\frac{25z^2}{2} - \frac{10z^3}{3} + \frac{z^4}{4} \right) \Big|_0^5 = \frac{625}{12}\delta\pi.$$

  We divide this result by the total mass of the cone, which is $\left(\frac{1}{3}\pi 5^2 \cdot 5 \right)\delta$:

$$\bar{z} = \frac{\frac{625}{12}\delta\pi}{\frac{1}{3}\pi 5^3 \delta} = \frac{5}{4} = 1.25 \text{ cm.}$$

  As predicted, the center of mass is closer to the base of the cone than its top.

33. We slice time into small intervals. Since t is given in seconds, we convert the minute to 60 seconds. We consider water loss over the time interval $0 \le t \le 60$. We also need to convert inches into feet since the velocity is given in ft/sec. Since 1 inch $= 1/12$ foot, the square hole has area $1/144$ square feet. For water flowing through a hole with constant velocity v, the amount of water which has passed through in some time, Δt, can be pictured as the rectangular solid in Figure 8.31, which has volume

$$\text{Area} \cdot \text{Height} = \text{Area} \cdot \text{Velocity} \cdot \text{Time.}$$

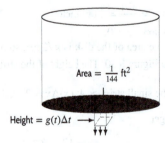

Figure 8.31: Volume of water
passing through hole

Over a small time interval of length Δt, starting at time t, water flows with a nearly constant velocity $v = g(t)$ through a hole $1/144$ square feet in area. In Δt seconds, we know that

$$\text{Water lost} \approx \left(\frac{1}{144} \text{ ft}^2\right)(g(t) \text{ ft/sec})(\Delta t \text{ sec}) = \left(\frac{1}{144}\right)g(t)\,\Delta t \text{ ft}^3.$$

Adding the water from all subintervals gives

$$\text{Total water lost} \approx \sum \frac{1}{144}g(t)\,\Delta t \text{ ft}^3.$$

As $\Delta t \to 0$, the sum tends to the definite integral:

$$\text{Total water lost} = \int_0^{60} \frac{1}{144}g(t)\,dt \text{ ft}^3.$$

Strengthen Your Understanding

37. The center of mass with density $\delta(x)$ for $0 \le x \le 10$ is given by

$$\text{Center of mass} = \frac{\int_0^{10} x\delta(x)\,dx}{\int_0^{10} \delta(x)\,dx}.$$

The correct formula for $\delta(x) = x^2$ is

$$\text{Center of mass} = \frac{\int_0^{10} x^3\,dx}{\int_0^{10} x^2\,dx}.$$

41. Any rod whose mass increases towards one end will skew the center of mass towards that end. As an example, if $\delta(x) = x$, then the total mass of the rod is

$$\text{Mass} = \int_0^2 x\,dx = 2$$

and the center of mass is

$$\text{Center of mass} = \frac{1}{2}\int_0^2 x \cdot \delta(x)\,dx = \frac{1}{2} \cdot \left.\frac{x^3}{3}\right|_0^2 = \frac{4}{3} \text{ cm}$$

which is not at the center of the rod.

45. False. Although the density is greater near the center, the area of the suburbs is much larger than the area of the inner city, and population is determined by both area and density. In fact, the population of the inner city:

$$\int_0^1 (10 - 3r)2\pi r\,dr = 2\pi(5r^2 - r^3)\Big|_0^1 = 8\pi$$

is less than the population of the suburbs:

$$\int_1^2 (10 - 3r)2\pi r\,dr = 2\pi(5r^2 - r^3)\Big|_1^2 = 16\pi.$$

49. False. If the density were constant this would be true, but suppose that all the mass on the left half is concentrated at $x = 0$ and all the mass on the right side is concentrated at $x = 3$. In order for the rod to balance at $x = 2$, the weight on the left side must be half the weight on the right side.

Solutions for Section 8.5

Exercises

1. This is in British units, so the distance raised, d, is first converted to feet: $d = 0.75$. feet.

$$\text{Work done} = F \cdot d = 40 \text{ lb} \cdot 0.75 \text{ ft} = 30 \text{ ft-lb.}$$

5. The work done is given by

$$W = \int_0^3 3x \, dx = \frac{3}{2} x^2 \Big|_0^3 = \frac{27}{2} \text{ joules.}$$

9. The only work is done by lifting the bucket initially, since the motion is parallel to the force of gravity, so the work is $2 \cdot 10 = 20$ ft-lb. When the child is walking and holding the bucket at a constant height, the force of gravity and the motion are at right angles, and the work done is zero. Thus, the total work done is 20 ft-lb.

Problems

13. When the anchor has been lifted through h feet, the length of chain in the water is $25 - h$ feet, so the total weight of the anchor and chain in the water is $50 + 3(25 - h)$ lb. Then

$$\text{Work to lift the anchor and chain } \Delta h \text{ higher} = \text{Weight} \cdot \text{Distance lifted}$$
$$= (100 + 3(25 - h))\Delta h.$$

To find the total work, we integrate from $h = 0$ to $h = 25$:

$$W = \int_0^{25} (100 + 3(25 - h)) dh = \int_0^{25} (175 - 3h) dh = \left(175h - \frac{3h^2}{2}\right) \Big|_0^{25} = 3437.5 \text{ft-lbs.}$$

17. We slice the water horizontally and find the work required to pump each horizontal slice of water over the top. See Figure 8.32. At a distance h ft above the bottom, a slice of thickness Δh has

$$\text{Volume} \approx 50 \cdot 20 \Delta h \text{ ft}^3.$$

Since the density of water is ρ lb/ft^3,

$$\text{Weight of the slice} \approx \rho(50 \cdot 20 \cdot \Delta h) \text{ lbs.}$$

The distance to lift the slice of water at height h ft is $10 - h$ ft, so

$$\text{Work to move one slice} = \rho \cdot \text{Volume} \cdot \text{Distance lifted}$$
$$\approx \rho(50 \cdot 20 \cdot \Delta h)(10 - h)$$
$$= 100\rho(10 - h)\Delta h \text{ ft-lb.}$$

The work done, W, to pump all the water is the sum of the work done on the pieces:

$$W \approx \sum 100\rho(10 - h)\Delta h.$$

As $\Delta h \to 0$, we obtain a definite integral. Since h varies from $h = 0$ to $h = 9$ and $\rho = 62.4$ lb/ft^3, the total work is:

$$W = \int_0^9 100\rho(10 - h) dh = 62400\left(10h - \frac{h^2}{2}\right)\Big|_0^9 = 62400(49.5) = 3,088,800.$$

The work to pump all the water out is 3,088,800 ft-lbs.

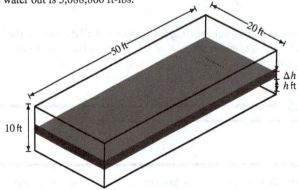

Figure 8.32

21.

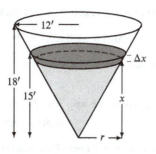

Figure 8.33

Let x be the depth of the water measured from the bottom of the tank. See Figure 8.33. It follows that $0 \leq x \leq 15$. Let r be the radius of the section of the cone with height x. By similar triangles, $\frac{r}{x} = \frac{12}{18}$, so $r = \frac{2}{3}x$. Then the work required to pump a layer of water with thickness of Δx at depth x over the top of the tank is $62.4\pi \left(\frac{2}{3}x\right)^2 \Delta x (18 - x)$. So the total work done by pumping the water over the top of the tank is

$$W = \int_0^{15} 62.4\pi \left(\frac{2}{3}x\right)^2 (18 - x)dx$$

$$= \frac{4}{9} \cdot 62.4\pi \int_0^{15} x^2(18 - x)dx$$

$$= \frac{4}{9} \cdot 62.4\pi \left(6x^3 - \frac{1}{4}x^4\right)\Big|_0^{15}$$

$$= \frac{4}{9} \cdot 62.4\pi(7593.75) \approx 661,619.41 \text{ ft-lb.}$$

25. Let h represent distance below the surface in feet. We slice the tank up into horizontal slabs of thickness Δh. From looking at Figure 8.34, we can see that the slabs will be rectangular. The length of any slab is 12 feet. The width w of a slab h units below the ground will equal $2x$, where $(14-h)^2 + x^2 = 16$, so $w = 2\sqrt{4^2 - (14 - h)^2}$. The volume of such a slab is therefore $12w\,\Delta h = 24\sqrt{16 - (14 - h)^2}\,\Delta h$ cubic feet; the slab weighs $42 \cdot 24\sqrt{16 - (14 - h)^2}\,\Delta h = 1008\sqrt{16 - (14 - h)^2}\,\Delta h$ pounds. So the total work done in pumping out all the gasoline is

$$\int_{10}^{18} 1008h\sqrt{16 - (14 - h)^2}\,dh = 1008 \int_{10}^{18} h\sqrt{16 - (14 - h)^2}\,dh.$$

Substitute $s = 14 - h$, $ds = -dh$. We get

$$1008 \int_{10}^{18} h\sqrt{16 - (14 - h)^2}\,dh = -1008 \int_4^{-4} (14 - s)\sqrt{16 - s^2}\,ds$$

$$= 1008 \cdot 14 \int_{-4}^4 \sqrt{16 - s^2}\,ds - 1008 \int_{-4}^4 s\sqrt{16 - s^2}\,ds.$$

The first integral represents the area of a semicircle of radius 4, which is 8π. The second is the integral of an odd function, over the interval $-4 \leq s \leq 4$, and is therefore 0. Hence, the total work is $1008 \cdot 14 \cdot 8\pi \approx 354,673$ foot-pounds.

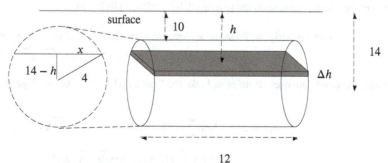

Figure 8.34

29. We divide the water against the dam into horizontal strips, each of thickness Δh and length 100.

$$\text{Area of each strip} \approx 100\Delta h \text{ ft}^2.$$

See Figure 8.35. The strip at height h ft from the bottom is at a water depth of $40 - h$, so, if δ lb/ft^3 is the density of water, we have:

$$\text{Force of one strip} = \delta \cdot \text{Depth} \cdot \text{Area}$$
$$\approx \delta(40 - h)(100\Delta h) \text{ lb.}$$

To find the total force, F, we integrate the force on a strip from $h = 0$ to $h = 40$, using $\delta = 62.4$ lb/ft^3:

$$F = \int_0^{40} \delta(40 - h)100\,dh = 100 \cdot 62.4 \left(40h - \frac{h^2}{2} \right)\Bigg|_0^{40} = 6240(800) = 4,992,000\text{lbs.}$$

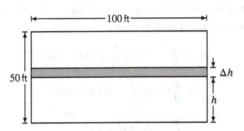

Figure 8.35

33. (a) Since water has density 62.4 lb/ft^3, at a depth of 12,500 feet,

$$\text{Pressure} = \text{Density} \times \text{Depth} = 62.4 \cdot 12,500 = 780,000 \text{ lb/square foot.}$$

To imagine this pressure, observe that it is equivalent to $780,000/144 \approx 5400$ pounds per square inch.

(b) To calculate the pressure on the porthole (window), we slice it into horizontal strips, as the pressure remains approximately constant along each one. See Figure 8.36. Since each strip is approximately rectangular

$$\text{Area of strip} \approx 2r\Delta h \text{ ft}^2.$$

To calculate r in terms of h, we use the Pythagorean Theorem:

$$r^2 + h^2 = 9$$
$$r = \sqrt{9 - h^2},$$

so

$$\text{Area of strip} \approx 2\sqrt{9 - h^2}\Delta h \text{ ft}^2.$$

The center of the porthole is at a depth of 12,500 feet below the surface, so the strip shown in Figure 8.36 is at a depth of $(12,500 - h)$ feet. Thus, pressure on the strip is $62.4(12,500 - h)$ lb/ft^2, so

$$\text{Force on strip} = \text{Pressure} \times \text{Area} \approx 62.4(12,500 - h)2\sqrt{9 - h^2}\Delta h \text{ lb}$$
$$= 124.8(12,500 - h)\sqrt{9 - h^2}\Delta h \text{ lb.}$$

To get the total force, we sum over all strips and take the limit as $\Delta h \to 0$. Since h ranges from -3 to 3, we get the integral

$$\text{Total force} = \lim_{\Delta h \to 0} \sum 124.8(12,500 - h)\sqrt{9 - h^2}\Delta h$$
$$= 124.8 \int_{-3}^{3} (12,500 - h)\sqrt{9 - h^2}\,dh \text{ lb.}$$

Evaluating the integral numerically, we obtain a total force of $2.2 \cdot 10^7$ pounds.

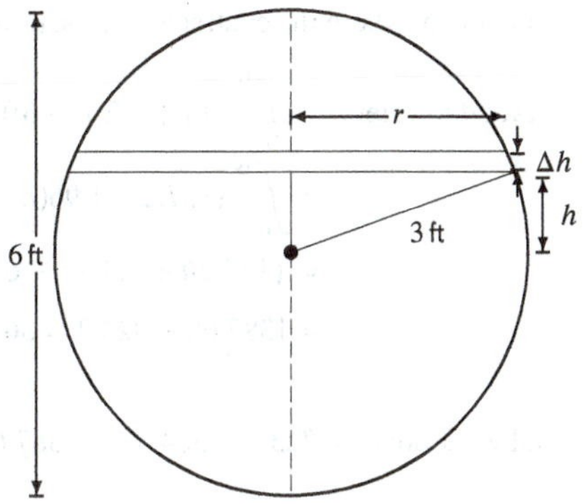

Figure 8.36: Center of circle is 12,500 ft below
the surface of ocean

37. Old houses may contain asbestos, now known to be dangerous; removal requires using a special vacuum. A contractor climbs a ladder and sucks up asbestos at a constant rate from a 10 m tall pipe covered by 0.2 kg/m using a vacuum weighing 14 kg with a 1.2 kg capacity.

    (a) Let h be the height of the vacuum from the ground. If the vacuum is empty at $h = 0$, find a formula for the mass of the vacuum and the asbestos inside as a function of h.
    (b) Approximate the work done by the contractor in lifting the vacuum from height h to $h + \Delta h$.
    (c) At what height does the vacuum fill up?
    (d) Find total work done lifting the vacuum from height $h = 0$ until the vacuum fills.
    (e) Assuming again an empty tank at $h = 0$, find the work done lifting the vacuum when removing the remaining asbestos.

    (a) Since the vacuum weighs 14 kg and gains mass at a constant rate of 0.2 kg/m, the mass, $m(h)$, of the vacuum at height h is

    $$m(h) = 14 + 0.2h \text{ kg}.$$

    (b) Assuming the vacuum has a constant mass of $m(h) = 14 + 0.2h$ kg between height h and $h + \Delta h$, we have:

    $$\text{Force due to gravity } = F = m(h) \cdot g = (14 + 0.2h) \text{ kg} \cdot 9.8 \text{ m/sec}^2,$$

    so the work done in moving the vacuum between h and $h + \Delta h$ is

    $$W = F \cdot d = m(h) \cdot g \cdot \Delta h = (14 + 0.2h) \text{ kg} \cdot 9.8 \text{ m/sec}^2 \cdot \Delta h \text{ m} = (137.2 + 1.96h) \, \Delta h \text{ joules}.$$

    (c) The vacuum has a capacity of 1.2 kg, and the pipe has 0.2 kg/m of insulation, so it fills up at a height of $1.2/0.2 = 6$ m.
    (d) By part (b), the work done in lifting the vacuum Δh m from height h can be approximated by $(137.2 + 1.96h) \, \Delta h$. Summing over Δh and taking $\Delta h \to 0$ gives an integral with lower limit $h = 0$ and upper limit $h = 6$, the height at which the vacuum reaches capacity:

    $$\text{Total work } = \lim_{\Delta h \to 0} \sum (137.2 + 1.96h) \, \Delta h = \int_0^6 (137.2 + 1.96h) \, dh.$$

    Evaluating the integral gives

    $$\text{Total work } = \int_0^6 (137.2 + 1.96h) \, dh$$
    $$= \left(137.2h + 0.98h^2\right)\Big|_0^6$$
    $$= 858.48 \text{ joules}.$$

    (e) First, the contractor moves the empty vacuum up a distance of 6 m; the work done for this is

    $$\text{Work to 6 meters } = 14 \cdot 9.8 \cdot 6 = 823.2 \text{ joules}.$$

    Following this, the contractor again starts removing the asbestos, so we use an integral to calculate the work done. Since the vacuum is empty at height $h = 6$, we modify the formula for mass to be

    $$m(h) = 14 + 0.2(h - 6) \text{ kg}$$

and the limits of the integral are now $h = 6$, where the contractor starts, to $h = 10$, the top of the pipe. Thus,

$$
\begin{aligned}
\text{Work beyond 6 meters} &= \int_6^{10} (14 + 0.2(h - 6)) \, 9.8 \, dh \\
&= \int_6^{10} (137.2 + 1.96(h - 6)) \, dh \\
&= \left. \left(137.2h + 0.98(h - 6)^2 \right) \right|_6^{10} \\
&= 1387.68 - 823.2 = 564.48 \text{ joules.}
\end{aligned}
$$

Thus,

$$
\text{Total work done} = 823.2 + 564.48 = 1387.68 \text{ joules.}
$$

Strengthen Your Understanding

41. The calculation given is correct when the entire rope is lifted 20 m. But the portions of the rope nearer the top are raised a shorter distance, so less work is required.

45. Let tank A be a cylinder of radius 1 meter and height 4 meters and tank B be a cylinder of radius 2 meters and height 1 meter as shown in Figure 8.37. The volume of each tank is $4\pi = 12.56$ meters^3. All the water in tank B has to be pumped at least 9 meters, whereas the water in the top part of tank A has to be pumped a shorter distance. Therefore, it takes less work to pump the water from tank A to a height of 10 meters than the water from tank B.

Tank A

1 m

4 m

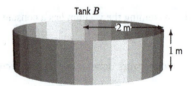

Tank B

2 m

1 m

Figure 8.37

49. True. Since pressure increases with depth and we want the pressure to be approximately constant on each strip, we use horizontal strips.

Additional Problems (online only)

53. This time, let's split the second rod into small slices of length dr. See Figure 8.38. Each slice is of mass $\frac{M_2}{l_2} \, dr$, since the density of the second rod is $\frac{M_2}{l_2}$. Since the slice is small, we can treat it as a particle at distance r away from the end of the first rod, as in Problem 52. By that problem, the force of attraction between the first rod and particle is

$$
\frac{GM_1 \frac{M_2}{l_2} \, dr}{(r)(r + l_1)}.
$$

So the total force of attraction between the rods is

$$
\begin{aligned}
\int_a^{a+l_2} \frac{GM_1 \frac{M_2}{l_2} \, dr}{(r)(r + l_1)} &= \frac{GM_1 M_2}{l_2} \int_a^{a+l_2} \frac{dr}{(r)(r + l_1)} \\
&= \frac{GM_1 M_2}{l_2} \int_a^{a+l_2} \frac{1}{l_1} \left(\frac{1}{r} - \frac{1}{r + l_1} \right) \, dr.
\end{aligned}
$$

$$= \frac{GM_1M_2}{l_1l_2} \left(\ln|r| - \ln|r + l_1| \right) \Big|_a^{a+l_2}$$

$$= \frac{GM_1M_2}{l_1l_2} \left[\ln|a + l_2| - \ln|a + l_1 + l_2| - \ln|a| + \ln|a + l_1| \right]$$

$$= \frac{GM_1M_2}{l_1l_2} \ln\left[\frac{(a + l_1)(a + l_2)}{a(a + l_1 + l_2)} \right].$$

This result is symmetric: if you switch l_1 and l_2 or M_1 and M_2, you get the same answer. That means it's not important which rod is "first," and which is "second."

Figure 8.38

Solutions for Section 8.6

Exercises

1. The future value, in dollars, is

$$C(1 + 0.02)^{20}.$$

5. The future value, in dollars, is

$$\int_0^{15} C e^{0.02(15-t)} \, dt.$$

9. We compute the future value first: we have

$$\text{Future value} = \int_0^5 5000e^{0.04(5-t)}dt = \$27{,}675.34.$$

We can compute the present value using an integral and the income stream or using the future value. We compute the present value, P, from the future value:

$$27{,}675.34 = Pe^{0.04(5)} \quad \text{so} \quad P = 22{,}658.65.$$

The future value of this income stream is $\$27{,}675.34$ and the present value of this income stream is $\$22{,}658.65$.

13. Since we want to have $\$20{,}000$ in 20 years starting with a single initial investment of $\$10{,}000$, we have

$$20{,}000 = 10{,}000e^{r(20)}.$$

Solving for r, we get:

$$2 = e^{r(20)}$$
$$r = \frac{\ln(2)}{20}$$
$$r = 0.03466 \quad \text{or} \quad r = 3.466\% \text{ per year.}$$

17. A constant income stream that pays 20,000 in 20 years has a yearly rate amount of $20,000/20 = 1000$ dollars/year. Thus, the future value, B, of the stream at the end of 20 years is

$$B = \int_0^{20} 1000 \, e^{0.02(20-t)} \, dt = 1000 \left. \frac{-e^{0.02(20-t)}}{0.02} \right|_0^{20} = 50,000(e^{0.4} - 1) = 24,591.23 \text{ dollars.}$$

Problems

21. (a) The lump sum payment has a present value of 120 million dollars . We compute the present value of the other option at 6% and 3% per year. An award of $195 million paid out continuously over 20 years works out to an income stream of 9.75 million dollars per year.

If the interest rate is 6%, compounded continuously, we have

$$\text{Present value at 6\%} = \int_0^{20} 9.75 e^{-0.06t} dt = \frac{9.75}{-0.06}(e^{-0.06 \cdot 20} - 1) = 113.556 \text{ million dollars.}$$

Since this amount is less than the lump sum payment of 120 million dollars, the lump sum payment is preferable if the interest rate is 6%.
If the interest rate is 3%, we have

$$\text{Present value at 3\%} = \int_0^{20} 9.75 e^{-0.03t} dt = \frac{9.75}{-0.03}(e^{-0.03 \cdot 20} - 1) = 146.636 \text{ million dollars.}$$

Since this amount is greater than the lump sum payment of 120 million dollars, taking payments continuously over 20 years is the better option if the interest rate is 3%.
(b) The assumption is that the interest rates would stay relatively high (closer to 6%). The interest rate at which the decision changes is the solution to the equation

$$\frac{9.75}{-x}(e^{-20x} - 1) = 120.$$

Using a graph or numerical methods gives $x = 5.3\%$. Thus if Mr. Nabors chooses the lump sum, he is assuming interests rates stay above 5.3%.

25. You should choose the payment which gives you the highest present value. The immediate lump-sum payment of $2800 obviously has a present value of exactly $2800, since you are getting it now. We can calculate the present value of the installment plan as:

$$PV = 1000e^{-0.03(0)} + 1000e^{-0.03(1)} + 1000e^{-0.03(2)}$$
$$\approx \$2912.21.$$

Since the installment payments offer a higher present value, you should accept the installment option.

29. We want to reach a target of $18,000 in 20 years, so we have:

$$18,000 = \int_0^{20} 300 \, e^{r(20-t)} \, dt.$$

In order to solve for r, we integrate, then solve the resulting equation graphically or numerically. We have:

$$\frac{18,000}{300} = \left. \frac{-e^{r(20-t)}}{r} \right|_0^{20}$$
$$60 = \frac{e^{20r} - 1}{r}.$$

We can solve for r by graphing the functions on the left- and right-hand sides of the last equation, and finding the intersection point. This yields $r = 0.09519$, or $r = 9.519\%$ per year.

33. Price in future $= P(1 + 20\sqrt{t})$.
The present value V of price satisfies $V = P(1 + 20\sqrt{t})e^{-0.05t}$.

We want to maximize V. To do so, we find the critical points of $V(t)$ for $t \geq 0$. (Recall that $\sqrt{t}$ is nondifferentiable at $t = 0$.)

$$\frac{dV}{dt} = P\left[\frac{20}{2\sqrt{t}}e^{-0.05t} + (1 + 20\sqrt{t})(-0.05e^{-0.05t})\right]$$

$$= Pe^{-0.05t}\left[\frac{10}{\sqrt{t}} - 0.05 - \sqrt{t}\right].$$

Setting $\frac{dV}{dt} = 0$ gives $\frac{10}{\sqrt{t}} - 0.05 - \sqrt{t} = 0$. Using a calculator, we find $t \approx 10$ years. Since $V'(t) > 0$ for $0 < t < 10$ and $V'(t) < 0$ for $t > 10$, we confirm that this is a maximum. Thus, the best time to sell the wine is in 10 years.

37. The supply curve, $S(q)$, represents the minimum price p per unit that the suppliers will be willing to supply some quantity q of the good for. See Figure 8.39. If the suppliers have q^* of the good and q^* is divided into subintervals of size Δq, then if the consumers could offer the suppliers for each Δq a price increase just sufficient to induce the suppliers to sell an additional Δq of the good, the consumers' total expenditure on q^* goods would be

$$p_1\Delta q + p_2\Delta q + \cdots = \sum p_i\Delta q.$$

As $\Delta q \to 0$ the Riemann sum becomes the integral $\displaystyle\int_0^{q^*} S(q)\,dq$. Thus $\displaystyle\int_0^{q^*} S(q)\,dq$ is the amount the consumers would pay if suppliers could be forced to sell at the lowest price they would be willing to accept.

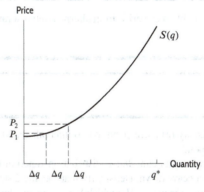

Figure 8.39

41.

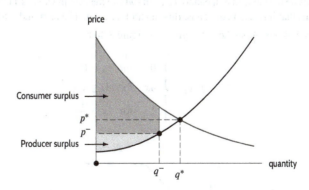

(a) The producer surplus is the area on the graph between p^- and the supply function. Lowering the price also lowers the producer surplus.

(b) Note that the consumer surplus—the area between the line p^- and the supply curve—increases or decreases depending on the functions describing the supply and demand and on the lowered price. (For example, the consumer surplus seems to be increased in the graph above, but if the price were brought down to \$0 then the consumer surplus would be zero, and hence clearly less than the consumer surplus at equilibrium.)

(c) The graph above shows that the total gains from the trade are decreased.

Strengthen Your Understanding

45. Producer surplus is measured in dollars.

49. Suppose there is an annual interest rate of 10%. Then the present value, A, of the $10,000 is:

$$A = \frac{10,000}{(1+0.10)^{10}} = 3855.$$

Thus, a deposit of roughly $3855 is needed now to have 10,000 dollars in 10 years, assuming a 10% interest rate.

There are two ways to think about the remaining values in the table. We can ask how much the amount needed now, $3855, will grow to in t years. This means the remaining values in the table arise by calculating:

$$3855(1+0.10)^t \quad \text{for} \quad t = 1, 2, 3, 4.$$

Alternatively, we realize that a single deposit made t years from now would have $10 - t$ years to grow to reach our target investment of $10,000 in 10 years. This means the remaining values in the table arise by calculating:

$$\frac{10,000}{(1+0.10)^{10-t}}, \quad \text{for} \quad t = 1, 2, 3, 4.$$

If we ignore rounding errors, we see that two approaches yield the same table values:

t (years from now)	0	1	2	3	4
$ (dollars)	3855	4241	4665	5131	5644

Other possible answers to this problem can arise from alternate interest rate choices.

Solutions for Section 8.7

Exercises

1. The two humps of probability in density (a) correspond to two intervals on which its cumulative distribution function is increasing. Thus (a) and (II) correspond.

A density function increases where its cumulative distribution function is concave up, and it decreases where its cumulative distribution function is concave down. Density (b) matches the distribution with both concave up and concave down sections, which is (I). Density (c) matches (III) which has a concave down section but no interval over which it is concave up.

5. Since the function takes on the value of 4, it cannot be a cdf (whose maximum value is 1). In addition, the function decreases for $x > c$, which means that it is not a cdf. Thus, this function is a pdf. The area under a pdf is 1, so $4c = 1$ giving $c = \frac{1}{4}$.

The pdf is $p(x) = 4$ for $0 \leq x \leq \frac{1}{4}$, so the cdf is given in Figure 8.40 by

$$P(x) = \begin{cases} 0 & \text{for} \quad x < 0 \\ 4x & \text{for} \quad 0 \leq x \leq \frac{1}{4} \\ 1 & \text{for} \quad x > \frac{1}{4} \end{cases}$$

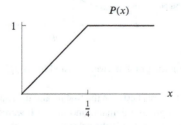

Figure 8.40

9. This function increases and levels off to c. The area under the curve is not finite, so it is not 1. Thus, the function must be a cdf, not a pdf, and $3c = 1$, so $c = 1/3$.

The pdf, $p(x)$ is the derivative, or slope, of the function shown, so, using $c = 1/3$,

$$p(x) = \begin{cases} 0 & \text{for } x < 0 \\ (1/3 - 0)/(2 - 0) = 1/6 & \text{for } 0 \le x \le 2 \\ (1 - 1/3)/(4 - 2) = 1/3 & \text{for } 2 < x \le 4 \\ 0 & \text{for } x > 4. \end{cases}$$

See Figure 8.41.

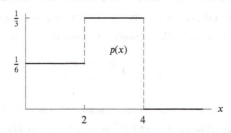

Figure 8.41

Problems

13. For a given energy E, Figure 8.42 shows that the area under the graph to the right of E is larger for graph B than it is for graph A. Therefore graph B has more molecules at higher kinetic energies, so it is the hotter gas. So graph A corresponds to 300 kelvins and graph B corresponds to 500 kelvins.

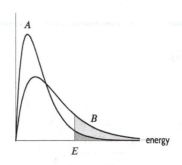

Figure 8.42

17. For a small interval Δx around 68, the fraction of the population of American men with heights in this interval is about $(0.2)\Delta x$. For example, taking $\Delta x = 0.1$, we can say that approximately $(0.2)(0.1) = 0.02 = 2\%$ of American men have heights between 68 and 68.1 inches.

21. (a) The percentage of calls lasting from 1 to 2 minutes is given by the integral

$$\int_1^2 p(x)\,dx \int_1^2 0.4e^{-0.4x}\,dx = e^{-0.4} - e^{-0.8} \approx 22.1\%.$$

(b) A similar calculation (changing the limits of integration) gives the percentage of calls lasting 1 minute or less as

$$\int_0^1 p(x)\,dx = \int_0^1 0.4e^{-0.4x}\,dx = 1 - e^{-0.4} \approx 33.0\%.$$

(c) The percentage of calls lasting 3 minutes or more is given by the improper integral

$$\int_3^\infty p(x)\,dx = \lim_{b\to\infty}\int_3^b 0.4e^{-0.4x}\,dx = \lim_{b\to\infty}(e^{-1.2} - e^{-0.4b}) = e^{-1.2} \approx 30.1\%.$$

(d) The cumulative distribution function is the integral of the probability density; thus,

$$C(h) = \int_0^h p(x)\,dx = \int_0^h 0.4e^{-0.4x}\,dx = 1 - e^{-0.4h}.$$

Strengthen Your Understanding

25. The fact that $p(1) = 0.02$ tells us that the probability that x falls in a small interval of length Δx around 1 is $0.02\Delta x$.

29. As $x \to \infty$ the function $P(x) = x^2 e^x$ grows without bound, whereas a cumulative distribution function must approach 1.

33. We can use
$$P(t) = \begin{cases} 0, & t < 0 \\ t, & 0 \le t \le 1 \\ 1, & t > 1. \end{cases}$$
It is a cumulative distribution function because $P(t)$ is an increasing function that increases from 0 to 1.

37. False. It is true that $p(x) \ge 0$ for all x, but we also need $\int_{-\infty}^\infty p(x)dx = 1$. Since $p(x) = 0$ for $x \le 0$, we need only check the integral from 0 to ∞. We have

$$\int_0^\infty xe^{-x^2}\,dx = \lim_{b\to\infty}\left(-\frac{1}{2}e^{-x^2}\right)\Big|_0^b = \frac{1}{2}.$$

Solutions for Section 8.8

Exercises

1.

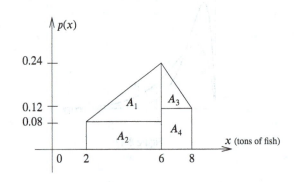

Splitting the figure into four pieces, we see that

$$\text{Area under the curve} = A_1 + A_2 + A_3 + A_4$$
$$= \frac{1}{2}(0.16)4 + 4(0.08) + \frac{1}{2}(0.12)2 + 2(0.12)$$
$$= 1.$$

We expect the area to be 1, since $\int_{-\infty}^\infty p(x)\,dx = 1$ for any probability density function, and $p(x)$ is 0 except when $2 \le x \le 8$.

5. Since a negative snowfall does not make sense,

$$\text{Mean annual snowfall} = \int_0^\infty xp(x)\,dx.$$

From part (IV), the mean snowfall is 2.65 m.

Problems

9. The shaded area represents the percentage of winters seeing snowfall totals of 3 m or less. Thus, the percentage of winters seeing more than 3 meters of total snowfall is 20%.

13. The median is the value of T such that $P(T) = 0.5$, so we solve

$$T - \frac{T^2}{4} = \frac{1}{2}$$

to get $T = 2 \pm \sqrt{2}$. Since the median is between 0 and 2 we discard the larger solution, so the median is $T = 2 - \sqrt{2}$. To find the mean, we first calculate the probability density

$$\text{Density} = p(x) = P'(x) = 1 - \frac{x}{2}$$

and then evaluate the integral

$$\text{Mean} = \int_0^2 xp(x)\,dx = \int_0^2 x - \frac{x^2}{2}\,dx = \frac{2}{3}.$$

So the mean is $2/3$.

17. (a) The cumulative distribution function

$$P(t) = \int_0^t p(x)dx = \text{Area under graph of density function } p(x) \text{ for } 0 \le x \le t$$

$$= \text{Fraction of population who survive } t \text{ years or less after treatment}$$

$$= \text{Fraction of population who survive up to } t \text{ years after treatment.}$$

(b) The probability that a randomly selected person survives for at least t years is the probability that he lives t years or longer, so

$$S(t) = \int_t^\infty p(x)\,dx = \lim_{b\to\infty} \int_t^b Ce^{-Ct}\,dx$$

$$= \lim_{b\to\infty} -e^{-Ct}\Big|_t^b = \lim_{b\to\infty} -e^{-Cb} - (-e^{-Ct}) = e^{-Ct},$$

or equivalently,

$$S(t) = 1 - \int_0^t p(x)\,dx = 1 - \int_0^t Ce^{-Ct}\,dx = 1 + e^{-Ct}\Big|_0^t = 1 + (e^{-Ct} - 1) = e^{-Ct}.$$

(c) The probability of surviving at least two years is

$$S(2) = e^{-C(2)} = 0.70$$

so

$$\ln e^{-C(2)} = \ln 0.70$$

$$-2C = \ln 0.7$$

$$C = -\frac{1}{2}\ln 0.7 \approx 0.178.$$

21. (a) First, we find the critical points of $p(x)$:

$$\frac{d}{dx}p(x) = \frac{1}{\sigma\sqrt{2\pi}}\left[\frac{-2(x-\mu)}{2\sigma^2}\right]e^{-\frac{(x-\mu)^2}{2\sigma^2}}$$

$$= -\frac{(x-\mu)}{\sigma^3\sqrt{2\pi}}e^{-\frac{(x-\mu)^2}{2\sigma^2}}.$$

This implies $x = \mu$ is the only critical point of $p(x)$.

To confirm that $p(x)$ is maximized at $x = \mu$, we rely on the first derivative test. As $-\frac{1}{\sigma^3\sqrt{2\pi}}e^{-\frac{(x-\mu)^2}{2\sigma^2}}$ is always negative, the sign of $p'(x)$ is the opposite of the sign of $(x-\mu)$; thus $p'(x) > 0$ when $x < \mu$, and $p'(x) < 0$ when $x > \mu$.

(b) To find the inflection points, we need to find where $p''(x)$ changes sign; that will happen only when $p''(x) = 0$. As

$$\frac{d^2}{dx^2}p(x) = -\frac{1}{\sigma^3\sqrt{2\pi}}e^{-\frac{(x-\mu)^2}{2\sigma^2}}\left[-\frac{(x-\mu)^2}{\sigma^2}+1\right],$$

$p''(x)$ changes sign when $\left[-\frac{(x-\mu)^2}{\sigma^2}+1\right]$ does, since the sign of the other factor is always negative. This occurs when

$$-\frac{(x-\mu)^2}{\sigma^2}+1 = 0,$$

$$-(x-\mu)^2 = -\sigma^2,$$

$$x - \mu = \pm\sigma.$$

Thus, $x = \mu + \sigma$ or $x = \mu - \sigma$. Since $p''(x) > 0$ for $x < \mu - \sigma$ and $x > \mu + \sigma$ and $p''(x) < 0$ for $\mu - \sigma \le x \le \mu + \sigma$, these are in fact points of inflection.

(c) μ represents the mean of the distribution, while σ is the standard deviation. In other words, σ gives a measure of the "spread" of the distribution, i.e., how tightly the observations are clustered about the mean. A small σ tells us that most of the data are close to the mean; a large σ tells us that the data is spread out.

25. (a) Let the $p(r)$ be the density function. Then $P(r) = \int_0^r p(x)\,dx$, and from the Fundamental Theorem of Calculus, $p(r) = \frac{d}{dr}P(r) = \frac{d}{dr}(1 - (2r^2 + 2r + 1)e^{-2r}) = -(4r + 2)e^{-2r} + 2(2r^2 + 2r + 1)e^{-2r}$, or $p(r) = 4r^2e^{-2r}$.

We have that $p'(r) = 8r(e^{-2r}) - 8r^2e^{-2r} = e^{-2r} \cdot 8r(1 - r)$, which is zero when $r = 0$ or $r = 1$, negative when $r > 1$, and positive when $r < 1$. Thus $p(1) = 4e^{-2} \approx 0.54$ is a relative maximum.

Here are sketches of $p(r)$ and the cumulative position $P(r)$:

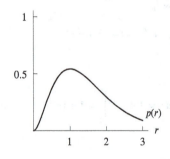
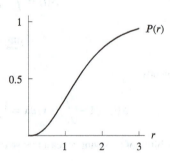

(b) The median distance is the distance r such that $P(r) = 1 - (2r^2 + 2r + 1)e^{-2r} = 0.5$, or equivalently, $(2r^2 + 2r + 1)e^{-2r} = 0.5$.

By experimentation with a calculator, we find that $r \approx 1.33$ Bohr radii is the median distance.

The mean distance is equal to the value of the integral $\int_0^\infty rp(r)\,dr = \lim_{x\to\infty}\int_0^x rp(r)\,dr$. We have that $\int_0^x rp(r)\,dr = \int_0^x 4r^3e^{-2r}\,dr$. Using the integral table, we get

$$\int_0^x 4r^3e^{-2r}\,dr = \left[\left(-\frac{1}{2}\right)4r^3 - \frac{1}{4}(12r^2) - \frac{1}{8}(24r) - \frac{1}{16}(24)\right]e^{-2x}\Big|_0^x$$

$$= \frac{3}{2} - \left[2x^3 + 3x^2 + 3x + \frac{3}{2}\right]e^{-2x}.$$

Taking the limit of this expression as $x \to \infty$, we see that all terms involving (powers of x or constants) $\cdot\, e^{-2x}$ have limit 0, and thus the mean distance is 1.5 Bohr radii.

The most likely distance is obtained by maximizing $p(r) = 4r^2 e^{-2r}$; as we have already seen this corresponds to $r = 1$ Bohr unit.

(c) Because it is the most likely distance of the electron from the nucleus.

Strengthen Your Understanding

29. We can take a constant, or uniform distribution

$$p(x) = \begin{cases} 0 & \text{for} \quad x < 0 \\ 1 & \text{for} \quad 0 \le x \le 1 \\ 0 & \text{for} \quad x > 1. \end{cases}$$

Since

$$\int_0^1 x\,dx = \frac{1}{2}, \quad \text{the mean is } \frac{1}{2}.$$

Since

$$\int_0^{1/2} p(x)\,dx = \int_0^{1/2} 1\,dx = \frac{1}{2}, \quad \text{the median is also } \frac{1}{2}.$$

33. False. Note that p is the density function for the population, not the cumulative density function. Thus $p(10) = p(20)$ means that x values near 10 are as likely as x values near 20.

Solutions for Chapter 8 Review

Exercises

1. Vertical slices are circular. Horizontal slices would be similar to ellipses in cross-section, or at least ovals (a word derived from *ovum*, the Latin word for egg).

Figure 8.43

5. (a) The equation $y = -x^2 + 6x$ can be solved for x as

$$x^2 - 6x + y = 0$$

$$x = \frac{6 \pm \sqrt{36 - 4y}}{2} = 3 \pm \sqrt{9 - y}.$$

The left end of the strip is given by $x = 3 - \sqrt{9 - y}$, and the right end is given by $x = 3 + \sqrt{9 - y}$. Thus,

$$\text{Area of region} \approx \sum ((3 + \sqrt{9 - y}) - (3 - \sqrt{9 - y}))\Delta y = \sum 2\sqrt{9 - y}\,\Delta y.$$

(b) We have

$$\text{Area} = \int_0^9 2\sqrt{9-y}\,dy = \left.\frac{2(9-y)^{3/2}}{-3/2}\right|_0^9 = 36.$$

9. We slice the region vertically. Each rotated slice is approximately a cylinder with radius $y = 4 - x^2$ and thickness Δx. See Figure 8.44. The volume, V, of a typical slice is $\pi(4 - x^2)^2\Delta x$. The volume of the object is the sum of the volumes of the slices:

$$V \approx \sum \pi(4 - x^2)^2\Delta x.$$

As $\Delta x \to 0$ we obtain an integral. Since the region lies between $x = -2$ and $x = 2$, we have:

$$V = \int_{-2}^2 \pi(4 - x^2)^2 dx = \pi\int_{-2}^2 (16 - 8x^2 + x^4)dx = \pi\left(16x - \frac{8x^3}{3} + \frac{x^5}{5}\right)\Bigg|_{-2}^2 = \frac{512\pi}{15} = 107.233.$$

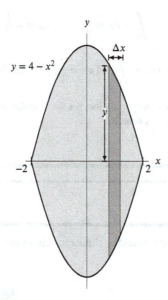

Figure 8.44

13. The region is bounded by $y = 2$, the y-axis and $y = x^{1/3}$. The two functions $y = 2$ and $y = x^{1/3}$ intersect at $(8, 2)$. We slice the volume with planes that are perpendicular to the y-axis. This divides the solid into thin cylinders with

$$\text{Volume} \approx \pi r^2\Delta y.$$

The radius is the distance from the y-axis to the curve $x = y^3$. Integrating from $y = 0$ to $y = 2$ we have

$$V = \int_0^2 \pi(y^3)^2\,dy.$$

17. The region is bounded by $x = 4x$ and $x = y^3$. The two functions intersect at $(0, 0)$ and $(8, 2)$. We slice the volume with planes that are perpendicular to the line $x = -3$. This divides the solid into thin washers with

$$\text{Volume} = \pi r_{out}^2\Delta y - \pi r_{in}^2\Delta y.$$

The inner radius is the distance from the line $x = -3$ to $x = y^3$ and the outer radius is the distance from the line $x = -3$ to the line $x = 4y$. Integrating from $y = 0$ to $y = 2$ we have

$$V = \int_0^2 \left[\pi(4y + 3)^2 - \pi(y^3 + 3)^2\right]\,dx.$$

21. We'll find the arc length of the top half of the ellipse, and multiply that by 2. In the top half of the ellipse, the equation $(x^2/a^2) + (y^2/b^2) = 1$ implies

$$y = +b\sqrt{1 - \frac{x^2}{a^2}}.$$

Differentiating $(x^2/a^2) + (y^2/b^2) = 1$ implicitly with respect to x gives us

$$\frac{2x}{a^2} + \frac{2y}{b^2}\frac{dy}{dx} = 0,$$

so

$$\frac{dy}{dx} = \frac{\frac{-2x}{a^2}}{\frac{2y}{b^2}} = -\frac{b^2 x}{a^2 y}.$$

Substituting this into the arc length formula, we get

$$\text{Arc Length} = \int_{-a}^{a} \sqrt{1 + \left(-\frac{b^2 x}{a^2 y}\right)^2}\, dx$$

$$= \int_{-a}^{a} \sqrt{1 + \left(\frac{b^4 x^2}{a^4(b^2)(1 - \frac{x^2}{a^2})}\right)}\, dx$$

$$= \int_{-a}^{a} \sqrt{1 + \left(\frac{b^2 x^2}{a^2(a^2 - x^2)}\right)}\, dx.$$

Hence the arc length of the entire ellipse is

$$2\int_{-a}^{a} \sqrt{1 + \left(\frac{b^2 x^2}{a^2(a^2 - x^2)}\right)}\, dx.$$

25. The arc length is given by

$$L = \int_{1}^{2} \sqrt{1 + e^{2x}}\, dx \approx 4.785.$$

Note that $\sqrt{1 + e^{2x}}$ does not have an obvious elementary antiderivative, so we use an approximation method to find an approximate value for L.

29.
- $\int_{0}^{1} f(x)\, dx$ gives the area under the graph of f from 0 to 1.
- The graph of f is concave up and passes through the points $(0, 0)$ and $(1, 1)$, so it lies below the line $y = x$.
- The area under $y = x$ from 0 to 1 is half the area of a square of side 1, or $1/2$. Thus, $\int_{0}^{1} f(x)\, dx < \frac{1}{2}$.

33.
- $\int_{0}^{1} \sqrt{1 + (f'(x))^2}\, dx$ gives the arc length of the graph of f from $x = 0$ to $x = 1$.
- The graph of f is concave up and contains $(0, 0)$ and $(1, 1)$, so it lies below the line $y = x$ on $0 < x < 1$. The arc length of the line between $(0, 0)$, and $(1, 1)$ is $\sqrt{1^2 + 1^2} = \sqrt{2}$.
- The line segment between $(0, 0)$ and $(1, 1)$ is shorter than the arc length of f, so $\int_{0}^{1} \sqrt{1 + (f'(x))^2}\, dx > \sqrt{2}$.

Problems

37. The two functions intersect at $(0, 0)$ and $(5, 25)$. We slice the volume with planes perpendicular to the vertical line $x = 8$. This divides the solid into thin washers with volume

$$\text{Volume of slice } = \pi((r_{out})^2 - (r_{in})^2)\Delta y.$$

The outer radius is the horizontal distance from the line $x = 8$ to the curve $x = y/5$, so $r_{out} = 8 - y/5$. Similarly, the inner radius is the horizontal distance from the line $x = 8$ to the curve $x = \sqrt{y}$, so $r_{in} = 8 - \sqrt{y}$. Integrating from $y = 0$ to $y = 25$ we have

$$V = \int_{0}^{25} \pi((8 - y/5)^2 - (8 - \sqrt{y})^2)\, dy.$$

41. We divide the region into vertical strips of thickness Δx. As a slice is rotated about the x-axis, it creates a disk of radius r_{out} from which has been removed a disk of radius r_{in}. We see in Figure 8.45 that $r_{\text{out}} = 5 + 2x$ and $r_{\text{in}} = 5$. Thus,

$$\text{Volume of a slice} \approx \pi(r_{\text{out}})^2 \Delta x - \pi(r_{\text{in}})^2 \Delta x = \pi(5 + 2x)^2 \Delta x - \pi(5)^2 \Delta x.$$

To find the total volume, V, we integrate this quantity between $x = 0$ and $x = 4$:

$$V = \int_0^4 (\pi(5+2x)^2 - \pi(5)^2) dx = \pi \int_0^4 ((5+2x)^2 - 25)\, dx = \pi \left(\frac{4}{3}x^3 + 10x^2\right)\Big|_0^4 = \frac{736\pi}{3} = 770.737.$$

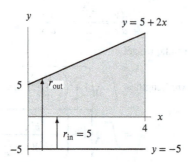

Figure 8.45

45. Slice the object into rings horizontally, as in Figure 8.46. A typical ring has thickness Δy, outer radius 1, and inner radius $1 - x = 1 - \sqrt{1 - y^2}$. Thus,

$$\text{Volume of ring} \approx \pi 1^2 \Delta y - \pi(1 - \sqrt{1 - y^2})^2 \Delta y = \pi(2\sqrt{1 - y^2} - (1 - y^2))\, \Delta y.$$

$$
\begin{aligned}
\text{Volume of solid} &= \int_0^1 \pi(2\sqrt{1 - y^2} - 1 + y^2)\, dy \\
&= 2\pi \int_0^1 \sqrt{1 - y^2}\, dy - \pi \int_0^1 1\, dy + \pi \int_0^1 y^2\, dy \\
&= 2\pi \cdot \frac{1}{2}\left(y\sqrt{1 - y^2}\Big|_0^1 + 1^2 \int_0^1 \frac{1}{\sqrt{1 - y^2}}\, dy\right) - \pi y\Big|_0^1 + \frac{\pi y^3}{3}\Big|_0^1 \\
&= \pi y\sqrt{1 - y^2} + \pi \arcsin y - \pi y + \frac{\pi y^3}{3}\Big|_0^1 \\
&= 0 + \frac{\pi^2}{2} - \pi + \frac{\pi}{3} - 0 - 0 + 0 - 0 \\
&= \frac{\pi^2}{2} - \frac{2\pi}{3} = 2.840.
\end{aligned}
$$

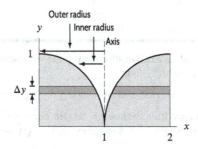

Figure 8.46: Cross-section of solid

49. The curve $y = x(x-3)^2$ has x-intercepts at $x = 0, 3$ and lies above the x-axis on this interval.

Thus, $\int_0^3 x(x-3)^2\, dx$ gives the area under the graph of f from $x = 0$ to $x = 3$.

53. (a) The line $y = ax$ must pass through (l, b). Hence $b = al$, so $a = b/l$.

(b) Cut the cone into N slices, slicing perpendicular to the x–axis. Each piece is almost a cylinder. The radius of the ith cylinder is $r(x_i) = \dfrac{bx_i}{l}$, so the volume

$$V \approx \sum_{i=1}^{N} \pi \left(\frac{bx_i}{l}\right)^2 \Delta x.$$

Therefore, as $N \to \infty$, we get

$$V = \int_0^l \pi b^2 l^{-2} x^2 dx$$

$$= \pi \frac{b^2}{l^2}\left[\frac{x^3}{3}\right]_0^l = \left(\pi \frac{b^2}{l^2}\right)\left(\frac{l^3}{3}\right) = \frac{1}{3}\pi b^2 l.$$

57. The arc length of the curve $y = f(t)$ from $t = 3$ to $t = 8$ is $\int_3^8 \sqrt{1 + (f'(t))^2}\, dt$. Thus, we want a function f such that

$$\int_3^8 \sqrt{1 + (f'(t))^2}\, dt = \int_3^8 \sqrt{1 + e^{6t}}\, dt.$$

Thus, we have

$$\left(f'(t)\right)^2 = e^{6t}.$$

One possibility is

$$f'(t) = e^{3t}$$
$$f(t) = \frac{1}{3}e^{3t} + C.$$

For any constant C, the original integral is the arc length of the curve $y = \frac{1}{3}e^{3t} + C$ from $t = 3$ to $t = 8$.

Another solution to $\left(f'(t)\right)^2 = e^{6t}$ is $f'(t) = -e^{3t}$, which gives $f(t) = -\frac{1}{3}e^{3t} + C$.

61. See Figure 8.47. The circles meet where

$$2a \cos\theta = a$$
$$\cos\theta = \frac{1}{2}$$
$$\theta = \pm\frac{\pi}{3}.$$

The area is obtained by subtraction:

$$\text{Area} = \int_{-\pi/3}^{\pi/3} \left(\frac{1}{2}(2a\cos\theta)^2 - \frac{1}{2}a^2\right) d\theta$$

$$= \int_{-\pi/3}^{\pi/3} \left(2a^2 \cos^2\theta - \frac{1}{2}a^2\right) d\theta$$

$$= \left(2a^2\left(\frac{1}{2}\cos\theta\sin\theta + \frac{\theta}{2}\right) - \frac{a^2}{2}\theta\right)\Bigg|_{-\pi/3}^{\pi/3}$$

$$= \left(\frac{\pi}{3} + \frac{\sqrt{3}}{2}\right)a^2.$$

Since

$$\frac{\left(\pi/3 + \sqrt{3}/2\right)a^2}{\pi a^2} = 61\%$$

the shaded region covers 61% of circle C.

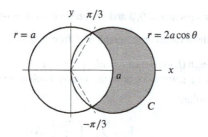

Figure 8.47

65. The total mass is 12 gm, so the center of mass is located at $\overline{x} = \frac{1}{12}(-5 \cdot 3 - 3 \cdot 3 + 2 \cdot 3 + 7 \cdot 3) = \frac{1}{4}$.

69. Let x be the distance from the bucket to the surface of the water. It follows that $0 \leq x \leq 40$. At x feet, the bucket weighs $(30 - \frac{1}{4}x)$, where the $\frac{1}{4}x$ term is due to the leak. When the bucket is x feet from the surface of the water, the work done by raising it Δx feet is $(30 - \frac{1}{4}x)\,\Delta x$. So the total work required to raise the bucket to the top is

$$W = \int_0^{40}(30 - \frac{1}{4}x)dx$$
$$= \left(30x - \frac{1}{8}x^2\right)\Big|_0^{40}$$
$$= 30(40) - \frac{1}{8}40^2 = 1000 \text{ ft-lb.}$$

73. (a) Divide the wall into horizontal strips, each of height Δh. See Figure 8.48. The area of each strip is $1000\Delta h$, and the pressure at depth h is $62.4h$, so

$$\text{Force on strip} \approx 1000(62.4h)\Delta h$$
$$\text{Force on dam} \approx \sum 1000(62.4h)\Delta h.$$

(b) The force on the dam is given by the integral

$$\text{Force on dam} = \int_0^{50} 1000(62.4h)\,dh = 62400\frac{h^2}{2}\Big|_0^{50} = 78{,}000{,}000 \text{ pounds.}$$

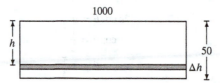

Figure 8.48

77. One good way to approach the problem is in terms of present values. In 1980, the present value of Germany's loan was 20 billion DM. Now let's figure out the rate that the Soviet Union would have to give money to Germany to pay off 10% interest on the loan by using the formula for the present value of a continuous stream. Since the Soviet Union sends gas at a constant rate, the rate of deposit, $P(t)$, is a constant c. Since they don't start sending the gas until after 5 years have passed, the present value of the loan is given by:

$$\text{Present Value} = \int_5^{\infty} P(t)e^{-rt}\,dt.$$

We want to find c so that

$$20{,}000{,}000{,}000 = \int_5^{\infty} ce^{-rt}\,dt = c\int_5^{\infty} e^{-rt}\,dt$$
$$= c\lim_{b\to\infty}(-10e^{-0.10t})\Big|_5^b = ce^{-0.10(5)}$$
$$\approx 6.065c.$$

Dividing, we see that c should be about 3.3 billion DM per year. At 0.10 DM per m^3 of natural gas, the Soviet Union must deliver gas at the constant, continuous rate of about 33 billion m^3 per year.

81. The statement $P(70) = 0.92$ means that 92% of the population has an age less than 70.

85. (a) Divide the cross-section of the blood into rings of radius r, width Δr. See Figure 8.49.

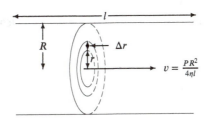

Figure 8.49

Then
$$\text{Area of ring} \approx 2\pi r \Delta r.$$

The velocity of the blood is approximately constant throughout the ring, so
$$\text{Rate blood flows through ring} \approx \text{Velocity} \cdot \text{Area}$$
$$= \frac{P}{4\eta l}(R^2 - r^2) \cdot 2\pi r \Delta r.$$

Thus, summing over all rings, we find the total blood flow:
$$\text{Rate blood flowing through blood vessel} \approx \sum \frac{P}{4\eta l}(R^2 - r^2) 2\pi r \Delta r.$$

Taking the limit as $\Delta r \to 0$, we get
$$\text{Rate blood flowing through blood vessel} = \int_0^R \frac{\pi P}{2\eta l}(R^2 r - r^3)dr$$

$$= \frac{\pi P}{2\eta l}\left(\frac{R^2 r^2}{2} - \frac{r^4}{4}\right)\Bigg|_0^R = \frac{\pi P R^4}{8\eta l}.$$

(b) Since
$$\text{Rate of blood flow} = \frac{\pi P R^4}{8\eta l},$$

if we take $k = \pi P/(8\eta l)$, then we have
$$\text{Rate of blood flow} = kR^4,$$

that is, rate of blood flow is proportional to R^4, in accordance with Poiseuille's Law.

89. Any small piece of mass ΔM on either of the two spheres has kinetic energy $\frac{1}{2}v^2\Delta M$. Since the angular velocity of the two spheres is the same, the actual velocity of the piece ΔM will depend on how far away it is from the axis of revolution. The further away a piece is from the axis, the faster it must be moving and the larger its velocity v. This is because if ΔM is at a distance r from the axis, in one revolution it must trace out a circular path of length $2\pi r$ about the axis. Since every piece in either sphere takes 1 minute to make 1 revolution, pieces farther from the axis must move faster, as they travel a greater distance.

Thus, since the thin spherical shell has more of its mass concentrated farther from the axis of rotation than does the solid sphere, the bulk of it is traveling faster than the bulk of the solid sphere. So, it has the higher kinetic energy.

CAS Challenge Problems

93. (a) The expression for arc length in terms of a definite integral gives
$$A(t) = \int_0^t \sqrt{1 + \left(\frac{1}{2\sqrt{x}}\right)^2}\, dx = \frac{2\sqrt{t}\sqrt{1 + 4t} + \operatorname{arcsinh}(2\sqrt{t})}{4}.$$

The integral was evaluated using a computer algebra system; different systems may give the answer in different forms. Some may involve ln instead of arcsinh, which is the inverse function of the hyperbolic sine function.

(b) Figure 8.51 shows that the graphs of $A(t)$ and the graph of $y = t$ look very similar. This suggests that $A(t) \approx t$.

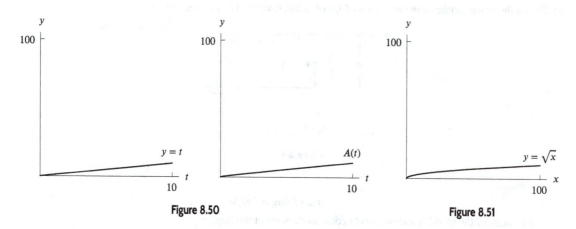

Figure 8.50   Figure 8.51

(c) The graph in Figure 8.51 is approximately horizontal and close to the x-axis. Thus, if we measure the arc length up to a certain x-value, the answer is approximately the same as if we had measured the length straight along the x-axis. Hence

$$A(t) \approx x = t.$$

So

$$A(t) \approx t.$$

CHAPTER NINE

Solutions for Section 9.1

Exercises

1. The first term is $2^1 + 1 = 3$. The second term is $2^2 + 1 = 5$. The third term is $2^3 + 1 = 9$, the fourth is $2^4 + 1 = 17$, and the fifth is $2^5 + 1 = 33$. The first five terms are $3, 5, 9, 17, 33$.

5. The first term is $(-1)^2(1/2)^0 = 1$. The second term is $(-1)^3(1/2)^1 = -1/2$. The first five terms are

$$1, -1/2, 1/4, -1/8, 1/16.$$

9. We observe that if we subtract 1 from each term of the sequence, we get $1, 4, 9, 16, 25, \ldots$, namely the squares $1^2, 2^2, 3^2, 4^2, 5^2, \ldots$. Thus $s_n = n^2 + 1$.

Problems

13. (a) matches (IV), since the sequence increases toward 1.
    (b) matches (III), since the odd terms increase toward 1 and the even terms decrease toward 1.
    (c) matches (II), since the sequence decreases toward 0.
    (d) matches (I), since the sequence decreases toward 1.

17. Since $\lim_{n \to \infty} x^n = 0$ if $|x| < 1$ and $|0.2| < 1$, we have $\lim_{n \to \infty}(0.2)^n = 0$, so the sequence converges to 0

21. Since $\lim_{n \to \infty} x^n = 0$ if $|x| < 1$ and $\left|\frac{2}{3}\right| < 1$, we have $\lim_{n \to \infty}\left(\frac{2^n}{3^n}\right) = \lim_{n \to \infty}\left(\frac{2}{3}\right)^n = 0$, so the sequence converges to 0.

25. As n increases, the term $2n$ is much larger in magnitude than $(-1)^n 5$ and the term $4n$ is much larger in magnitude than $(-1)^n 3$. Thus dividing the numerator and denominator by n and using the fact that $\lim_{n \to \infty} 1/n = 0$, we have

$$\lim_{n \to \infty} \frac{2n + (-1)^n 5}{4n - (-1)^n 3} = \lim_{n \to \infty} \frac{2 + (-1)^n 5/n}{4 - (-1)^n 3/n} = \frac{1}{2}.$$

Thus, the sequence converges to $1/2$.

29. We have $s_2 = s_1 + 2 = 3$ and $s_3 = s_2 + 3 = 6$. Continuing, we get

$$1, 3, 6, 10, 15, 21.$$

33. We have

$$a_2 = a_{2-1} + 3 \cdot 2 = a_1 + 6 \quad = 8 + 6 \quad = 14 \qquad b_2 = b_{2-1} + a_{2-1} = b_1 + a_1 = 5 + 8 \quad = 13$$
$$a_3 = a_{3-1} + 3 \cdot 3 = a_2 + 9 \quad = 14 + 9 \quad = 23 \qquad b_3 = b_{3-1} + a_{3-1} = b_2 + a_2 = 13 + 14 = 27$$
$$a_4 = a_{4-1} + 3 \cdot 4 = a_3 + 12 = 23 + 12 = 35 \qquad b_4 = b_{4-1} + a_{4-1} = b_3 + a_3 = 27 + 23 = 50$$
$$b_5 = b_{5-1} + a_{5-1} = b_4 + a_4 = 50 + 35 = 85.$$

37. Each term is twice the previous term minus one, so a recursive definition is $s_n = 2s_{n-1} - 1$ for $n > 1$ and $s_1 = 3$. We also notice that the differences of consecutive terms are powers of 2, so $s_2 = s_1 + 2$, $s_3 = s_2 + 2^2$, and so on. Thus another recursive definition is $s_n = s_{n-1} + 2^{n-1}$ for $n > 1$ and $s_1 = 3$.

41. For $n > 1$, if $s_n = 3n - 2$, then $s_{n-1} = 3(n-1) - 2 = 3n - 5$, so

$$s_n - s_{n-1} = (3n - 2) - (3n - 5) = 3,$$

giving

$$s_n = s_{n-1} + 3.$$

In addition, $s_1 = 3 \cdot 1 - 2 = 1$.

45. We have

$$s_3 = \sum_{k=0}^{3} 3 \cdot 2^k = 3 \cdot 2^0 + 3 \cdot 2^1 + 3 \cdot 2^2 + 3 \cdot 2^3 = 3\,(1 + 2 + 4 + 8) = 45.$$

49. The first 6 terms of the sequence for the sampling is

$$\cos 0.5, \ \cos 1.0, \ \cos 1.5, \ \cos 2.0, \ \cos 2.5, \ \cos 3.0$$
$$= 0.878, \ 0.540, \ 0.071, \ -0.416, \ -0.801, \ -0.990.$$

53. The first smoothing gives

$$1.5, 2, 3, 4, 5, 6, 7 \ldots$$

The second smoothing gives

$$1.75, 2.17, 3, 4, 5, 6 \ldots$$

Terms which are already the same as their average with their neighbors are not changed.

57. The sequence converges to 1. By the tenth term, it stabilizes to three decimal places at 1.000.

61. (a) The bottom row contains k cans, the next one contains $(k-1)$ cans, then $(k-2)$ and so on. Thus, there are k rows. Since the top row contains 1 can, the second contains 2 cans, etc, we have $a_n = n$.
   (b) Since the nth row contains n cans, $a_n = n$,

$$T_n = T_{n-1} + a_n$$

   gives

$$T_n = T_{n-1} + n, \quad \text{for } n > 1.$$

   In addition, $T_1 = 1$.
   (c) If $T_n = \frac{1}{2}n(n+1)$, then $T_{n-1} = \frac{1}{2}(n-1)n$, so

$$T_n - T_{n-1} = \frac{1}{2}n(n+1) - \frac{1}{2}n(n-1) = \frac{n}{2}(n+1-(n-1)) = n.$$

   In addition, $T_1 = \frac{1}{2} \cdot 1(2) = 1$.

Strengthen Your Understanding

65. A decreasing sequence does not have to converge to 0; in fact, it does not have to converge at all (consider the sequence $s_n = -n$, for example). In this case, the limit of the sequence is

$$\lim_{n\to\infty} \frac{3n+10}{7n+3} = \lim_{n\to\infty} \frac{3+10/n}{7+3/n} = \frac{3+0}{7+0} = \frac{3}{7}.$$

69. One example is the sequence $s_n = n$. This sequence is increasing and therefore monotone, but it does not converge because the terms do not get closer and closer to any specific finite value.

73. True. If there is only a finite number of terms greater than a million, then we can choose the largest of them to be an upper bound M for the sequence. Thus the sequence is bounded by $0 < s_n \leq M$ for all n.

77. False. The decreasing sequence $-1, -2, -3, \ldots$ has all terms less than a million, but it has no lower bound. Thus it is unbounded.

Additional Problems (online only)

81. We use Theorem 9.1, so we must show that s_n is bounded. Since t_n converges, it is bounded so there is a number M, such that $t_n \leq M$ for all n. Therefore $s_n \leq t_n \leq M$ for all n. Since s_n is increasing, $s_1 \leq s_n$ for all n. Thus if we let $K = s_1$, we have $K \leq s_n \leq M$ for all n, so s_n is bounded. Therefore, s_n converges.

Solutions for Section 9.2

Exercises

1. Series, because the terms are added together.

5. Series, because the terms are added together.

9. Yes, $a = 2$, ratio $= 1/2$.

13. No. Ratio between successive terms is not constant: $\dfrac{6z^2}{3z} = 2z$, while $\dfrac{9z^3}{6z^2} = \dfrac{3}{2}z$.

17. No. Ratio between successive terms is not constant: $\dfrac{-z^4}{z^2} = -z^2$, while $\dfrac{z^8}{-z^4} = -z^4$.

21. Since there is no common ratio between successive terms of the series, this series is not geometric.

25. The series has 10 terms. The first term is $a = 0.2$ and the constant ratio is $x = 0.1$, so

$$\text{Sum} = \frac{0.2(1 - x^{10})}{(1 - x)} = \frac{0.2(1 - (0.1)^{10})}{0.9} = 0.222.$$

29. We have

$$\text{Sum} = -810 + 540 - 360 + 240 - 160 + \cdots$$
$$= -810 + (-810) \cdot \left(-\frac{2}{3}\right) + (-810) \cdot \left(-\frac{2}{3}\right)^2 + (-810) \cdot \left(-\frac{2}{3}\right)^3 + \cdots$$
$$= \frac{-810}{1 - (-2/3)} = -486.$$

33. The geometric series has first term $a = 1/2$ and common ratio $r = 1 - x/2$. Hence,

$$\sum_{n=0}^{\infty} \frac{1}{2}\left(1 - \frac{x}{2}\right)^n = \frac{a}{1 - r} = \frac{1/2}{1 - (1 - x/2)} = \frac{1}{x}.$$

The series converges for $-1 < r < 1$, which is

$$-1 < 1 - \frac{x}{2} < 1,$$

that is,

$$0 < x < 4.$$

Problems

37. This is a geometric series with first term y and ratio $-y$:

$$y - y^2 + y^3 - y^4 + \cdots = \frac{y}{1 - (-y)} = \frac{y}{1 + y}.$$

This series converges for $|-y| < 1$, that is for $-1 < y < 1$.

41. This is a geometric series with first term 8 and ratio $x^2 - 5$, so

$$8 + 8\left(x^2 - 5\right) + 8\left(x^2 - 5\right)^2 + 8\left(x^2 - 5\right)^3 + \cdots = \frac{8}{1 - (x^2 - 5)}.$$

This series converges for $|x^2 - 5| < 1$. Since $x^2 - 5 < 1$ when $-\sqrt{6} < x < \sqrt{6}$ and $x^2 - 5 > -1$ when $x > 2$ or $x < -2$, the infinite series converges over the intervals $-\sqrt{6} < x < -2$ and $2 < x < \sqrt{6}$.

45. At $x = \pi$ the geometric series has first term 3 and ratio $\cos \pi = -1$. Since the ratio equals -1, the series does not converge.

49. (a) $0.232323\ldots = 0.23 + 0.23(0.01) + 0.23(0.01)^2 + \cdots$, which is a geometric series with $a = 0.23$ and $x = 0.01$.

(b) The sum is $\dfrac{0.23}{1 - 0.01} = \dfrac{0.23}{0.99} = \dfrac{23}{99}$.

53. (a) In 2014, the quantity mined is predicted to be $17.9(1.025)$ million tonnes.
In 2015, the quantity predicted is $17.9(1.025)^2$ million tonnes, and n years after 2013, the quantity is $17.9(1.025)^n$. Thus,

$$\text{Total quantity mined} = 17.9(1.025) + 17.9(1.025)^2 + \ldots + 17.9(1.025)^n \text{ million tonnes.}$$

To sum, we factor out $17.9(1.025)$, giving

$$\text{Total quantity mined} = 17.9(1.025)\left(1 + (1.025) + \ldots + (1.025)^{n-1}\right)$$

$$= 17.9(1.025)\left(\frac{1.025^n - 1}{1.025 - 1}\right) = 733.9\left(1.025^n - 1\right) \text{ million tonnes.}$$

(b) The reserves are exhausted when

$$733.9\,(1.025^n - 1) = 690,$$

that is, when

$$1.025^n = \frac{690}{733.9} + 1$$

$$n = \frac{\ln((690/733.9) + 1)}{\ln 1.025} = 26.8 \text{ years.}$$

Thus, the copper is predicted to be exhausted in the year 2040.

(c) The sum is a right-hand Riemann sum approximating the integral $\int_0^n 17.9(1.025)^t \, dt$; see Figure 9.1. That is, if $\Delta t = 1$, the interval $0 \le t \le n$ can be divided into n subintervals and the right-hand Riemann sum is

$$\text{RIGHT}(n) = \left(17.9(1.025) + 17.9(1.025)^2 + \ldots + 17.9(1.025)^n\right) \cdot 1$$

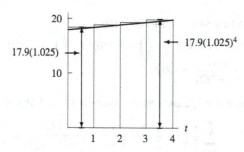

Figure 9.1: Right Riemann sum for $n = 4$

57.

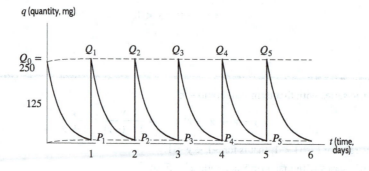

61. (a)

$$\text{Present value of first coupon} = \frac{50}{1.05}$$

$$\text{Present value of second coupon} = \frac{50}{(1.05)^2}, \text{etc.}$$

$$\text{Total present value} = \underbrace{\frac{50}{1.05} + \frac{50}{(1.05)^2} + \cdots + \frac{50}{(1.05)^{10}}}_{\text{coupons}} + \underbrace{\frac{1000}{(1.05)^{10}}}_{\text{principal}}$$

$$= \frac{50}{1.05}\left(1 + \frac{1}{1.05} + \cdots + \frac{1}{(1.05)^9}\right) + \frac{1000}{(1.05)^{10}}$$

$$= \frac{50}{1.05}\left(\frac{1 - \left(\frac{1}{1.05}\right)^{10}}{1 - \frac{1}{1.05}}\right) + \frac{1000}{(1.05)^{10}}$$

$$= 386.087 + 613.913$$

$$= \$1000$$

(b) When the interest rate is 5%, the present value equals the principal.

(c) When the interest rate is more than 5%, the present value is smaller than it is when interest is 5% and must therefore be less than the principal. Since the bond will sell for around its present value, it will sell for less than the principal; hence the description *trading at discount*.

(d) When the interest rate is less than 5%, the present value is more than the principal. Hence the bond will be selling for more than the principal, and is described as *trading at a premium*.

Strengthen Your Understanding

65. The formula for the sum of an infinite geometric series does not apply because the common ratio is not between -1 and 1.

69. One way to find an example is to start with a geometric series with four distinct terms, and then rescale the terms so that their sum is 10.

For example, we might start with the series

$$1 + \frac{1}{2} + \frac{1}{4} + \frac{1}{8},$$

whose sum is 15/8. Since we want a sum of 10, we scale these terms up by a factor of

$$\frac{10}{15/8} = \frac{16}{3}.$$

This gives the series

$$\frac{16}{3} + \frac{8}{3} + \frac{4}{3} + \frac{2}{3} = 10.$$

This series is geometric with common ratio $1/2$.

73. (c). The common ratio in series (I) and (IV) is between -1 and 1. The common ratio in (II) and (III) is greater than one.

Solutions for Section 9.3

Exercises

1. The series is $1 + 2 + 3 + 4 + 5 + \cdots$. The sequence of partial sums is

$$S_1 = 1, \quad S_2 = 1 + 2, \quad S_3 = 1 + 2 + 3, \quad S_4 = 1 + 2 + 3 + 4, \quad S_5 = 1 + 2 + 3 + 4 + 5, \ldots$$

which is

$$1, \quad 3, \quad 6, \quad 10, \quad 15 \ldots.$$

5. We use the integral test with $f(x) = x/(x^2 + 1)$ to determine whether this series converges or diverges. We determine whether the corresponding improper integral $\int_1^\infty \frac{x}{x^2 + 1} dx$ converges or diverges:

$$\int_1^\infty \frac{x}{x^2 + 1} dx = \lim_{b \to \infty} \int_1^b \frac{x}{x^2 + 1} dx = \lim_{b \to \infty} \frac{1}{2} \ln(x^2 + 1) \Big|_1^b = \lim_{b \to \infty} \left(\frac{1}{2} \ln(b^2 + 1) - \frac{1}{2} \ln 2 \right) = \infty.$$

Since the integral $\int_1^\infty \frac{x}{x^2 + 1} dx$ diverges, we conclude from the integral test that the series $\sum_{n=1}^\infty \frac{n}{n^2 + 1}$ diverges.

9. The improper integral $\int_0^\infty \frac{1}{x^2 + 1} dx$ converges to $\frac{\pi}{2}$, since

$$\int_0^b \frac{1}{x^2 + 1} dx = \arctan x \Big|_0^b = \arctan b - \arctan 0 = \arctan b,$$

and $\lim_{b\to\infty} \arctan b = \frac{\pi}{2}$. The terms of the series $\sum_{n=1}^{\infty} \frac{1}{n^2+1}$ form a right hand sum for the improper integral; each term represents the area of a rectangle of width 1 fitting completely under the graph of the function $\frac{1}{x^2+1}$. (See Figure 9.2.) Thus the sequence of partial sums is bounded above by $\frac{\pi}{2}$. Since the partial sums are increasing (every new term added is positive), the series is guaranteed to converge to some number less than or equal to $\pi/2$ by Theorem 9.1.

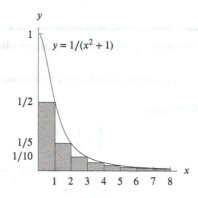

Figure 9.2

Problems

13. Using the integral test, we compare the series with

$$\int_0^\infty \frac{3}{x+2}\,dx = \lim_{b\to\infty} \int_0^b \frac{3}{x+2}\,dx = 3\ln|x+2|\Big|_0^b .$$

Since $\ln(b+2)$ is unbounded as $b \to \infty$, the integral diverges and therefore so does the series.

17. Since the terms in the series are positive and decreasing, we can use the integral test. We calculate the corresponding improper integral using the substitution $w = 1 + x^2$:

$$\int_0^\infty \frac{2x}{(1+x^2)^2}\,dx = \lim_{b\to\infty} \int_0^b \frac{2x}{(1+x^2)^2}\,dx = \lim_{b\to\infty} \frac{-1}{(1+x^2)}\Big|_0^b = \lim_{b\to\infty} \left(\frac{-1}{1+b^2}+1\right) = 1.$$

Since the limit exists, the integral converges, so the series $\sum_{n=0}^{\infty} \frac{2n}{(1+n^2)^2}$ converges.

21. Using the integral test, we compare the series with

$$\int_0^\infty \frac{3}{x^2+4}\,dx = \lim_{b\to\infty} \int_0^b \frac{3}{x^2+4}\,dx = \frac{3}{2}\lim_{b\to\infty} \arctan\left(\frac{x}{2}\right)\Big|_0^b = \frac{3}{2}\lim_{b\to\infty} \arctan\left(\frac{b}{2}\right) = \frac{3\pi}{4},$$

by integral table V-24. Since the integral converges so does the series.

25. Both $\sum_{n=1}^{\infty} \left(\frac{1}{2}\right)^n$ and $\sum_{n=1}^{\infty} \left(\frac{2}{3}\right)^n$ are convergent geometric series. Therefore, by Property 1 of Theorem 9.2, the series $\sum_{n=1}^{\infty} \left(\left(\frac{1}{2}\right)^n + \left(\frac{2}{3}\right)^n\right)$ converges.

29. Since the terms in the series are positive and decreasing, we can use the integral test. We calculate the corresponding improper integral using the substitution $w = 1 + \ln x$:

$$\int_1^\infty \frac{1}{x(1+\ln x)}\,dx = \lim_{b\to\infty} \int_1^b \frac{1}{x(1+\ln x)}\,dx = \lim_{b\to\infty} \ln(1+\ln x)\Big|_1^b = \lim_{b\to\infty} \ln(1+\ln b).$$

Since the limit does not exist, the integral diverges, so the series $\sum_{n=1}^{\infty} \frac{1}{n(1+\ln n)}$ diverges.

33. Using $\ln(2^n) = n \ln 2$, we see that

$$\sum \frac{1}{\ln(2^n)} = \sum \frac{1}{(\ln 2)n}.$$

The series on the right is the harmonic series multiplied by $1/\ln 2$. Since the harmonic series diverges, $\sum_{n=1}^{\infty} 1/\ln(2^n)$ diverges.

37. (a) A common denominator is $k(k+1)$ so

$$\frac{1}{k} - \frac{1}{k+1} = \frac{k+1}{k(k+1)} - \frac{k}{k(k+1)} = \frac{k+1-k}{k(k+1)} = \frac{1}{k(k+1)}.$$

(b) Using the result of part (a), the partial sum can be written as

$$S_3 = \frac{1}{1 \cdot 2} + \frac{1}{2 \cdot 3} + \frac{1}{3 \cdot 4} = \frac{1}{1} - \frac{1}{2} + \frac{1}{2} - \frac{1}{3} + \frac{1}{3} - \frac{1}{4} = 1 - \frac{1}{4}.$$

All of the intermediate terms cancel out, leaving only the first and last terms. Thus $S_{10} = 1 - \frac{1}{11}$ and $S_n = 1 - \frac{1}{n+1}$.

(c) The limit of S_n as $n \to \infty$ is $\lim_{n \to \infty} \left(1 - \frac{1}{n+1}\right) = 1 - 0 = 1$. Thus the series $\displaystyle\sum_{k=1}^{\infty} \frac{1}{k(k+1)}$ converges to 1.

41. We have $a_n = S_n - S_{n-1}$. If $\sum a_n$ converges, then $S = \lim_{n \to \infty} S_n$ exists. Hence $\lim_{n \to \infty} S_{n-1}$ exists and is equal to S also. Thus

$$\lim_{n \to \infty} a_n = \lim_{n \to \infty} (S_n - S_{n-1}) = \lim_{n \to \infty} S_n - \lim_{n \to \infty} S_{n-1} = S - S = 0.$$

Strengthen Your Understanding

45. If the terms of a series do not approach zero, the series does not converge. But just because the terms approach zero does not mean the series converges. For example, $\sum(1/n)$ diverges even though the terms approach zero.

49. The series $\sum_{n=1}^{\infty} 1/n$ is an example. The terms of the series converge to zero, but the series is a p-series with $p \le 1$, and therefore the series diverges.

53. False. For example, if $a_n = 1/n$ and $b_n = -1/n$, then $|a_n + b_n| = 0$, so $\sum |a_n + b_n|$ converges. However $\sum |a_n|$ and $\sum |b_n|$ are the harmonic series, which diverge.

57. False. If $a_n = b_n = 1/n$, then $\sum a_n$ and $\sum b_n$ do not converge. However, $a_n b_n = 1/n^2$, so $\sum a_n b_n$ does converge.

Additional Problems (online only)

61. (a) Since for $x > 0$,

$$\int \frac{1}{x \ln x} dx = \ln(\ln x) + C$$

we have

$$\int_2^{\infty} \frac{1}{x \ln x} dx = \lim_{b \to \infty} \int_2^b \frac{1}{x \ln x} dx = \lim_{b \to \infty} (\ln(\ln b) - \ln(\ln 2)) = \infty.$$

The series diverges by the integral test.

(b) The terms in each group are decreasing so we can bound each group as follows:

$$\frac{1}{3 \ln 3} + \frac{1}{4 \ln 4} > \frac{1}{4 \ln 4} + \frac{1}{4 \ln 4} = \frac{1}{2 \ln 4}$$

and

$$\frac{1}{5 \ln 5} + \frac{1}{6 \ln 6} + \frac{1}{7 \ln 7} + \frac{1}{8 \ln 8} > 4 \frac{1}{8 \ln 8} = \frac{1}{2 \ln 8}.$$

Similarly, the group whose final term is $1/(2^n \ln(2^n))$ is greater than $1/(2 \ln(2^n)) = 1/(2(\ln 2)n)$. Thus

$$\sum_{n=2}^{2^N} \frac{1}{n \ln n} > \sum_{n=1}^{N} \frac{1}{2(\ln 2)n}.$$

The series on the right is the harmonic series multiplied by the constant $1/(2 \ln 2)$. Since the harmonic series diverges, $\sum 1/(n \ln n)$ diverges.

65. (a) The right-hand sum for $\int_0^N x^N dx$ with $\Delta x = 1$ is the sum $1^5 \cdot 1 + 2^5 \cdot 1 + 3^5 \cdot 1 + \cdots + N^5 \cdot 1 = S_N$. This sum is greater than the integral because the integrand x^5 is increasing on the interval $0 < x < N$. Since $\int_0^N x^5 dx = N^6/6$, we have $S_N > N^6/6$.

(b) The left-hand sum for $\int_1^{N+1} x^N dx$ with $\Delta x = 1$ is the sum $1^5 \cdot 1 + 2^5 \cdot 1 + 3^5 \cdot 1 + \cdots + N^5 \cdot 1 = S_N$. This sum is less than the integral because the integrand x^5 is increasing on the interval $1 < x < N+1$. Since $\int_1^{N+1} x^5 dx = ((N+1)^6 - 1)/6$, we have $S_N < ((N+1)^6 - 1)/6$.

(c) By parts (a) and (b) we have

$$\frac{N^6/6}{N^6/6} = 1 < \frac{S_N}{N^6/6} < \frac{((N+1)^6 - 1)/6}{N^6/6} = \left(1 + \frac{1}{N}\right)^6 - \frac{1}{N^6}.$$

Since both $\lim_{N\to\infty} 1 = 1$ and $\lim_{N\to\infty}\left(\left(1 + \frac{1}{N}\right)^6 - \frac{1}{N^6}\right) = 1$, we conclude that the limit in the middle also equals 1, $\lim_{N\to\infty} S_N/(N^6/6) = 1$.

Solutions for Section 9.4

Exercises

1. Let $a_n = 1/(n - 3)$, for $n \geq 4$. Since $n - 3 < n$, we have $1/(n-3) > 1/n$, so

$$a_n > \frac{1}{n}.$$

The harmonic series $\sum_{n=4}^{\infty} \frac{1}{n}$ diverges, so the comparison test tells us that the series $\sum_{n=4}^{\infty} \frac{1}{n-3}$ also diverges.

5. As n gets large, polynomials behave like the leading term, so for large n,

$$\frac{n^3 + 1}{n^4 + 2n^3 + 2n} \quad \text{behaves like} \quad \frac{n^3}{n^4} = \frac{1}{n}.$$

Since the series $\sum_{n=1}^{\infty} 1/n$ diverges, we predict that the given series will diverge.

9. Let $a_n = n^2/(n^4 + 1)$. Since $n^4 + 1 > n^4$, we have $\frac{1}{n^4 + 1} < \frac{1}{n^4}$, so

$$a_n = \frac{n^2}{n^4 + 1} < \frac{n^2}{n^4} = \frac{1}{n^2},$$

therefore

$$0 < a_n < \frac{1}{n^2}.$$

Since the p-series $\sum_{n=1}^{\infty} \frac{1}{n^2}$ converges, the comparison test tells us that the series $\sum_{n=1}^{\infty} \frac{n^2}{n^4 + 1}$ converges also.

13. We know that $|\sin n| < 1$, so

$$\frac{n \sin^2 n}{n^3 + 1} \leq \frac{n}{n^3 + 1} < \frac{n}{n^3} = \frac{1}{n^2}.$$

Since the p-series $\sum_{n=1}^{\infty} \frac{1}{n^2}$ converges, comparison gives that $\sum_{n=1}^{\infty} \frac{n \sin^2 n}{n^3 + 1}$ converges.

17. Since $a_n = 1/(ne^n)$, replacing n by $n + 1$ gives $a_{n+1} = 1/((n + 1)e^{n+1})$. Thus

$$\frac{|a_{n+1}|}{|a_n|} = \frac{\frac{1}{(n+1)e^{n+1}}}{\frac{1}{ne^n}} = \frac{ne^n}{(n+1)e^{n+1}} = \left(\frac{n}{n+1}\right)\frac{1}{e}.$$

Therefore

$$L = \lim_{n\to\infty} \frac{|a_{n+1}|}{|a_n|} = \frac{1}{e} < 1.$$

Since $L < 1$, the ratio test tells us that $\sum_{n=1}^{\infty} \frac{1}{ne^n}$ converges.

21. Since $a_n = 2^n/(n^3 + 1)$, replacing n by $n + 1$ gives $a_{n+1} = 2^{n+1}/((n + 1)^3 + 1)$. Thus

$$\frac{|a_{n+1}|}{|a_n|} = \frac{\dfrac{2^{n+1}}{(n + 1)^3 + 1}}{\dfrac{2^n}{n^3 + 1}} = \frac{2^{n+1}}{(n + 1)^3 + 1} \cdot \frac{n^3 + 1}{2^n} = 2\frac{n^3 + 1}{(n + 1)^3 + 1},$$

so

$$L = \lim_{n \to \infty} \frac{|a_{n+1}|}{|a_n|} = 2.$$

Since $L > 1$ the ratio test tells us that the series $\displaystyle\sum_{n=0}^{\infty} \frac{2^n}{n^3 + 1}$ diverges.

25. The nth term $a_n = 1/(n^4 - 7)$ behaves like $1/n^4$ for large n, so we take $b_n = 1/n^4$. We have

$$\lim_{n \to \infty} \frac{a_n}{b_n} = \lim_{n \to \infty} \frac{1/(n^4 - 7)}{1/n^4} = \lim_{n \to \infty} \frac{n^4}{n^4 - 7} = 1.$$

The limit comparison test applies with $c = 1$. The p-series $\displaystyle\sum_{n=1}^{\infty} \frac{1}{n^4}$ converges because $p = 4 > 1$. Therefore

$\displaystyle\sum_{n=1}^{\infty} \frac{1}{n^4 - 7}$ also converges.

29. The nth term $a_n = 1/(2\sqrt{n} + \sqrt{n + 2})$ behaves like $1/(3\sqrt{n})$ for large n, so we take $b_n = 1/(3\sqrt{n})$. We have

$$\lim_{n \to \infty} \frac{a_n}{b_n} = \lim_{n \to \infty} \frac{1/(2\sqrt{n} + \sqrt{n + 2})}{1/(3\sqrt{n})} = \lim_{n \to \infty} \frac{3\sqrt{n}}{2\sqrt{n} + \sqrt{n + 2}}$$

$$= \lim_{n \to \infty} \frac{3\sqrt{n}}{\sqrt{n}\left(2 + \sqrt{1 + 2/n}\right)}$$

$$= \lim_{n \to \infty} \frac{3}{2 + \sqrt{1 + 2/n}} = \frac{3}{2 + \sqrt{1 + 0}}$$

$$= 1.$$

The limit comparison test applies with $c = 1$. The series $\displaystyle\sum_{n=1}^{\infty} \frac{1}{3\sqrt{n}}$ diverges because it is a multiple of a p-series with

$p = 1/2 < 1$. Therefore $\displaystyle\sum_{n=1}^{\infty} \frac{1}{2\sqrt{n} + \sqrt{n + 2}}$ also diverges.

33. Even though the first term is negative, the terms alternate in sign, so it is an alternating series.

37. Let $a_n = 1/\sqrt{n}$. Then replacing n by $n + 1$ we have $a_{n+1} = 1/\sqrt{n + 1}$. Since $\sqrt{n + 1} > \sqrt{n}$, we have $\dfrac{1}{\sqrt{n + 1}} < \dfrac{1}{\sqrt{n}}$,

hence $a_{n+1} < a_n$. In addition, $\lim_{n \to \infty} a_n = 0$ so $\displaystyle\sum_{n=1}^{\infty} \frac{(-1)^{n-1}}{\sqrt{n}}$ converges by the alternating series test.

41. Let $a_n = 1/(n^2 + 2n + 1) = 1/(n + 1)^2$. Then replacing n by $n + 1$ gives $a_{n+1} = 1/(n + 2)^2$. Since $n + 2 > n + 1$, we have

$$\frac{1}{(n + 2)^2} < \frac{1}{(n + 1)^2}$$

so

$$0 < a_{n+1} < a_n.$$

We also have $\lim_{n \to \infty} a_n = 0$. Therefore, the alternating series test tells us that the series $\displaystyle\sum_{n=1}^{\infty} \frac{(-1)^{n-1}}{n^2 + 2n + 1}$ converges.

45. The series $\displaystyle\sum_{n=1}^{\infty} \frac{(-1)^n}{2n}$ converges by the alternating series test. However $\displaystyle\sum_{n=1}^{\infty} \frac{1}{2n}$ diverges because it is a multiple of the

harmonic series. Thus $\displaystyle\sum_{n=1}^{\infty} \frac{(-1)^n}{2n}$ is conditionally convergent.

49. We first check absolute convergence by deciding whether $\sum_{n=2}^{\infty} 1/(n \ln n)$ converges by using the integral test. Since

$$\int_2^{\infty} \frac{dx}{x \ln x} = \lim_{b \to \infty} \int_2^b \frac{dx}{x \ln x} = \lim_{b \to \infty} \ln(\ln(x)) \Big|_2^b = \lim_{b \to \infty} (\ln(\ln(b)) - \ln(\ln(2))),$$

and since this limit does not exist, $\displaystyle\sum_{n=2}^{\infty} \frac{1}{n \ln n}$ diverges.

We now check conditional convergence. The original series is alternating so we check whether $a_{n+1} < a_n$. Consider $a_n = f(n)$, where $f(x) = 1/(x \ln x)$. Since

$$\frac{d}{dx}\left(\frac{1}{x \ln x}\right) = \frac{-1}{x^2 \ln x}\left(1 + \frac{1}{\ln x}\right)$$

is negative for $x > 1$, we know that a_n is decreasing for $n \geq 2$. Thus, for $n \geq 2$

$$a_{n+1} = \frac{1}{(n+1)\ln(n+1)} < \frac{1}{n \ln n} = a_n.$$

Since $1/(n \ln n) \to 0$ as $n \to \infty$, we see that $\displaystyle\sum_{n=2}^{\infty} \frac{(-1)^{n-1}}{n \ln n}$ is conditionally convergent.

Problems

53. The comparison test requires that $a_n = (-1)^n/n^2$ be positive. It is not.

57. The sequence $a_n = n$ does not satisfy either $a_{n+1} < a_n$ or $\lim_{n \to \infty} a_n = 0$.

61. The partial sums are $S_1 = 1$, $S_2 = -1$, $S_3 = 2$, $S_{10} = -5$, $S_{11} = 6$, $S_{100} = -50$, $S_{101} = 51$, $S_{1000} = -500$, $S_{1001} = 501$, which appear to be oscillating further and further from 0. This series does not converge.

65. We take $a_n = \dfrac{1}{4n+3}$. Because a_n is proportional to $1/n$ as $n \to \infty$, we take $b_n = 1/n$. Since

$$\lim_{n \to \infty} \frac{a_n}{b_n} = \lim_{n \to \infty} \frac{1}{4n+3} \cdot \frac{n}{1} = \lim_{n \to \infty} \frac{n}{4n+3} = \frac{1}{4},$$

the limit comparison test applies with $c = 1/4$. Since $\sum_{n=1}^{\infty} 1/n$ diverges, the limit comparison test shows that $\displaystyle\sum_{n=1}^{\infty} \frac{1}{4n+3}$ also diverges.

69. As $n \to \infty$, we see that

$$\frac{n+2}{n^2-1} \to \frac{n}{n^2} = \frac{1}{n}.$$

Since $\sum(1/n)$ diverges, we expect our series to have the same behavior.

More precisely, for all $n \geq 2$, we have

$$0 \leq \frac{1}{n} = \frac{n}{n^2} \leq \frac{n+2}{n^2-1},$$

so $\displaystyle\sum_{n=2}^{\infty} \frac{n+2}{n^2-1}$ diverges by comparison with the divergent series $\displaystyle\sum \frac{1}{n}$.

73. First, observe that

$$\frac{1+3^n}{4^n} = \frac{1}{4^n} + \frac{3^n}{4^n} = \left(\frac{1}{4}\right)^n + \left(\frac{3}{4}\right)^n.$$

Since $\sum_{n=1}^{\infty}(1/4)^n$ and $\sum_{n=1}^{\infty}(3/4)^n$ are geometric series with common ratios of $1/4$ and $3/4$, respectively, both of these series converge. It therefore follows from part 1 of Theorem 9.2 that $\sum_{n=1}^{\infty}(1+3^n)/(4^n)$ converges.

77. The first few terms of the series may be written

$$1 + e^{-1} + e^{-2} + e^{-3} + \cdots;$$

this is a geometric series with $a = 1$ and $x = e^{-1} = 1/e$. Since $|x| < 1$, the geometric series converges to $S = \dfrac{1}{1-x} = \dfrac{1}{1-e^{-1}} = \dfrac{e}{e-1}$.

81. Since the exponential, 2^n, grows faster than the power, n^2, the terms are growing in size. Thus, $\lim_{n\to\infty} a_n \neq 0$. We conclude that this series diverges.

85. For $n \geq 1$, we know that $n + 1 \geq n$ and $3n^2 - 2 \leq 3n^2$, so

$$\frac{n+1}{3n^2 - 2} \geq \frac{n}{3n^2} = \frac{1}{3n} \geq 0.$$

Thus, every term in the series $\sum_{n=1}^{\infty} \frac{n+1}{3n^2 - 2}$ is greater than or equal to the corresponding term in $\sum_{n=1}^{\infty} 1/(3n)$. Since $\sum_{n=1}^{\infty} 1/n$

diverges as a p-series, the series $\sum_{n=1}^{\infty} 1/(3n)$ also diverges, so it follows from the comparison test that $\sum_{n=1}^{\infty} \frac{n+1}{3n^2 - 2}$ diverges.

89. Since $0 \leq |\sin n| \leq 1$ for all n, we may be able to compare with $1/n^2$. We have $0 \leq |\sin n/n^2| \leq 1/n^2$ for all n. So $\sum |\sin n/n^2|$ converges by comparison with the convergent series $\sum(1/n^2)$. Therefore $\sum(\sin n/n^2)$ also converges, since absolute convergence implies convergence by Theorem 9.6.

93. (a) Since

$$\lim_{n\to\infty} \frac{e^n}{1 + e^{2n}} = \lim_{n\to\infty} \frac{e^n/e^{2n}}{(1/e^{2n}) + (e^{2n}/e^{2n})} = \lim_{n\to\infty} \frac{e^{-n}}{e^{-2n} + 1} = 0,$$

the sequence $\left\{ \frac{e^n}{1 + e^{2n}} \right\}_{n=1}^{\infty}$ converges to 0.

(b) For $n \geq 1$, we have $1 + e^{2n} \geq 1$, so

$$\frac{e^n}{1 + e^{2n}} \leq \frac{e^n}{e^{2n}} = e^{-n}.$$

Since $\sum_{n=1}^{\infty} e^{-n}$ is a geometric series with a common ratio of $1/e < 1$, it converges. It therefore follows from the comparison test that $\sum_{n=1}^{\infty} \frac{e^n}{1 + e^{2n}}$ converges.

97. (a) Since

$$\{a_n\}_{n=1}^{\infty} = 1, 3, 1, 3, 1, 3, \ldots,$$

we see that $\lim_{n\to\infty} a_n$ does not exist. Therefore, $\{2 + (-1)^n\}_{n=1}^{\infty}$ diverges.

(b) From part (a), we know that $\lim_{n\to\infty} a_n$ does not exist. Therefore, by Property 3 of Theorem 9.2, the series $\sum_{n=1}^{\infty} 2 + (-1)^n$ diverges.

101. The series $\sum_{n=2}^{\infty} \frac{(-1)^{n+1}}{\ln n}$ converges by the alternating series test. However, since $\ln n \leq n$ for $n \geq 2$, we have $1/\ln n \geq 1/n$, so

the series $\sum 1/\ln n$ diverges by comparison with the harmonic series, $\sum 1/n$. Thus, $\sum_{n=1}^{\infty} \frac{(-1)^{n+1}}{\ln n}$ is conditionally convergent

which means its sum can be affected by reordering the terms.

105. For $a > 0$, the terms of the series are positive and eventually decreasing. We use the integral test and calculate the corresponding improper integral:

$$\int_1^{\infty} \frac{\ln x}{x^a} \, dx = \lim_{b\to\infty} \int_1^b \frac{\ln x}{x^a} \, dx.$$

For $a \neq 1$, use integration by parts with $u = \ln x$ and $v' = x^{-a}$:

$$\int \frac{\ln x}{x^a} \, dx = \frac{x^{-a+1}}{-a+1} \ln x - \int \frac{x^{-a}}{-a+1} \, dx = \frac{x^{-a+1}}{-a+1} \ln x - \frac{x^{-a+1}}{(-a+1)^2} = \frac{x^{-a+1}}{-a+1} \left(\ln x - \frac{1}{-a+1} \right).$$

Thus,

$$\lim_{b\to\infty} \int_1^b \frac{\ln x}{x^a} \, dx = \lim_{b\to\infty} \left[\frac{b^{-a+1}}{-a+1} \left(\ln b - \frac{1}{-a+1} \right) \right] + \frac{1}{(-a+1)^2} = \frac{1}{-a+1} \lim_{b\to\infty} \frac{\ln b - 1/(-a+1)}{b^{a-1}} + \frac{1}{(-a+1)^2}.$$

Use l'Hopital's Rule to obtain

$$\lim_{b\to\infty} \left(\frac{\ln b - 1/(-a+1)}{b^{a-1}} \right) = \lim_{b\to\infty} \left(\frac{1/b}{(a-1)b^{a-2}} \right) = \lim_{b\to\infty} \frac{1}{(a-1)b^{a-1}}.$$

This limit exists for $a - 1 > 0$ and does not exist for $a - 1 < 0$. Thus the series converges for $a > 1$ and diverges for $0 < a < 1$.

For $a = 1$, the series $\sum_{n=1}^{\infty} \dfrac{\ln n}{n}$ diverges because $\int_1^{\infty} \dfrac{\ln x}{x}\, dx = \lim_{b \to \infty} \left. \dfrac{(\ln x)^2}{2} \right|_1^b$, and this limit diverges. For $a \le 0$, $\lim_{n \to \infty} \dfrac{\ln n}{n^a}$

does not exist, so the series diverges by Property 3 of Theorem 9.2. Thus $\sum_{n=1}^{\infty} \dfrac{\ln n}{n}$ converges for $a > 1$ and diverges for $a \le 1$.

109. The nth partial sum of the series is given by

$$S_n = \frac{1}{2} - \frac{1}{24} + \frac{1}{720} - \cdots + \frac{(-1)^{n-1}}{(2n)!},$$

so the absolute value of the first term omitted is $1/(2n+2)!$. By Theorem 9.9, we know that the value, S, of the sum differs from S_n by less than $1/(2n + 2)!$. Thus, we want to choose n large enough so that $1/(2n + 2)! \le 0.01$. Substituting $n = 2$ into the expression $1/(2n+2)!$ yields $1/720$ which is less than 0.01. We therefore take 2 or more terms in our partial sum.

Strengthen Your Understanding

113. The series is not alternating since $(-1)^{2n} = 1$ for all $n > 0$, so we cannot use the alternating series test. The series is the same as $\sum_{n=1}^{\infty} 1/n^2$ which converges by the p-test with $p = 2$.

117. The following series is alternating:

$$\sum a_n = \sum (-1)^n n,$$

but since the terms do not tend to 0 as $n \to \infty$, the series does not converge.

121. True. This is one of the statements of the comparison test.

125. True, since if we write out the terms of the series, using the fact that $\cos(2\pi n) = 1$ for all n, we have

$$(-1)^0 \cos 0 + (-1)^1 \cos(2\pi) + (-1)^2 \cos(4\pi) + (-1)^3 \cos(6\pi) + \cdots$$
$$= 1 \cdot 1 - 1 \cdot 1 + 1 \cdot 1 - 1 \cdot 1 + \cdots$$
$$= 1 - 1 + 1 - 1 + \cdots.$$

This is an alternating series.

129. This statement is false. The statement is true if the series converges by the alternating series test, but not in general. Consider, for example, the alternating series

$$S = 10 - 0.01 + 0.8 - 0.7 - 0 + 0 - 0 + \cdots.$$

Since the later terms are all 0, we can find the sum exactly:

$$S = 10.69.$$

If we approximated the sum by the first term, $S_1 = 10$, the magnitude of the first term omitted would be 0.01. Thus, if the statement in this problem were true, we would say that the true value of the sum lay between $10 + 0.01 = 10.01$ and $10 - 0.01 = 9.99$ which it does not.

133. False. The alternating harmonic series $\sum \dfrac{(-1)^n}{n}$ is conditionally convergent because it converges by the Alternating Series test, but the harmonic series $\sum \left| \dfrac{(-1)^n}{n} \right| = \sum \dfrac{1}{n}$ is divergent. The alternating harmonic series is not absolutely convergent.

Additional Problems (online only)

137. (a) Show that the sum of positive numbers in each group of fractions is at least $1/4$.
(b) Show that the sum of each group of fractions is at least $1/12$.
(c) Explain why this shows that this particular rearrangement of the alternating harmonic series diverges.
(a) For the first two groups we have $1 > 1/4$ and $1/3 > 1/4$. For the remaining groups, observe that

$$\frac{1}{5} + \frac{1}{7} > \frac{1}{8} + \frac{1}{8} = \frac{1}{4}$$

$$\frac{1}{9} + \frac{1}{11} + \frac{1}{13} + \frac{1}{15} > 4 \cdot \frac{1}{16} = \frac{1}{4}$$

$$\frac{1}{17} + \cdots + \frac{1}{31} > 8 \cdot \frac{1}{32} = \frac{1}{4}$$

$$\frac{1}{33} + \cdots + \frac{1}{63} > 16 \cdot \frac{1}{64} = \frac{1}{4}.$$

Continuing in this fashion, we see the sum of positive numbers in each group of fractions is at least $1/4$.

(b) Using our solution to part (a), we have:

$$1 - \frac{1}{2} = \frac{1}{2} > \frac{1}{12}$$

$$\frac{1}{3} - \frac{1}{4} = \frac{1}{12}$$

$$\frac{1}{5} + \frac{1}{7} - \frac{1}{6} > \frac{1}{4} - \frac{1}{6} = \frac{1}{12}$$

$$\frac{1}{9} + \cdots + \frac{1}{15} - \frac{1}{8} > \frac{1}{4} - \frac{1}{8} > \frac{1}{12}$$

$$\frac{1}{17} + \cdots + \frac{1}{31} - \frac{1}{10} > \frac{1}{4} - \frac{1}{10} > \frac{1}{12}.$$

Since in each consecutive group we subtract off a smaller number from $1/4$ as we did in the previous group, we see that the sum of the fractions in each grouping is no less than $1/12$.

(c) Since the sum of the first n groups is at least $n/12$, it follows that the partial sums are not bounded. Thus, after this particular rearrangement, the alternating harmonic series diverges.

141. Since $\lim_{n \to \infty} a_n/b_n = \infty$, for large enough n we have $a_n/b_n > 1$ and thus $a_n > b_n$. By the comparison test applied to $\sum a_n$ and $\sum b_n$, the series $\sum a_n$ diverges.

Solutions for Section 9.5

Exercises

1. Yes.

5. The general term can be written as $\dfrac{1 \cdot 3 \cdot 5 \cdots (2n - 1)}{2^n \cdot n!} x^n$ for $n \geq 1$. Other answers are possible.

9. The general term can be written as $\dfrac{(x - a)^n}{2^{n-1} \cdot n!}$ for $n \geq 1$. Other answers are possible.

13. Since $C_n = n^3$, replacing n by $n + 1$ gives $C_{n+1} = (n + 1)^3$. Using the ratio test, with $a_n = n^3 x^n$, we have

$$\frac{|a_{n+1}|}{|a_n|} = |x| \frac{|C_{n+1}|}{|C_n|} = |x| \frac{(n + 1)^3}{n^3} = |x| \left(\frac{n + 1}{n}\right)^3.$$

We have

$$\lim_{n \to \infty} \frac{|a_{n+1}|}{|a_n|} = |x|.$$

Thus the radius of convergence is $R = 1$.

17. We use the ratio test:

$$\left|\frac{a_{n+1}}{a_n}\right| = \left|\frac{(x - 3)^{n+1}}{(n + 1)2^{n+1}} \cdot \frac{n2^n}{(x - 3)^n}\right| = |x - 3| \frac{n}{2(n + 1)}.$$

Therefore, we have

$$\lim_{n \to \infty} \left|\frac{a_{n+1}}{a_n}\right| = |x - 3| \lim_{n \to \infty} \frac{n}{2(n + 1)} = \frac{|x - 3|}{2}.$$

Thus by the ratio test, the series converges if $|x - 3|/2 < 1$, that is $|x - 3| < 2$. The radius of convergence is $R = 2$.

21. Here $C_n = (2n)!/(n!)^2$. We have:

$$\left| \frac{a_{n+1}}{a_n} \right| = \left| \frac{(2(n+1))!/((n+1)!)^2 x^{n+1}}{(2n)!/(n!)^2 x^n} \right| = \frac{(2(n+1))!}{(2n)!} \cdot \frac{(n!)^2}{((n+1)!)^2} |x|$$

$$= \frac{(2n+2)(2n+1)|x|}{(n+1)^2} \to 4|x| \text{ as } n \to \infty.$$

Thus, the radius of convergence is $R = 1/4$.

Problems

25. (a) The general term of the series is x^n/n if n is odd and $-x^n/n$ if n is even, so $C_n = (-1)^{n-1}/n$, and we can use the ratio test. We have

$$\lim_{n \to \infty} \frac{|a_{n+1}|}{|a_n|} = |x| \lim_{n \to \infty} \frac{|(-1)^n/(n+1)|}{|(-1)^{n-1}/n|} = |x| \lim_{n \to \infty} \frac{n}{n+1} = |x|.$$

Therefore the radius of convergence is $R = 1$. This tells us that the power series converges for $|x| < 1$ and does not converge for $|x| > 1$. Notice that the radius of convergence does not tell us what happens at the endpoints, $x = \pm 1$.

(b) The endpoints of the interval of convergence are $x = \pm 1$. At $x = 1$, we have the series

$$1 - \frac{1}{2} + \frac{1}{3} - \frac{1}{4} + \cdots + \frac{(-1)^{n-1}}{n} + \cdots$$

This is an alternating series with $a_n = 1/n$, so by the alternating series test, it converges. At $x = -1$, we have the series

$$-1 - \frac{1}{2} - \frac{1}{3} - \frac{1}{4} - \cdots - \frac{1}{n} - \cdots$$

This is the negative of the harmonic series, so it does not converge. Therefore the right endpoint is included, and the left endpoint is not included in the interval of convergence, which is $-1 < x \le 1$.

29. We use the ratio test:

$$\left| \frac{a_{n+1}}{a_n} \right| = \left| \frac{(n+1)^2 x^{2(n+1)}}{2^{2(n+1)}} \cdot \frac{2^{2n}}{n^2 x^{2n}} \right| = \left(\frac{n+1}{n} \right)^2 \cdot \frac{x^2}{4}.$$

Since $(n+1)/n \to 1$ as $n \to \infty$, we have

$$\lim_{n \to \infty} \left| \frac{a_{n+1}}{a_n} \right| = \frac{x^2}{4}.$$

We have $x^2/4 < 1$ when $|x| < 2$. The radius of convergence is 2 and the series converges for $-2 < x < 2$.

We check the endpoints. For $x = -2$, we have

$$\sum_{n=1}^{\infty} \frac{n^2 x^{2n}}{2^{2n}} = \sum_{n=1}^{\infty} \frac{n^2 (-2)^{2n}}{2^{2n}} = \sum_{n=1}^{\infty} n^2,$$

which diverges. Similarly, for $x = 2$, we have

$$\sum_{n=1}^{\infty} \frac{n^2 x^{2n}}{2^{2n}} = \sum_{n=1}^{\infty} \frac{n^2 2^{2n}}{2^{2n}} = \sum_{n=1}^{\infty} n^2,$$

which diverges. The series diverges at both endpoints, so the interval of convergence is $-2 < x < 2$.

33. We use the ratio test:

$$\left| \frac{a_{n+1}}{a_n} \right| = \left| \frac{(5x)^{n+1}/\sqrt{n+1}}{(5x)^n/\sqrt{n}} \right| = 5|x| \sqrt{\frac{n}{n+1}}.$$

Since $\sqrt{\dfrac{n}{n+1}} \to 1$ as $n \to \infty$, we have

$$\lim_{n \to \infty} \left| \frac{a_{n+1}}{a_n} \right| = 5|x|.$$

We have $5|x| < 1$ when $|x| < 1/5$. The radius of convergence is $1/5$ and the series converges for $-1/5 < x < 1/5$.

We check the endpoints. For $x = -1/5$, we have

$$\sum_{n=1}^{\infty} \frac{(5x)^n}{\sqrt{n}} = \sum_{n=1}^{\infty} \frac{(-1)^n}{\sqrt{n}}.$$

This is an alternating series. Since we have $0 < \dfrac{1}{\sqrt{n+1}} < \dfrac{1}{\sqrt{n}}$ for all $n \geq 1$ and $\lim\limits_{n\to\infty} \dfrac{1}{\sqrt{n}} = 0$, by the alternating series test, the series converges at the endpoint $x = -1/5$. For $x = 1/5$, we have

$$\sum_{n=1}^{\infty} \frac{(5x)^n}{\sqrt{n}} = \sum_{n=1}^{\infty} \frac{1}{\sqrt{n}}.$$

This is a p-series with $p = 1/2$ and it diverges. Therefore, the interval of convergence is $-1/5 \leq x < 1/5$.

37. The expression $3/(1 - z/2)$ has the form $a/(1 - x)$, which is the sum of a geometric series with $a = 3$, $x = z/2$. The power series is

$$3 + 3(z/2) + 3(z/2)^2 + 3(z/2)^3 + \cdots = \sum_{n=0}^{\infty} 3(z/2)^n.$$

This series converges for $|z/2| < 1$, that is, for $-2 < z < 2$.

41. The radius of convergence of the series, R, is at least 4 but no larger than 7.

(a) False. Since $10 > R$ the series diverges.
(b) True. Since $3 < R$ the series converges.
(c) False. Since $1 < R$ the series converges.
(d) Not possible to determine since the radius of convergence may be more or less than 6.

45. Taking the hint, we have:

$$\pi = 3 + 0.1 + 0.04 + 0.001 + \cdots$$
$$= 3 + 1(0.1) + 4(0.1)^2 + 1(0.1)^3 + \cdots.$$

This means $k = 0.1$ and $C_0 = 3, C_1 = 1, C_2 = 4, C_3 = 1$.

49. (a) Since only odd powers are involved in the series for $g(x)$,

$$g(x) = x - \frac{x^3}{3!} + \frac{x^5}{5!} - \frac{x^7}{7!} + \cdots,$$

we see that $g(x)$ is odd. Substituting $x = 0$ gives $g(0) = 0$.

(b) Differentiating term by term gives

$$g'(x) = 1 - 3\frac{x^2}{3!} + 5\frac{x^4}{5!} - 7\frac{x^6}{7!} + \cdots$$
$$= 1 - \frac{x^2}{2!} + \frac{x^4}{4!} - \frac{x^6}{6!} + \cdots.$$

$$g''(x) = 0 - 2\frac{x}{2!} + 4\frac{x^3}{4!} - 6\frac{x^5}{6!} + \cdots$$
$$= -x + \frac{x^3}{3!} - \frac{x^5}{5!} + \cdots.$$

So we see $g''(x) = -g(x)$.

(c) We guess $g(x) = \sin x$ since then $g'(x) = \cos x$ and $g''(x) = -\sin x = g(x)$. We check $g(0) = 0 = \sin 0$ and $g'(0) = 1 = \cos 0$.

Strengthen Your Understanding

53. In order to get a power series that does not converge at $x = 0$, we need to construct a power series about a point other than $x = 0$. We'll use $a = 2$, and the series

$$\sum_{n=0}^{\infty} \frac{(x-2)^n}{n}.$$

If we let $x = 0$ and use the ratio test to determine the convergence or divergence of the series, we get

$$\lim_{n\to\infty} \left| \frac{(-2)^{n+1}/(n+1)}{(-2)^n/n} \right| = \lim_{n\to\infty} \left| \frac{-2n}{n+1} \right| = 2.$$

Since the limit is greater than 1, the series diverges at $x = 0$ by the ratio test.

57. False. Writing out terms, we have

$$(x-1) + (x-2)^2 + (x-3)^3 + \cdots.$$

A power series is a sum of powers of $(x - a)$ for constant a. In this case, the value of a changes from term to term, so it is not a power series.

61. True. The radius of convergence, R, is given by $\lim_{n\to\infty} |C_{n+1}|/|C_n| = 1/R$, if this limit exists, and since these series have the same coefficients, C_n, the radii of convergence are the same.

65. True. Since the power series converges at $x = 10$, the radius of convergence is at least 10. Thus, $x = -9$ must be within the interval of convergence.

69. True. The interval of convergence is centered on $x = a$, so $a = (-11 + 1)/2 = -5$.

Solutions for Chapter 9 Review

Exercises

1. $3 + \dfrac{3}{2} + \dfrac{3}{4} + \dfrac{3}{8} \cdots + \dfrac{3}{2^{10}} = 3\left(1 + \dfrac{1}{2} + \cdots + \dfrac{1}{2^{10}}\right) = \dfrac{3\left(1 - \frac{1}{2^{11}}\right)}{1 - \frac{1}{2}} = \dfrac{3\left(2^{11} - 1\right)}{2^{10}}$

5. If $b = 1$, then the sum is 6. If $b \neq 1$, we use the formula for the sum of a finite geometric series. We can write the series as

$$b^5 + b^5 \cdot b + b^5 \cdot b^2 + b^5 \cdot b^3 + b^5 \cdot b^4 + b^5 \cdot b^5.$$

This is a six-term geometric series ($n = 6$) with initial term $a = b^5$ and constant ratio $x = b$:

$$\text{Sum} = \frac{a(1 - x^n)}{1 - x} = \frac{b^5(1 - b^6)}{1 - b}.$$

9. We have

$$S_1 = 36.$$
$$S_2 = 36 + 36\left(\frac{1}{3}\right) = 48.$$
$$S_3 = 36 + 36\left(\frac{1}{3}\right) + 36\left(\frac{1}{3}\right)^2 = 52.$$
$$S_4 = 36 + 36\left(\frac{1}{3}\right) + 36\left(\frac{1}{3}\right)^2 + 36\left(\frac{1}{3}\right)^3 = 53.333.$$

Here we have $a = 36$ and $x = 1/3$, so

$$S_n = \frac{a(1 - x^n)}{1 - x} = \frac{36(1 - (1/3)^n)}{1 - 1/3}.$$

As $n \to \infty$, we see that $S_n \to 36/(2/3) = 54$.

13. As n increases, the term $4n$ is much larger than 3 and $7n$ is much larger than 5. Thus dividing the numerator and denominator by n and using the fact that $\lim_{n\to\infty} 1/n = 0$, we have

$$\lim_{n\to\infty} \frac{3 + 4n}{5 + 7n} = \lim_{n\to\infty} \frac{(3/n) + 4}{(5/n) + 7} = \frac{4}{7}.$$

Thus, the sequence converges to $4/7$.

17. We use the integral test with $f(x) = 1/x^3$ to determine whether this series converges or diverges. To do so we determine whether the corresponding improper integral $\int_1^\infty \dfrac{1}{x^3} dx$ converges or diverges:

$$\int_1^\infty \frac{1}{x^3} dx = \lim_{b\to\infty} \int_1^b \frac{1}{x^3} dx = \lim_{b\to\infty} \left.\frac{-1}{2x^2}\right|_1^b = \lim_{b\to\infty} \left(\frac{-1}{2b^2} + \frac{1}{2}\right) = \frac{1}{2}.$$

Since the integral $\int_1^\infty \dfrac{1}{x^3} dx$ converges, we conclude from the integral test that the series $\sum_{n=1}^\infty \dfrac{1}{n^3}$ converges.

21. Since $a_n = 1/(2^n n!)$, replacing n by $n+1$ gives $a_{n+1} = 1/(2^{n+1}(n+1)!)$. Thus

$$\frac{|a_{n+1}|}{|a_n|} = \frac{\dfrac{1}{2^{n+1}(n+1)!}}{\dfrac{1}{2^n n!}} = \frac{2^n n!}{2^{n+1}(n+1)!} = \frac{1}{2(n+1)},$$

so

$$L = \lim_{n \to \infty} \frac{|a_{n+1}|}{|a_n|} = \lim_{n \to \infty} \frac{1}{2n+2} = 0.$$

Since $L < 1$, the ratio test tells us that $\displaystyle\sum_{n=1}^{\infty} \frac{1}{2^n n!}$ converges.

25. Let $a_n = 1/\sqrt{n^2+1}$. Then replacing n by $n+1$ we have $a_{n+1} = 1/\sqrt{(n+1)^2+1}$. Since $\sqrt{(n+1)^2+1} > \sqrt{n^2+1}$, we have

$$\frac{1}{\sqrt{(n+1)^2+1}} < \frac{1}{\sqrt{n^2+1}},$$

so

$$0 < a_{n+1} < a_n.$$

In addition, $\lim_{n \to \infty} a_n = 0$ so $\displaystyle\sum_{n=0}^{\infty} \frac{(-1)^n}{\sqrt{n^2+1}}$ converges by the alternating series test.

29. Since

$$\lim_{n \to \infty} a_n = \lim_{n \to \infty} \frac{1}{\arctan n} = \frac{2}{\pi} \neq 0$$

we know that $\displaystyle\sum \frac{(-1)^{n-1}}{\arctan n}$ diverges by Property 3 of Theorem 9.2.

33. The nth term $a_n = (n^3 - 2n^2 + n + 1)/(n^5 - 2)$ behaves like $n^3/n^5 = 1/n^2$ for large n, so we take $b_n = 1/n^2$. We have

$$\lim_{n \to \infty} \frac{a_n}{b_n} = \lim_{n \to \infty} \frac{(n^3 - 2n^2 + n + 1)/(n^5 - 2)}{1/n^2} = \lim_{n \to \infty} \frac{n^5 - 2n^4 + n^3 + n^2}{n^5 - 2} = 1.$$

The limit comparison test applies with $c = 1$. The p-series $\sum 1/n^2$ converges because $p = 2 > 1$. Therefore the series $\sum (n^3 - 2n^2 + n + 1)/(n^5 - 2)$ also converges.

37. This is a p-series with $p > 1$, so it converges.

41. We use the integral test to determine whether this series converges or diverges. To do so we determine whether the corresponding improper integral $\displaystyle\int_1^{\infty} \frac{x^2}{x^3+1}\,dx$ converges or diverges:

$$\int_1^{\infty} \frac{x^2}{x^3+1}\,dx = \lim_{b \to \infty} \int_1^b \frac{x^2}{x^3+1}\,dx = \lim_{b \to \infty} \frac{1}{3} \ln|x^3 + 1| \Big|_1^b = \lim_{b \to \infty} \left(\frac{1}{3}\ln(b^3 + 1) - \frac{1}{3}\ln 2 \right).$$

Since the limit does not exist, the integral $\displaystyle\int_1^{\infty} \frac{x^2}{x^3+1}\,dx$ diverges an so we conclude from the integral test that the series $\displaystyle\sum_{n=1}^{\infty} \frac{n^2}{n^3+1}$ diverges. The limit comparison test with $b_n = 1/n$ can also be used.

45. We use the ratio test:

$$\left| \frac{a_{n+1}}{a_n} \right| = \left| \frac{3^{2n+2}}{(2n+2)!} \cdot \frac{(2n)!}{3^{2n}} \right| = 3^2 \frac{1}{(2n+2)(2n+1)}.$$

Therefore we have

$$\lim_{n \to \infty} \left| \frac{a_{n+1}}{a_n} \right| = 3^2 \lim_{n \to \infty} \frac{1}{(2n+2)(2n+1)} = 0.$$

Thus by the ratio test, the series converges.

49. The series can be written as

$$\sum_{n=0}^{\infty} \frac{2 + 3^n}{5^n} = \sum_{n=0}^{\infty} \left(\frac{2}{5^n} + \frac{3^n}{5^n} \right) = \sum_{n=0}^{\infty} \left(2\left(\frac{1}{5}\right)^n + \left(\frac{3}{5}\right)^n \right).$$

The series $\sum_{n=0}^{\infty} \left(\frac{1}{5}\right)^n$ is a geometric series which converges because $|\frac{1}{5}| < 1$. Likewise, the geometric series $\sum_{n=0}^{\infty} \left(\frac{3}{5}\right)^n$ converges because $|\frac{3}{5}| < 1$. Since both series converge, Property 1 of Theorem 9.2 tells us that the series $\sum_{n=0}^{\infty} \frac{2+3^n}{5^n}$ also converges.

53. The nth term $a_n = 1/(n^3 - 3)$ behaves like $1/n^3$ for large n, so we take $b_n = 1/n^3$. We have

$$\lim_{n\to\infty} \frac{a_n}{b_n} = \lim_{n\to\infty} \frac{1/(n^3-3)}{1/n^3} = \lim_{n\to\infty} \frac{n^3}{n^3-3} = 1.$$

The limit comparison test applies with $c = 1$. The p-series $\sum 1/n^3$ converges because $p = 3 > 1$. Therefore $\sum 1/(n^3 - 3)$ also converges.

57. Since $\ln n$ grows much more slowly than n, we suspect that $(\ln n)^2 < n$ for large n. This can be confirmed with L'Hopital's rule.

$$\lim_{n\to\infty} \frac{(\ln n)^2}{n} = \lim_{n\to\infty} \frac{2(\ln n)/n}{1} = \lim_{n\to\infty} \frac{2(\ln n)}{n} = 0.$$

Therefore, for large n, we have $(\ln n)^2/n < 1$, and hence for large n,

$$\frac{1}{n} < \frac{1}{(\ln n)^2}.$$

Thus $\sum_{n=2}^{\infty} 1/(\ln n)^2$ diverges by comparison with the divergent harmonic series $\sum 1/n$.

61. Here the coefficient of the nth term is $C_n = n/(2n + 1)$. Now we have

$$\left|\frac{a_{n+1}}{a_n}\right| = \left|\frac{((n+1)/(2n+3))x^{n+1}}{(n/(2n+1))x^n}\right| = \frac{(n+1)(2n+1)}{n(2n+3)}|x| \to |x| \text{ as } n \to \infty.$$

Thus, by the ratio test, the radius of convergence is $R = 1$.

65. We use the ratio test to find the radius of convergence:

$$\left|\frac{a_{n+1}}{a_n}\right| = \left|\frac{x^{n+1}}{(n+1)!} \cdot \frac{n!}{x^n}\right| = \left|\frac{x}{n+1}\right|.$$

Since $\lim_{n\to\infty} |x|/(n+1) = 0$ for all x, the radius of convergence is $R = \infty$. There are no endpoints to check. The interval of convergence is all real numbers $-\infty < x < \infty$.

Problems

69. We have

$$\begin{aligned}
a_2 &= a_1 + 2 \cdot 2 = 5 + 4 &= 9 \qquad \text{because } a_1 = 5 \\
a_3 &= a_2 + 2 \cdot 3 = 9 + 6 &= 15 \\
a_4 &= a_3 + 2 \cdot 4 = 15 + 8 &= 23.
\end{aligned}$$

73. (a) Using an argument similar to Example 7 in Section 9.5, we take

$$a_n = (-1)^n \frac{t^{2n}}{(2n)!},$$

so, replacing n by $n + 1$,

$$a_{n+1} = (-1)^{n+1} \frac{t^{2(n+1)}}{(2(n+1))!} = (-1)^{n+1} \frac{t^{2n+2}}{(2n+2)!}.$$

Thus,

$$\frac{|a_{n+1}|}{|a_n|} = \frac{|(-1)^{n+1} t^{2n+2}/(2n+2)!|}{|(-1)^n t^{2n}/(2n)!|} = \frac{t^2}{(2n+2)(2n+1)},$$

so

$$\lim_{n\to\infty} \frac{|a_{n+1}|}{|a_n|} = \lim_{n\to\infty} \frac{t^2}{(2n+2)(2n+1)} = 0.$$

The radius of convergence is therefore ∞, so the series converges for all t. Therefore the domain of h is all real numbers.

(b) Since h involves only even powers,

$$h(t) = 1 - \frac{t^2}{2!} + \frac{t^4}{4!} - \frac{t^6}{6!} + \cdots,$$

h is an even function.

(c) Differentiating term by term, we have

$$h'(t) = 0 - 2\frac{t}{2!} + 4\frac{t^3}{4!} - 6\frac{t^6}{6!} + \cdots$$

$$= -t + \frac{t^3}{3!} - \frac{t^5}{5!} + \cdots.$$

$$h''(t) = -1 + 3\frac{t^2}{3!} - 5\frac{t^4}{5!} + \cdots$$

$$= -1 + \frac{t^2}{2!} - \frac{t^4}{4!} + \cdots.$$

So we see $h''(t) = -h(t)$.

77.

$$\text{Present value of first coupon} = \frac{50}{1.06}$$

$$\text{Present value of second coupon} = \frac{50}{(1.06)^2}, \text{ etc.}$$

$$\text{Total present value} = \underbrace{\frac{50}{1.06} + \frac{50}{(1.06)^2} + \cdots + \frac{50}{(1.06)^{10}}}_{\text{coupons}} + \underbrace{\frac{1000}{(1.06)^{10}}}_{\text{principal}}$$

$$= \frac{50}{1.06}\left(1 + \frac{1}{1.06} + \cdots + \frac{1}{(1.06)^9}\right) + \frac{1000}{(1.06)^{10}}$$

$$= \frac{50}{1.06}\left(\frac{1 - \left(\frac{1}{1.06}\right)^{10}}{1 - \frac{1}{1.06}}\right) + \frac{1000}{(1.06)^{10}}$$

$$= 368.004 + 558.395$$

$$= \$926.40$$

81. A person should expect to pay the present value of the bond on the day it is bought.

$$\text{Present value of first payment} = \frac{10}{1.04}$$

$$\text{Present value of second payment} = \frac{10}{(1.04)^2}, \text{ etc.}$$

Therefore,

$$\text{Total present value} = \frac{10}{1.04} + \frac{10}{(1.04)^2} + \frac{10}{(1.04)^3} + \cdots.$$

This is a geometric series with $a = \frac{10}{1.04}$ and $x = \frac{1}{1.04}$, so

$$\text{Total present value} = \frac{\frac{10}{1.04}}{1 - \frac{1}{1.04}} = \text{£}250.$$

85. This series converges by the alternating series test, so we can use Theorem 9.9. The nth partial sum of the series is given by

$$S_n = 1 - \frac{1}{6} + \frac{1}{120} - \cdots + \frac{(-1)^{n-1}}{(2n-1)!},$$

so the absolute value of the first term omitted is $1/(2n + 1)!$. By Theorem 9.9, we know that the true value of the sum differs from S_n by less than $1/(2n + 1)!$. Thus, we want to choose n large enough so that $1/(2n + 1)! \le 0.01$. Substituting $n = 2$ into the expression $1/(2n + 1)!$ yields $1/720$ which is less than 0.01, so $S_2 = 1 - (1/6) = 5/6$ approximates the sum to within 0.01 of the actual sum.

89. We have $0 \le a_n/n \le a_n$ for all $n \ge 1$. Therefore, since $\sum a_n$ converges, $\sum a_n/n$ converges by the Comparison Test.

93. Since $\sum a_n$ converges, we know that $\lim_{n\to\infty} a_n = 0$. Therefore, we can choose a positive integer N large enough so that $|a_n| \le 1$ for all $n \ge N$, so we have $0 \le a_n^2 \le a_n$ for all $n \ge N$. Thus, by Property 2 of Theorem 9.2, $\sum a_n^2$ converges by comparison with the convergent series $\sum a_n$.

97. We want to estimate $\sum_{k=1}^{100,000} \frac{1}{k}$ using left and right Riemann sum approximations to $f(x) = 1/x$ on the interval $1 \le x \le$ 100,000. Figure 9.3 shows a left Riemann sum approximation with 99,999 terms. Since $f(x)$ is decreasing, the left Riemann sum overestimates the area under the curve. Figure 9.3 shows that the first term in the sum is $f(1) \cdot 1$ and the last is $f(99,999) \cdot 1$, so we have

$$\int_1^{100,000} \frac{1}{x}\, dx < \text{LHS} = f(1) \cdot 1 + f(2) \cdot 1 + \cdots + f(99,999) \cdot 1.$$

Since $f(x) = 1/x$, the left Riemann sum is

$$\text{LHS} = \frac{1}{1} \cdot 1 + \frac{1}{2} \cdot 1 + \cdots + \frac{1}{99,999} \cdot 1 = \sum_{k=1}^{99,999} \frac{1}{k},$$

so

$$\int_1^{100,000} \frac{1}{x}\, dx < \sum_{k=1}^{99,999} \frac{1}{k}.$$

Since we want the sum to go $k = 100,000$ rather than $k = 99,999$, we add $1/100,000$ to both sides:

$$\int_1^{100,000} \frac{1}{x}\, dx + \frac{1}{100,000} < \sum_{k=1}^{99,999} \frac{1}{k} + \frac{1}{100,000} = \sum_{k=1}^{100,000} \frac{1}{k}.$$

The left Riemann sum has therefore given us an underestimate for our sum. We now use the right Riemann sum in Figure 9.4 to get an overestimate for our sum.

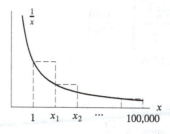

Figure 9.3

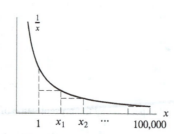

Figure 9.4

The right Riemann sum again has 99,999 terms, but this time the sum underestimates the area under the curve. Figure 9.4 shows that the first rectangle has area $f(2) \cdot 1$ and the last $f(100,000) \cdot 1$, so we have

$$\text{RHS} = f(2) \cdot 1 + f(3) \cdot 1 + \cdots + f(100,000) \cdot 1 < \int_1^{100,000} \frac{1}{x}\, dx.$$

Since $f(x) = 1/x$, the right Riemann sum is

$$\text{RHS} = \frac{1}{2} \cdot 1 + \frac{1}{3} \cdot 1 + \cdots + \frac{1}{100,000} \cdot 1 = \sum_{k=2}^{100,000} \frac{1}{k}.$$

So

$$\sum_{k=2}^{100,000} \frac{1}{k} < \int_1^{100,000} \frac{1}{x}\, dx.$$

Since we want the sum to start at $k = 1$, we add 1 to both sides:

$$\sum_{k=1}^{100,000} \frac{1}{k} = \frac{1}{1} + \sum_{k=2}^{100,000} \frac{1}{k} < 1 + \int_1^{100,000} \frac{1}{x}\, dx.$$

Putting these under- and overestimates together, we have

$$\int_1^{100,000} \frac{1}{x}\, dx + \frac{1}{100,000} < \sum_{k=1}^{100,000} \frac{1}{k} < 1 + \int_1^{100,000} \frac{1}{x}\, dx.$$

Since $\int_1^{100,000} \frac{1}{x}\, dx = \ln 100,000 - \ln 1 = 11.513$, we have

$$11.513 < \sum_{k=1}^{100,000} \frac{1}{k} < 12.513.$$

Therefore we have $\displaystyle\sum_{k=1}^{100,000} \frac{1}{k} \approx 12.$

CHAPTER TEN

Solutions for Section 10.1

Exercises

1. Let $f(x) = \dfrac{1}{1-x} = (1-x)^{-1}$. Then $f(0) = 1$.

$$f'(x) = 1!(1-x)^{-2} \quad f'(0) = 1!,$$
$$f''(x) = 2!(1-x)^{-3} \quad f''(0) = 2!,$$
$$f'''(x) = 3!(1-x)^{-4} \quad f'''(0) = 3!,$$
$$f^{(4)}(x) = 4!(1-x)^{-5} \quad f^{(4)}(0) = 4!,$$
$$f^{(5)}(x) = 5!(1-x)^{-6} \quad f^{(5)}(0) = 5!,$$
$$f^{(6)}(x) = 6!(1-x)^{-7} \quad f^{(6)}(0) = 6!,$$
$$f^{(7)}(x) = 7!(1-x)^{-8} \quad f^{(7)}(0) = 7!.$$

$$P_3(x) = 1 + x + x^2 + x^3,$$
$$P_5(x) = 1 + x + x^2 + x^3 + x^4 + x^5,$$
$$P_7(x) = 1 + x + x^2 + x^3 + x^4 + x^5 + x^6 + x^7.$$

5. Let $f(x) = \cos x$. Then $f(0) = \cos(0) = 1$, and

$$f'(x) = -\sin x \quad f'(0) = 0,$$
$$f''(x) = -\cos x \quad f''(0) = -1,$$
$$f'''(x) = \sin x \quad f'''(0) = 0,$$
$$f^{(4)}(x) = \cos x \quad f^{(4)}(0) = 1,$$
$$f^{(5)}(x) = -\sin x \quad f^{(5)}(0) = 0,$$
$$f^{(6)}(x) = -\cos x \quad f^{(6)}(0) = -1.$$

Thus,

$$P_2(x) = 1 - \frac{x^2}{2!},$$
$$P_4(x) = 1 - \frac{x^2}{2!} + \frac{x^4}{4!},$$
$$P_6(x) = 1 - \frac{x^2}{2!} + \frac{x^4}{4!} - \frac{x^6}{6!}.$$

9. Let $f(x) = \dfrac{1}{\sqrt{1+x}} = (1+x)^{-1/2}$. Then $f(0) = 1$.

$$f'(x) = -\tfrac{1}{2}(1+x)^{-3/2} \quad f'(0) = -\tfrac{1}{2},$$
$$f''(x) = \tfrac{3}{2^2}(1+x)^{-5/2} \quad f''(0) = \tfrac{3}{2^2},$$
$$f'''(x) = -\tfrac{3 \cdot 5}{2^3}(1+x)^{-7/2} \quad f'''(0) = -\tfrac{3 \cdot 5}{2^3},$$
$$f^{(4)}(x) = \tfrac{3 \cdot 5 \cdot 7}{2^4}(1+x)^{-9/2} \quad f^{(4)}(0) = \tfrac{3 \cdot 5 \cdot 7}{2^4}$$

Then,

$$P_2(x) = 1 - \frac{1}{2}x + \frac{1}{2!}\frac{3}{2^2}x^2 = 1 - \frac{1}{2}x + \frac{3}{8}x^2,$$

$$P_3(x) = P_2(x) - \frac{1}{3!}\frac{3\cdot 5}{2^3}x^3 = 1 - \frac{1}{2}x + \frac{3}{8}x^2 - \frac{5}{16}x^3,$$

$$P_4(x) = P_3(x) + \frac{1}{4!}\frac{3\cdot 5\cdot 7}{2^4}x^4 = 1 - \frac{1}{2}x + \frac{3}{8}x^2 - \frac{5}{16}x^3 + \frac{35}{128}x^4.$$

13. Let $f(x) = \frac{1}{1+x} = (1+x)^{-1}$. Then $f'(x) = -(1+x)^{-2}$, $f''(x) = 2(1+x)^{-3}$, $f'''(x) = -6(1+x)^{-4}$, $f^{(4)}(x) = 24(1+x)^{-5}$.
So $f(2) = \frac{1}{3}$, $f'(2) = -\frac{1}{3^2}$, $f''(2) = \frac{2}{3^3}$, $f'''(2) = -\frac{6}{3^4}$, and $f^{(4)}(2) = \frac{24}{3^5}$. Therefore,

$$P_4(x) = \frac{1}{3} - \frac{1}{3^2}(x-2) + \frac{2}{3^3}\frac{1}{2!}(x-2)^2 - \frac{6}{3^4}\frac{1}{3!}(x-2)^3 + \frac{24}{3^5}\frac{1}{4!}(x-2)^4$$

$$= \frac{1}{3}\left(1 - \frac{x-2}{3} + \frac{(x-2)^2}{3^2} - \frac{(x-2)^3}{3^3} + \frac{(x-2)^4}{3^4}\right).$$

Problems

17. Using the fact that

$$f(x) \approx P_3(x) = f(0) + f'(0)x + \frac{f''(0)}{2!}x^2 + \frac{f'''(0)}{3!}x^3$$

and identifying coefficients with those given for $P_3(x)$, we obtain the following:
(a) $f(0) =$ constant term which equals 2,   so $f(0) = 2$.
(b) $f'(0) =$ coefficient of x which equals -1,   so $f'(0) = -1$.
(c) $\frac{f''(0)}{2!} =$ coefficient of x^2 which equals $-1/3$,   so $f''(0) = -2/3$.
(d) $\frac{f'''(0)}{3!} =$ coefficient of x^3 which equals 2,   so $f'''(0) = 12$.

21. The third degree Taylor polynomial of $f(x)$ will have the same terms as the seventh degree polynomial but only up to the x^3 term. So the third degree Taylor polynomial of $f(x)$ is given by

$$P_3(x) = 1 - \frac{x}{3} + \frac{5x^2}{7} + 8x^3.$$

25. (a) The graph of $f(x) = \ln(2x+4)$ is in Figure 10.1. Since the slope of function is positive at $x = 0$, the tangent line approximation has a positive slope.
(b) Evaluating the function and the first derivative at $x = 0$ gives $f(0) = \ln 4$ and

$$f'(x) = \frac{2}{2x+4} = \frac{1}{x+2} \qquad \text{so} \qquad f'(0) = \frac{1}{2}.$$

The tangent line approximation is the first Taylor approximation, namely

$$y = P_1(x) = \ln 4 + \frac{1}{2}x.$$

(c) Evaluating the second and third derivatives at $x = 0$, we have

$$f''(x) = -\frac{1}{(x+2)^2} \qquad \text{so} \qquad f''(0) = -\frac{1}{4},$$

$$f'''(x) = \frac{2}{(x+2)^3} \qquad \text{so} \qquad f'''(0) = \frac{1}{4}.$$

Thus the second and third Taylor approximations about 0 are

$$P_2(x) = \ln 4 + \frac{1}{2}x - \frac{1}{4}\cdot\frac{1}{2!}x^2 = \ln 4 + \frac{1}{2}x - \frac{1}{8}x^2$$

$$P_3(x) = \ln 4 + \frac{1}{2}x - \frac{1}{8}x^2 + \frac{1}{4}\cdot\frac{1}{3!}x^3 = \ln 4 + \frac{1}{2}x - \frac{1}{8}x^2 + \frac{1}{24}x^3.$$

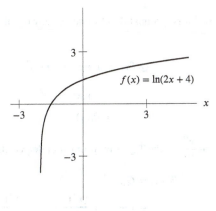

Figure 10.1

29. As we can see from Problem 26, a is the y-intercept of $f(x)$, b is the slope of the tangent line to $f(x)$ at $x = 0$ and c tells us the concavity of $f(x)$ near $x = 0$.
So $a < 0$, $b < 0$ and $c > 0$.

33. The third-degree Taylor polynomial has a fourth derivative of zero, no matter what the fourth derivative of $f(x)$ is, so we cannot find $f^{(4)}(1)$ knowing $P_3(x)$.

37.

$$\lim_{x \to 0} \frac{\sin x}{x} = \lim_{x \to 0} \frac{x - \frac{x^3}{3!}}{x} = \lim_{x \to 0} \left(1 - \frac{x^2}{3!}\right) = 1.$$

41. (a) Since the coefficient of the x-term of each f is 1, we know $f_1'(0) = f_2'(0) = f_3'(0) = 1$. Thus, each of the fs slopes upward near 0, and are in the second figure.

The coefficient of the x-term in g_1 and in g_2 is 1, so $g_1'(0) = g_2'(0) = 1$. For g_3 however, $g_3'(0) = -1$. Thus, g_1 and g_2 slope up near 0, but g_3 slopes down. The gs are in the first figure.

(b) Since $g_1(0) = g_2(0) = g_3(0) = 1$, the point A is $(0, 1)$.
Since $f_1(0) = f_2(0) = f_3(0) = 2$, the point B is $(0, 2)$.

(c) Since g_3 slopes down, g_3 is I. Since the coefficient of x^2 for g_1 is 2, we know

$$\frac{g_1''(0)}{2!} = 2 \qquad \text{so} \qquad g_1''(0) = 4.$$

By similar reasoning $g_2''(0) = 2$. Since g_1 and g_2 are concave up, and g_1 has a larger second derivative, g_1 is III and g_2 is II.

Calculating the second derivatives of the fs from the coefficients x^2, we find

$$f_1''(0) = 4 \qquad f_2''(0) = -2 \qquad f_3''(0) = 2.$$

Thus, f_1 and f_3 are concave up, with f_1 having the larger second derivative, so f_1 is III and f_3 is II. Then f_2 is concave down and is I.

45. (a) $f(x) = e^{x^2}$.
$f'(x) = 2xe^{x^2}$, $f''(x) = 2(1 + 2x^2)e^{x^2}$, $f'''(x) = 4(3x + 2x^3)e^{x^2}$,
$f^{(4)}(x) = 4(3 + 6x^2)e^{x^2} + 4(3x + 2x^3)2xe^{x^2}$.
The Taylor polynomial about $x = 0$ is

$$P_4(x) = 1 + \frac{0}{1!}x + \frac{2}{2!}x^2 + \frac{0}{3!}x^3 + \frac{12}{4!}x^4$$
$$= 1 + x^2 + \frac{1}{2}x^4.$$

(b) $f(x) = e^x$. The Taylor polynomial of degree 2 is

$$Q_2(x) = 1 + \frac{x}{1!} + \frac{x^2}{2!} = 1 + x + \frac{1}{2}x^2.$$

If we substitute x^2 for x in the Taylor polynomial for e^x of degree 2, we will get $P_4(x)$, the Taylor polynomial for e^{x^2} of degree 4:

$$Q_2(x^2) = 1 + x^2 + \frac{1}{2}(x^2)^2$$
$$= 1 + x^2 + \frac{1}{2}x^4$$
$$= P_4(x).$$

(c) Let $Q_{10}(x) = 1 + \frac{x}{1!} + \frac{x^2}{2!} + \cdots + \frac{x^{10}}{10!}$ be the Taylor polynomial of degree 10 for e^x about $x = 0$. Then

$$P_{20}(x) = Q_{10}(x^2)$$
$$= 1 + \frac{x^2}{1!} + \frac{(x^2)^2}{2!} + \cdots + \frac{(x^2)^{10}}{10!}$$
$$= 1 + \frac{x^2}{1!} + \frac{x^4}{2!} + \cdots + \frac{x^{20}}{10!}.$$

(d) Let $e^x \approx Q_5(x) = 1 + \frac{x}{1!} + \cdots + \frac{x^5}{5!}$. Then

$$e^{-2x} \approx Q_5(-2x)$$
$$= 1 + \frac{-2x}{1!} + \frac{(-2x)^2}{2!} + \frac{(-2x)^3}{3!} + \frac{(-2x)^4}{4!} + \frac{(-2x)^5}{5!}$$
$$= 1 - 2x + 2x^2 - \frac{4}{3}x^3 + \frac{2}{3}x^4 - \frac{4}{15}x^5.$$

49. (a) The graphs of $y = \cos x$ and $y = 1 - 0.1x$ cross at $x = 0$ and for another x-value just to the right of $x = 0$. (There are other crossings much further to the right.)

(b) Since

$$\cos x \approx 1 - \frac{x^2}{2}$$

the equation becomes

$$1 - \frac{x^2}{2} = 1 - 0.1x$$
$$\frac{x^2}{2} = 0.1x$$
$$x = 0, 0.2.$$

The solution $x = 0$ is an exact solution to the original equation; $x = 0.2$ is an approximate solution to the original equation.

Strengthen Your Understanding

53. The coefficient of the x term is given by $f'(0)$ and $f'(0) = 1$.

57. An example is $f(x) = 1 + x$ and $g(x) = 1 + x + x^2$. The Taylor polynomial approximation to both f and g near $x = 0$ is $1 + x$. Many other examples are possible.

61. False. $P_2(x) = f(5) + f'(5)(x - 5) + (f''(5)/2)(x - 5)^2 = e^5 + e^5(x - 5) + (e^5/2)(x - 5)^2$.

65. False. Since $f(-1) = g(-1)$ the graphs of f and g intersect at $x = -1$. Since $f'(-1) < g'(-1)$, the slope of f is less than the slope of g at $x = -1$. Thus $f(x) > g(x)$ for all x sufficiently close to -1 on the left, and $f(x) < g(x)$ for all x sufficiently close to -1 on the right.

Solutions for Section 10.2

Exercises

1. Differentiating $(1 + x)^{3/2}$:

$$f(x) = (1 + x)^{3/2} \qquad\qquad\qquad\qquad f(0) = 1,$$
$$f'(x) = (3/2)(1 + x)^{1/2} \qquad\qquad\qquad f'(0) = \tfrac{3}{2},$$
$$f''(x) = (1/2)(3/2)(1 + x)^{-1/2} = (3/4)(1 + x)^{-1/2} \qquad f''(0) = \tfrac{3}{4},$$
$$f'''(x) = (-1/2)(3/4)(1 + x)^{-3/2} = (-3/8)(1 + x)^{-3/2} \quad f'''(0) = -\tfrac{3}{8}.$$

$$f(x) = (1 + x)^{3/2} = 1 + \frac{3}{2} \cdot x + \frac{(3/4)x^2}{2!} + \frac{(-3/8)x^3}{3!} + \cdots$$
$$= 1 + \frac{3x}{2} + \frac{3x^2}{8} - \frac{x^3}{16} + \cdots$$

5.

$$f(x) = \tfrac{1}{1-x} = (1 - x)^{-1} \qquad\qquad\qquad f(0) = 1,$$
$$f'(x) = -(1 - x)^{-2}(-1) = (1 - x)^{-2} \qquad f'(0) = 1,$$
$$f''(x) = -2(1 - x)^{-3}(-1) = 2(1 - x)^{-3} \quad f''(0) = 2,$$
$$f'''(x) = -6(1 - x)^{-4}(-1) = 6(1 - x)^{-4} \quad f'''(0) = 6.$$

$$f(x) = \frac{1}{1 - x} = 1 + 1 \cdot x + \frac{2x^2}{2!} + \frac{6x^3}{3!} + \cdots$$
$$= 1 + x + x^2 + x^3 + \cdots$$

9. Differentiating $f(x) = \ln(5 + 2x)$, simplifying and evaluating at $x = 0$ gives:

$$f(x) = \ln(5 + 2x) \qquad \text{so} \quad f(0) = \ln 5,$$
$$f'(x) = \tfrac{2}{5+2x} \qquad \text{so} \quad f'(0) = \tfrac{2}{5},$$
$$f''(x) = -\tfrac{4}{(5+2x)^2} \qquad \text{so} \quad f''(0) = -\tfrac{4}{25},$$
$$f'''(x) = \tfrac{16}{(5+2x)^3} \qquad \text{so} \quad f'''(0) = \tfrac{16}{125}.$$

Therefore, the Taylor series about 0 is

$$\ln 5 + 2x = \ln 5 + \frac{(2/5)}{1!}x - \frac{(4/25)}{2!}x^2 + \frac{(16/25)}{3!}x^3 + \cdots$$
$$= \ln 5 + \frac{2}{5}x - \frac{2}{25}x^2 + \frac{8}{475}x^3 + \cdots.$$

13.

$$f(\theta) = \sin \theta \qquad f(-\tfrac{\pi}{4}) = -\tfrac{\sqrt{2}}{2},$$
$$f'(\theta) = \cos \theta \qquad f'(-\tfrac{\pi}{4}) = \tfrac{\sqrt{2}}{2},$$
$$f''(\theta) = -\sin \theta \qquad f''(-\tfrac{\pi}{4}) = \tfrac{\sqrt{2}}{2},$$
$$f'''(\theta) = -\cos \theta \qquad f'''(-\tfrac{\pi}{4}) = -\tfrac{\sqrt{2}}{2}.$$

$$\sin \theta = -\frac{\sqrt{2}}{2} + \frac{\sqrt{2}}{2}\left(\theta + \frac{\pi}{4}\right) + \frac{\sqrt{2}}{2}\frac{(\theta + \frac{\pi}{4})^2}{2!} - \frac{\sqrt{2}}{2}\frac{(\theta + \frac{\pi}{4})^3}{3!} + \cdots$$
$$= -\frac{\sqrt{2}}{2} + \frac{\sqrt{2}}{2}\left(\theta + \frac{\pi}{4}\right) + \frac{\sqrt{2}}{4}\left(\theta + \frac{\pi}{4}\right)^2 - \frac{\sqrt{2}}{12}\left(\theta + \frac{\pi}{4}\right)^3 + \cdots.$$

17. Using the derivatives from Problem 15, we have

$$f(-1) = -1, \quad f'(-1) = -1, \quad f''(-1) = -2, \quad f'''(-1) = -6.$$

Hence,

$$\frac{1}{x} = -1 - (x + 1) - \frac{2(x+1)^2}{2!} - \frac{6(x+1)^3}{3!} - \cdots$$
$$= -1 - (x + 1) - (x + 1)^2 - (x + 1)^3 - \cdots$$

21. The general term can be written as $(-1)^{n-1}x^n/n$ for $n \geq 1$.

25. The general term can be written as $(-1)^k x^{4k+2}/(2k)!$ for $k \geq 0$.

Problems

29. We know that the Taylor series for e^x around 0 is given by

$$e^x = 1 + x + \frac{x^2}{2!} + \frac{x^3}{3!} + \cdots.$$

Using the right hand side of the above equation for e^x in the expression $\dfrac{e^x - 1}{x}$, we have

$$\lim_{x \to 0} \frac{e^x - 1}{x} = \lim_{x \to 0} \frac{1 + x + \frac{x^2}{2!} + \frac{x^3}{3!} + \cdots - 1}{x}.$$

Simplifying we get

$$\lim_{x \to 0} \frac{x + \frac{x^2}{2!} + \frac{x^3}{3!} + \cdots}{x} = \lim_{x \to 0} 1 + \frac{x}{2!} + \frac{x^2}{3!} + \cdots = 1.$$

Hence this limit is equal to 1.

33. The Taylor series for $f(x)$ about $x = 0$ is

$$f(0) + f'(0)x + \frac{f''(0)}{2!}x^2 + \frac{f'''(0)}{3!}x^3 + \cdots + \frac{f^{(n)}(0)}{n!}x^n + \cdots.$$

Therefore, in order for all the coefficients to be positive, we must have $f^{(n)}(0) > 0$ for all n.

37. By looking at Figure 10.2 we can that the Taylor polynomials are reasonable approximations for the function $f(x) = \frac{1}{\sqrt{1+x}}$ between $x = -1$ and $x = 1$. Thus a good guess is that the interval of convergence is $-1 < x < 1$.

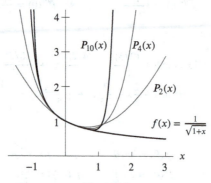

Figure 10.2

41. The Taylor series for $\ln(1 - x)$ is

$$\ln(1 - x) = -x - \frac{x^2}{2} - \frac{x^3}{3} - \cdots - \frac{x^n}{n} - \cdots,$$

so

$$\lim_{n \to \infty} \frac{|a_{n+1}|}{|a_n|} = |x| \lim_{n \to \infty} \frac{1/(n+1)}{1/n} = |x| \lim_{n \to \infty} \left| \frac{n}{n+1} \right| = |x|.$$

Thus the series converges for $|x| < 1$, and the radius of convergence is 1. Note: This series can be obtained from the series for $\ln(1 + x)$ by replacing x by $-x$ and has the same radius of convergence as the series for $\ln(1 + x)$.

45. This is the series for $1/(1-x)$ with x replaced by $1/4$, so the series converges to $1/(1-(1/4)) = 4/3$.

49. This is the series for e^x with $x = 3$ substituted. Thus

$$1 + 3 + \frac{9}{2!} + \frac{27}{3!} + \frac{81}{4!} + \cdots = 1 + 3 + \frac{3^2}{2!} + \frac{3^3}{3!} + \frac{3^4}{4!} + \cdots = e^3.$$

53. Since $1 + x + x^2 + x^3 + \cdots = \dfrac{1}{1-x}$, a geometric series, we solve $\dfrac{1}{1-x} = 5$ giving $\dfrac{1}{5} = 1 - x$, so $x = \dfrac{4}{5}$.

57. Looking at the first few terms of $C(x)$, we see that $C(x) = x +$ higher-powered terms. This means $C'(x) = 1 +$ higher-powered terms, so $C'(0) = 1$ since at $x = 0$ all higher-powered terms all vanish.

Strengthen Your Understanding

61. The left hand side of the equation is finite, namely -1, whereas the right hand side of the equation is infinite. The statement is wrong, since

$$\frac{1}{1-x} = 1 + x + x^2 + x^3 + \cdots$$

only for $-1 < x < 1$.

65. False. The Taylor series for $\sin x$ about $x = \pi$ is calculated by taking derivatives and using the formula

$$f(a) + f'(a)(x-a) + \frac{f''(a)}{2!}(x-a)^2 + \cdots.$$

The series for $\sin x$ about $x = \pi$ turns out to be

$$-(x - \pi) + \frac{(x-\pi)^3}{3!} - \frac{(x-\pi)^5}{5!} + \cdots.$$

69. True. For large x, the graph of $P_{10}(x)$ looks like the graph of its highest powered term, $x^{10}/10!$. But e^x grows faster than any power, so e^x gets further and further away from $x^{10}/10! \approx P_{10}(x)$.

Solutions for Section 10.3

Exercises

1. Substitute $y = -x$ into $e^y = 1 + y + \frac{y^2}{2!} + \frac{y^3}{3!} + \cdots$. We get

$$e^{-x} = 1 + (-x) + \frac{(-x)^2}{2!} + \frac{(-x)^3}{3!} + \cdots$$

$$= 1 - x + \frac{x^2}{2!} - \frac{x^3}{3!} + \cdots.$$

5. Since $\frac{d}{dx}(\arcsin x) = \frac{1}{\sqrt{1-x^2}} = 1 + \frac{1}{2}x^2 + \frac{3}{8}x^4 + \frac{5}{16}x^6 + \cdots$, integrating gives

$$\arcsin x = c + x + \frac{1}{6}x^3 + \frac{3}{40}x^5 + \frac{5}{112}x^7 + \cdots.$$

Since $\arcsin 0 = 0$, $c = 0$.

9.

$$\phi^3 \cos(\phi^2) = \phi^3 \left(1 - \frac{(\phi^2)^2}{2!} + \frac{(\phi^2)^4}{4!} - \frac{(\phi^2)^6}{6!} + \cdots \right)$$

$$= \phi^3 - \frac{\phi^7}{2!} + \frac{\phi^{11}}{4!} - \frac{\phi^{15}}{6!} + \cdots$$

13. Multiplying out gives $(1 + x)^3 = 1 + 3x + 3x^2 + x^3$. Since this polynomial equals the original function for all x, it must be the Taylor series. The general term is $0 \cdot x^n$ for $n \geq 4$.

17. Since we have

$$\ln(1 + x) = x + \frac{x^2}{2} + \frac{x^3}{3} + \frac{x^4}{4} + \frac{x^5}{5} + \frac{x^6}{6} + \frac{x^7}{7} + \cdots$$

we conclude that

$$\ln(1 + 0.5) = \ln\left(\frac{3}{2}\right) = 0.5 + \frac{0.5^2}{2} + \frac{0.5^3}{3} + \frac{0.5^4}{4} + \frac{0.5^5}{5} + \frac{0.5^6}{6} + \frac{0.5^7}{7} + \cdots.$$

Notice that the series for $\ln(1 + x)$ converges for all $|x| < 1$, including $x = 0.5$.

21. Using the binomial expansion for $(1 + x)^{-2}$ with $x = r/a$:

$$\frac{1}{(a + r)^2} = \frac{1}{\left(a + a\left(\frac{r}{a}\right)\right)^2} = \frac{1}{\left(a\left(1 + \frac{r}{a}\right)\right)^2} = \frac{1}{a^2}\left(1 + \left(\frac{r}{a}\right)\right)^{-2}$$

$$= \frac{1}{a^2}\left(1 + (-2)\left(\frac{r}{a}\right) + \frac{(-2)(-3)}{2!}\left(\frac{r}{a}\right)^2 + \frac{(-2)(-3)(-4)}{3!}\left(\frac{r}{a}\right)^3 + \cdots\right)$$

$$= \frac{1}{a^2}\left(1 - 2\left(\frac{r}{a}\right) + 3\left(\frac{r}{a}\right)^2 - 4\left(\frac{r}{a}\right)^3 + \cdots\right).$$

Problems

25.

$$e^t \cos t = \left(1 + t + \frac{t^2}{2!} + \frac{t^3}{3!} + \frac{t^4}{4!} + \cdots\right)\left(1 - \frac{t^2}{2!} + \frac{t^4}{4!} - \frac{t^6}{6!} + \cdots\right)$$

Multiplying out and collecting terms gives

$$e^t \cos t = 1 + t + \left(\frac{t^2}{2!} - \frac{t^2}{2!}\right) + \left(\frac{t^3}{3!} - \frac{t^3}{2!}\right) + \left(\frac{t^4}{4!} + \frac{t^4}{4!} - \frac{t^4}{(2!)^2}\right) + \cdots$$

$$= 1 + t - \frac{t^3}{3} - \frac{t^4}{6} + \cdots.$$

29. (a) Since $x = 2z + 1$, substituting $z = (x - 1)/2$ into $f(x)$ we get

$$\frac{1}{x + 1} = \frac{1}{2 + 2z} = \frac{1}{2}\left(\frac{1}{1 + z}\right) = h(z).$$

Using a binomial series:

$$h(z) = \frac{1}{2}\left(\frac{1}{1 + z}\right) = \frac{1}{2}(1 + z)^{-1}$$

$$= \frac{1}{2}\left(1 + (-1)z + \frac{(-1)(-2)}{2!}z^2 + \frac{(-1)(-2)(-3)}{3!}z^3 + \cdots + \frac{(-1)(-2)(-3)\cdots(-n)}{n!}z^n + \cdots\right)$$

$$= \frac{1}{2} - \frac{z}{2} + \frac{z^2}{2} - \frac{z^3}{2} + \cdots + \frac{(-1)^n z^n}{2} + \cdots$$

Substituting $z = (x - 1)/2$ for z we get

$$\frac{1}{x + 1} = \frac{1}{2} - \frac{(x - 1)}{4} + \frac{(x - 1)^2}{8} - \frac{(x - 1)^3}{16} + \cdots + (-1)^n\frac{(x - 1)^n}{2^{n+1}} + \cdots$$

(b) Observe that $(x - 1)^2$ is already written as a Taylor series around $x = 1$, so we have

$$\frac{(x - 1)^2}{x + 1} = (x - 1)^2\left(\frac{1}{x + 1}\right)$$

$$= (x - 1)^2\left(\frac{1}{2} - \frac{(x - 1)}{4} + \frac{(x - 1)^2}{8} - \frac{(x - 1)^3}{16} + \cdots + (-1)^n\frac{(x - 1)^n}{2^{n+1}} + \cdots\right)$$

$$= \frac{(x - 1)^2}{2} - \frac{(x - 1)^3}{4} + \frac{(x - 1)^4}{8} - \frac{(x - 1)^5}{16} + \cdots + (-1)^n\frac{(x - 1)^{n+2}}{2^{n+1}} + \cdots$$

33. (a) Since the Taylor series for e^x and e^{-x} are given by

$$e^x = 1 + x + \frac{x^2}{2!} + \frac{x^3}{3!} + \frac{x^4}{4!} + \frac{x^5}{5!} + \cdots$$

$$e^{-x} = 1 - x + \frac{x^2}{2!} - \frac{x^3}{3!} + \frac{x^4}{4!} - \frac{x^5}{5!} + \cdots,$$

we have

$$e^x - e^{-x} = 0 + 2x + 0\frac{x^2}{2!} + 2\frac{x^3}{3!} + 0\frac{x^4}{4!} + 2\frac{x^5}{5!} \cdots = 2x + \frac{x^3}{3} + \frac{x^5}{60} + \cdots .$$

(b) For x near 0, we can approximate $e^x - e^{-x}$ by its third degree Taylor polynomial, $P_3(x)$:

$$e^x - e^{-x} \approx P_3(x) = 2x + \frac{x^3}{3}.$$

The function $P_3(x)$ is a cubic polynomial whose graph is symmetric about the origin.

37. The Taylor series for $1 + \sin\theta$ near $\theta = 0$ is

$$1 + \sin\theta = 1 + \theta - \frac{\theta^3}{3!} + \frac{\theta^5}{5!} - \cdots$$

The Taylor series for e^θ is

$$e^\theta = 1 + \theta + \frac{\theta^2}{2!} + \frac{\theta^3}{3!} + \cdots$$

The Taylor series for $\frac{1}{\sqrt{1+\theta}}$ is

$$\frac{1}{\sqrt{1+\theta}} = 1 - \frac{\theta}{2} + \frac{3\theta^2}{8} - \frac{5\theta^3}{16} + \cdots$$

Substituting -2θ into this formula yields

$$\frac{1}{\sqrt{1-2\theta}} = 1 + \theta + \frac{3}{2}\theta^2 + \frac{5}{2}\theta^3 + \cdots$$

These three series are identical in the constant and linear terms, but they differ in their quadratic terms. For values of θ near zero, the quadratic terms dominate all of the subsequent terms, so we can use the approximations

$$1 + \sin\theta \approx 1 + \theta$$

$$e^\theta \approx 1 + \theta + \frac{1}{2}\theta^2$$

$$\frac{1}{\sqrt{1-2\theta}} \approx 1 + \theta + \frac{3}{2}\theta^2.$$

Clearly $1 + \sin\theta$ is smallest, because the θ^2 term is zero, and the θ^2 terms of the other two series are positive. The function $1/\sqrt{1-2\theta}$ is largest, because the coefficient of its θ^2 term is the greatest. Therefore, for θ near zero,

$$1 + \sin\theta \le e^\theta \le \frac{1}{\sqrt{1-2\theta}}.$$

41. The Taylor series for $\sin t$ about $t = 0$ is $\sin t = t - \frac{1}{3!}t^3 + \frac{1}{5!}t^5 - \frac{1}{7!}t^7 + \cdots$. This gives the Taylor series for $\sin\left(t^2\right)$ about $t = 0$:

$$\sin\left(t^2\right) = t^2 - \frac{1}{3!}\left(t^2\right)^3 + \frac{1}{5!}\left(t^2\right)^5 - \frac{1}{7!}\left(t^2\right)^7 + \cdots = t^2 - \frac{1}{3!}t^6 + \frac{1}{5!}t^{10} - \frac{1}{7!}t^{14} + \cdots,$$

so

$$f(x) = \int_0^x \sin\left(t^2\right)\,dt$$

$$= \int_0^x \left(t^2 - \frac{1}{3!}t^6 + \frac{1}{5!}t^{10} - \frac{1}{7!}t^{14} + \cdots\right)\,dt$$

$$= \frac{1}{3}x^3 - \frac{1}{7}\frac{1}{3!}x^7 + \frac{1}{11}\frac{1}{5!}x^{11} - \frac{1}{15}\frac{1}{7!}x^{15} + \cdots$$

$$= \frac{1}{3}x^3 - \frac{1}{42}x^7 + \frac{1}{1320}x^{11} - \frac{1}{75,600}x^{15} + \cdots$$

45. (a)
$$f(x) = (1 + ax)(1 + bx)^{-1} = (1 + ax)\left(1 - bx + (bx)^2 - (bx)^3 + \cdots\right)$$
$$= 1 + (a - b)x + (b^2 - ab)x^2 + \cdots$$

(b) $e^x = 1 + x + \frac{x^2}{2} + \cdots$
Equating coefficients:
$$a - b = 1,$$
$$b^2 - ab = \frac{1}{2}.$$

Solving gives $a = \frac{1}{2}, b = -\frac{1}{2}$.

49. (a) Since we have
$$\theta'' + \frac{g}{l}\sin\theta = 0,$$
we use $\sin\theta \approx \theta$ to get the approximate differential equation
$$\theta'' + \frac{g}{l}\theta = 0.$$

(b) To compare the two estimates of the period, we consider
$$\text{Percent difference} = \frac{\text{Accurate period} - \text{Approximate period}}{\text{Approximate period}} = \frac{2\pi\sqrt{l/g}(\theta_0^2/16)}{2\pi\sqrt{l/g}} = \frac{1}{16}\theta_0^2.$$

Since the Taylor series is based on the derivative formulas for the sine and cosine, which are valid in radians, not degrees, we convert $\theta_0 = 20°$ into radians:
$$20° = 20 \cdot \frac{\pi}{180} = \frac{\pi}{9}.$$

Substituting gives
$$\text{Percent difference} = \frac{1}{16}\left(\frac{\pi}{9}\right)^2 = 0.0076 = 0.76.\%$$

53. (a) To find when V takes on its minimum values, set $\frac{dV}{dr} = 0$. So
$$-V_0\frac{d}{dr}\left(2\left(\frac{r_0}{r}\right)^6 - \left(\frac{r_0}{r}\right)^{12}\right) = 0$$
$$-V_0\left(-12r_0^6 r^{-7} + 12r_0^{12} r^{-13}\right) = 0$$
$$12r_0^6 r^{-7} = 12r_0^{12} r^{-13}$$
$$r_0^6 = r^6$$
$$r = r_0.$$

Rewriting $V'(r)$ as $\frac{12r_0^6 V_0}{r^7}\left(1 - \left(\frac{r_0}{r}\right)^6\right)$, we see that $V'(r) > 0$ for $r > r_0$ and $V'(r) < 0$ for $r < r_0$. Thus, $V = -V_0(2(1)^6 - (1)^{12}) = -V_0$ is a minimum.
(Note: We discard the negative root $-r_0$ since the distance r must be positive.)

(b)

$$V(r) = -V_0\left(2\left(\frac{r_0}{r}\right)^6 - \left(\frac{r_0}{r}\right)^{12}\right)$$
$$V'(r) = -V_0(-12r_0^6 r^{-7} + 12r_0^{12} r^{-13})$$
$$V''(r) = -V_0(84r_0^6 r^{-8} - 156r_0^{12} r^{-14})$$

$$V(r_0) = -V_0$$
$$V'(r_0) = 0$$
$$V''(r_0) = 72V_0 r_0^{-2}$$

The Taylor series is thus:
$$V(r) = -V_0 + 72V_0 r_0^{-2} \cdot (r - r_0)^2 \cdot \frac{1}{2} + \cdots$$

(c) The difference between V and its minimum value $-V_0$ is
$$V - (-V_0) = 36V_0\frac{(r - r_0)^2}{r_0^2} + \cdots$$

which is approximately proportional to $(r - r_0)^2$ since terms containing higher powers of $(r - r_0)$ have relatively small values for r near r_0.

(d) From part (a) we know that $dV/dr = 0$ when $r = r_0$, hence $F = 0$ when $r = r_0$. Since, if we discard powers of $(r - r_0)$ higher than the second,

$$V(r) \approx -V_0 \left(1 - 36 \frac{(r - r_0)^2}{r_0^2} \right)$$

giving

$$F = -\frac{dV}{dr} \approx 72 \cdot \frac{r - r_0}{r_0^2} (-V_0) = -72 V_0 \frac{r - r_0}{r_o^2}.$$

So F is approximately proportional to $(r - r_0)$.

57. We first note that the Taylor series for $\cos x$ about $x = 0$ is $\cos x = 1 - x^2/2 + x^4/4! - x^6/6! + \cdots$. Therefore, we see that $1/a_3 = -1/6!$ so $a_3 = -6! = -720$ since substituting $\sqrt{x}$ into the Taylor series for $\cos x$ gives

$$\cos \sqrt{x} = 1 - \frac{1}{2} \left(\sqrt{x} \right)^2 + \frac{1}{4!} \left(\sqrt{x} \right)^4 - \frac{1}{6!} \left(\sqrt{x} \right)^6 + \cdots$$

$$= 1 - \frac{1}{2} x + \frac{1}{24} x^2 - \frac{1}{6!} x^3 + \cdots.$$

Strengthen Your Understanding

61. The order of operations has not been respected. In e^{-x} the function e^x is composed with the function $-x$. Therefore we have: $e^{-x} = 1 + (-x) + \frac{(-x)^2}{2!} + \frac{(-x)^3}{3!} + \cdots = 1 - x + \frac{x^2}{2!} - \frac{x^3}{3!} + \cdots$.

65. True. Since the Taylor series for $\cos x$ has only even powers, multiplying by x^3 gives only odd powers.

69. (c). The Taylor series for

$$3 \tan(x/3) = 3(x/3 + (x/3)^3/3 + 21(x/3)^5/120 + \cdots) = x + x^3/27 + 7x^5/3240 + \cdots.$$

Solutions for Section 10.4

Exercises

1. The error bound in approximating $e^{0.1}$ using the Taylor polynomial of degree 3 for $f(x) = e^x$ about $x = 0$ is:

$$|E_3| = |f(0.1) - P_3(0.1)| \leq \frac{M \cdot |0.1 - 0|^4}{4!} = \frac{M(0.1)^4}{24},$$

where $|f^{(4)}(x)| \leq M$ for $0 \leq x \leq 0.1$. Now, $f^{(4)}(x) = e^x$. Since e^x is increasing for all x, we see that $|f^{(4)}(x)|$ is maximized for x between 0 and 0.1 when $x = 0.1$. Thus,

$$|f^{(4)}| \leq e^{0.1},$$

so

$$|E_3| \leq \frac{e^{0.1} \cdot (0.1)^4}{24} = 0.00000460.$$

The Taylor polynomial of degree 3 is

$$P_3(x) = 1 + x + \frac{1}{2!} x^2 + \frac{1}{3!} x^3.$$

The approximation is $P_3(0.1)$, so the actual error is

$$E_3 = e^{0.1} - P_3(0.1) = 1.10517092 - 1.10516667 = 0.00000425,$$

which is slightly less than the bound.

5. The error bound in approximating $\ln(1.5)$ using the Taylor polynomial of degree 3 for $f(x) = \ln(1 + x)$ about $x = 0$ is:

$$|E_4| = |f(0.5) - P_3(0.5)| \le \frac{M \cdot |0.5 - 0|^4}{4!} = \frac{M(0.5)^4}{24},$$

where $|f^{(4)}(x)| \le M$ for $0 \le x \le 0.5$. Since

$$f^{(4)}(x) = \frac{-3!}{(1 + x)^4}$$

and the denominator attains its minimum when $x = 0$, we have $|f^{(4)}(x)| \le 3!$, so

$$|E_4| \le \frac{3! \, (0.5)^4}{24} = 0.0156.$$

The Taylor polynomial of degree 3 is

$$P_3(x) = 0 + x + (-1)\frac{x^2}{2!} + (-1)(-2)\frac{x^3}{3!}$$

$$= x - \frac{1}{2}x^2 + \frac{1}{3}x^3.$$

The approximation is $P_3(0.5)$, so the actual error is

$$E_3 = \ln(1.5) - P_3(0.5) = 0.4055 - 0.4167 = -0.0112$$

which is slightly less, in absolute value, than the bound.

9. The Taylor series about 0 for e^{-x} is

$$e^{-x} = 1 - x + \frac{x^2}{2!} - \frac{x^3}{3!} + \frac{x^4}{4!} - \frac{x^5}{5!} + \cdots.$$

Substituting $x = 1$ gives

$$e^{-1} = \underbrace{1 - 1 + \frac{1}{2!} - \frac{1}{3!} + \frac{1}{4!}}_{P_4(1)} - \frac{1}{5!} + \cdots.$$

The series for e^{-1} is alternating and the absolute value of the terms is monotonically decreasing toward zero. Thus the error is bounded by

$$E_4(1) = |f(1) - P_4(1)| \le \frac{1}{5!}.$$

13. At $x = 1$ the Taylor series is

$$f(1) = \sum_{n=0}^{\infty} (-1)^n \frac{1}{n^2 + 1}$$

The series is alternating and the absolute value of the terms is monotonically decreasing toward zero since $n^2 + 1$ increases monotonically without bound. Thus the error is bounded by

$$E_6(1) = |f(1) - P_6(1)| \le \frac{1}{7^2 + 1} = \frac{1}{50}.$$

Problems

17. (a) The third degree Taylor approximation of degree 3 of e^x around 0 is

$$P_3(x) = 1 + x + \frac{x^2}{2!} + \frac{x^3}{3!}.$$

Using the third-degree error bound, if $|f^{(4)}(x)| \le M$ for $-2 \le x \le 2$, then

$$|E_3(x)| = |f(x) - P_3(x)| \le \frac{M}{4!} \cdot |x|^4 \le \frac{M2^4}{4!}.$$

Since $|f^{(4)}(x)| = e^x$, and e^x is increasing on $[-2, 2]$,

$$f^{(4)}(x) \le e^2 \text{ for all } x \in [-2, 2].$$

So we can let $M = e^2$ and we get

$$|E_3(x)| < \frac{e^2 \cdot 2^4}{4!} \approx 5$$

(b) The actual maximum error is $|e^2 - P_3(2)| = 1.06$.

21. (a) The Taylor polynomial of degree 0 about $t = 0$ for $f(t) = e^t$ is simply $P_0(x) = 1$. Since $e^t \ge 1$ on $[0, 0.5]$, the approximation is an underestimate.

(b) Using the zero degree error bound, if $|f'(t)| \le M$ for $0 \le t \le 0.5$, then

$$|E_0| \le M \cdot |t| \le M(0.5).$$

Since $|f'(t)| = |e^t| = e^t$ is increasing on $[0, 0.5]$,

$$|f'(t)| \le e^{0.5} < \sqrt{4} = 2.$$

Therefore

$$|E_0| \le (2)(0.5) = 1.$$

(Note: By looking at a graph of $f(t)$ and its 0th degree approximation, it is easy to see that the greatest error occurs when $t = 0.5$, and the error is $e^{0.5} - 1 \approx 0.65 < 1$. So our error bound works.)

25. By the results of Problem 24, if we approximate $\cos 1$ using the nth degree polynomial, the error is at most $\frac{1}{(n+1)!}$. For the answer to be correct to four decimal places, the error must be less than 0.00005. Thus, the first n such that $\frac{1}{(n+1)!} < 0.00005$ will work. In particular, when $n = 7$, $\frac{1}{8!} = \frac{1}{40370} < 0.00005$, so the 7th degree Taylor polynomial will give the desired result. For six decimal places, we need $\frac{1}{(n+1)!} < 0.0000005$. Since $n = 9$ works, the 9th degree Taylor polynomial is sufficient.

29. (a) Since $4 \arctan 1 = \pi$, we approximate π by approximating $4 \arctan x$ by Taylor polynomials with $x = 1$. Let $f(x) = 4 \arctan x$. We find the Taylor polynomial of f about $x = 0$.

$$f(0) = 0$$
$$f'(x) = \frac{4}{1 + x^2} \quad \text{so} \quad f'(0) = 4$$
$$f''(x) = -\frac{8x}{(1 + x^2)^2} \quad \text{so} \quad f''(0) = 0$$
$$f'''(x) = -\frac{8}{(1 + x^2)^2} + \frac{32x^2}{(1 + x^2)^3} \quad \text{so} \quad f'''(0) = -8.$$

Thus, the third degree Taylor polynomial for f is

$$F_3(x) = \frac{4x}{1!} - \frac{8}{3!}x^3 = 4x - \frac{4}{3}x^3.$$

In particular,

$$F_3(1) = 4 - \frac{4}{3} = \frac{8}{3} \approx 2.67.$$

Note: If you already have the Taylor series for $1/(1 + x^2)$, the Taylor polynomial for $\arctan x$ can also be found by integration.

(b) We now approximate π by looking at $g(x) = 2 \arcsin x$ about $x = 0$ and substituting $x = 1$.

$$g(0) = 0$$
$$g'(x) = \frac{2}{\sqrt{1 - x^2}} \quad \text{so} \quad g'(0) = 2$$
$$g''(x) = \frac{2x}{(1 - x^2)^{\frac{3}{2}}} \quad \text{so} \quad g''(0) = 0$$
$$g'''(x) = \frac{2}{(1 - x^2)^{\frac{3}{2}}} + \frac{6x^2}{(1 - x^2)^{\frac{5}{2}}} \quad \text{so} \quad g'''(0) = 2.$$

Thus, the third degree Taylor polynomial for g is

$$G_3(x) = \frac{2x}{1!} + \frac{2x^3}{3!} = 2x + \frac{1}{3}x^3.$$

In particular,

$$G_3(1) = \frac{7}{3} \approx 2.33.$$

Note: If you already have the Taylor series for $1/\sqrt{1-x^2}$, the Taylor polynomial for $\arcsin x$ can also be found by integration.

(c) To estimate the maximum possible error, $|E_3|$, in the approximation using the arctangent, we need a bound on the fourth derivative of $f(x) = \arctan x$ on $0 \le x \le 1$. Since

$$f^{(4)}(x) = -\frac{192x^3}{(1+x^2)^4} + \frac{96x}{(1+x^2)^3},$$

now use a graphing calculator to see that the maximum value of $|f^{(4)}(x)|$, on $0 \le x \le 1$ is about 18.6. Thus,

$$|E_3| \le \frac{18.6}{4!} \approx 0.78.$$

(Notice that $\pi \approx 3.14$ is within 0.78 of 2.67.)

(d) To estimate the maximum possible error, $|E_n|$, in an approximation using the arcsine, we need a bound on the derivatives of $g(x) = \arcsin x$ on $0 \le x \le 1$. The derivatives of $\arcsin x$ contain terms of the form $(1 - x^2)^{-a}$, for some positive a. As x gets close to 1, the value of $(1 - x^2)^{-a}$ approaches ∞. Thus, we cannot get a bound on the derivatives of $\arcsin x$, so the error formula does not give us a bound on $|E_n|$.

Strengthen Your Understanding

33. This statement is not correct, since $f(x)$ and its Taylor approximation $P_n(x)$ around a have the same value at $x = a$. So $f(a) = P_n(a)$ for all n.

37. According to the error formula we need $\max |f^{(n+1)}| \le M$ on the interval between 0 and x. Since we are working with the second-degree Taylor polynomial we have $n = 2$. Therefore, we need a function and an interval such that the third derivative of the function on that interval is less or equal to 4. We can choose, for example, $f(x) = e^x$ and c small enough such that $f^{(3)}(x) = e^x \le 4$ on $[-c, c]$. There are many choices of c, for example $c = 1$ would work.

41. True. When $x = 1$,

$$\sum_{n=0}^{\infty} \frac{f^{(n)}(0)}{n!} x^n = \sum_{n=0}^{\infty} \frac{f^{(n)}(0)}{n!}.$$

Since $f^{(n)}(0) \ge n!$, the terms of this series are all greater than 1. So the series cannot converge

Solutions for Section 10.5

Exercises

1. No, a Fourier series has terms of the form $\cos nx$, not $\cos^n x$.

5.

$$a_0 = \frac{1}{2\pi} \int_{-\pi}^{\pi} f(x)\, dx = \frac{1}{2\pi} \left[\int_{-\pi}^{0} -1\, dx + \int_{0}^{\pi} 1\, dx \right] = 0$$

$$a_1 = \frac{1}{\pi} \int_{-\pi}^{\pi} f(x) \cos x\, dx = \frac{1}{\pi} \left[\int_{-\pi}^{0} -\cos x\, dx + \int_{0}^{\pi} \cos x\, dx \right]$$

$$= \frac{1}{\pi} \left[-\sin x \Big|_{-\pi}^{0} + \sin x \Big|_{0}^{\pi} \right] = 0.$$

Similarly, a_2 and a_3 are both 0. See Figure 10.3.

(In fact, notice $f(x)\cos nx$ is an odd function, so $\int_{-\pi}^{\pi} f(x)\cos nx = 0$.)

$$b_1 = \frac{1}{\pi}\int_{-\pi}^{\pi} f(x)\sin x\,dx = \frac{1}{\pi}\left[\int_{-\pi}^{0} -\sin x\,dx + \int_{0}^{\pi}\sin x\,dx\right]$$

$$= \frac{1}{\pi}\left[\cos x\Big|_{-\pi}^{0} + (-\cos x)\Big|_{0}^{\pi}\right] = \frac{4}{\pi}$$

$$b_2 = \frac{1}{\pi}\int_{-\pi}^{\pi} f(x)\sin 2x\,dx = \frac{1}{\pi}\left[\int_{-\pi}^{0} -\sin 2x\,dx + \int_{0}^{\pi}\sin 2x\,dx\right]$$

$$= \frac{1}{\pi}\left[\frac{1}{2}\cos 2x\Big|_{-\pi}^{0} + (-\frac{1}{2}\cos 2x)\Big|_{0}^{\pi}\right] = 0.$$

$$b_3 = \frac{1}{\pi}\int_{-\pi}^{\pi} f(x)\sin 3x\,dx = \frac{1}{\pi}\left[\int_{-\pi}^{0} -\sin 3x\,dx + \int_{0}^{\pi}\sin 3x\,dx\right]$$

$$= \frac{1}{\pi}\left[\frac{1}{3}\cos 3x\Big|_{-\pi}^{0} + (-\frac{1}{3}\cos 3x)\Big|_{0}^{\pi}\right] = \frac{4}{3\pi}.$$

Thus, $F_1(x) = F_2(x) = \frac{4}{\pi}\sin x$ and $F_3(x) = \frac{4}{\pi}\sin x + \frac{4}{3\pi}\sin 3x$. See Figure 10.4.

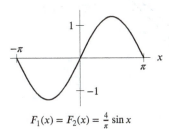

$F_1(x) = F_2(x) = \frac{4}{\pi}\sin x$

Figure 10.3

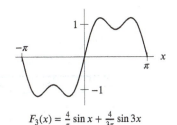

$F_3(x) = \frac{4}{\pi}\sin x + \frac{4}{3\pi}\sin 3x$

Figure 10.4

9.

$$a_0 = \frac{1}{2\pi}\int_{-\pi}^{\pi} h(x)\,dx = \frac{1}{2\pi}\int_{0}^{\pi} x\,dx = \frac{\pi}{4}$$

As in Problem 10, we use the integral table (III-15 and III-16) to find formulas for a_n and b_n.

$$a_n = \frac{1}{\pi}\int_{-\pi}^{\pi} h(x)\cos(nx)\,dx = \frac{1}{\pi}\int_{0}^{\pi} x\cos nx\,dx = \frac{1}{\pi}\left(\frac{x}{n}\sin(nx) + \frac{1}{n^2}\cos(nx)\right)\Big|_{0}^{\pi}$$

$$= \frac{1}{\pi}\left(\frac{1}{n^2}\cos(n\pi) - \frac{1}{n^2}\right)$$

$$= \frac{1}{n^2\pi}\left(\cos(n\pi) - 1\right).$$

Note that since $\cos(n\pi) = (-1)^n$, $a_n = 0$ if n is even and $a_n = -\frac{2}{n^2\pi}$ if n is odd.

$$b_n = \frac{1}{\pi}\int_{-\pi}^{\pi} h(x)\cos(nx)\,dx = \frac{1}{\pi}\int_{0}^{\pi} x\sin x\,dx$$

$$= \frac{1}{\pi}\left(-\frac{x}{n}\cos(nx) + \frac{1}{n^2}\sin(nx)\right)\Big|_{0}^{\pi}$$

$$= \frac{1}{\pi}\left(-\frac{\pi}{n}\cos(n\pi)\right)$$

$$= -\frac{1}{n}\cos(n\pi)$$

$$= \frac{1}{n}(-1)^{n+1} \quad \text{if } n \geq 1$$

We have that the nth Fourier polynomial for h (for $n \geq 1$) is

$$H_n(x) = \frac{\pi}{4} + \sum_{i=1}^{n} \left(\frac{1}{i^2 \pi} \Big(\cos(i\pi) - 1 \Big) \cdot \cos(ix) + \frac{(-1)^{i+1} \sin(ix)}{i} \right).$$

This can also be written as

$$H_n(x) = \frac{\pi}{4} + \sum_{i=1}^{n} \frac{(-1)^{i+1} \sin(ix)}{i} + \sum_{i=1}^{\left[\frac{n}{2}\right]} \frac{-2}{(2i-1)^2 \pi} \cos((2i-1)x)$$

where $\left[\frac{n}{2}\right]$ denotes the biggest integer smaller than or equal to $\frac{n}{2}$. In particular, we have the graphs in Figure 10.5.

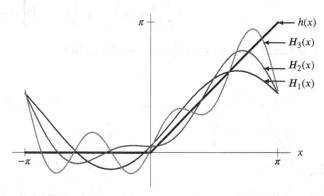

Figure 10.5

Problems

13. (a)   (i) The graph of $y = \sin x + \frac{1}{3}\sin 3x$ is in Figure 10.6.

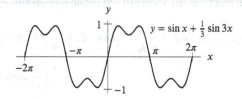

Figure 10.6

(ii) The graph of $y = \sin x + \frac{1}{3}\sin 3x + \frac{1}{5}\sin 5x$ is in Figure 10.6.

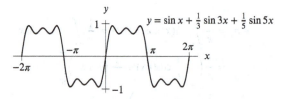

Figure 10.7

(b) Following the pattern, we add the term $\frac{1}{7}\sin 7x$ to get Figure 10.8.

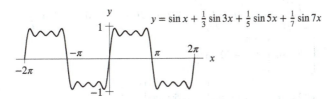

$$y = \sin x + \tfrac{1}{3}\sin 3x + \tfrac{1}{5}\sin 5x + \tfrac{1}{7}\sin 7x$$

Figure 10.8

(c) The equation is

$$f(x) = \begin{cases} \vdots & \vdots \\ 1 & -2\pi \le x < -\pi \\ -1 & -\pi \le x < 0 \\ 1 & 0 \le x < \pi \\ -1 & \pi \le x < 2\pi \\ \vdots & \vdots \end{cases}$$

The square wave function is not continuous at $x = 0,\ \pm\pi,\ \pm2\pi, \dots$. See Figure 10.9.

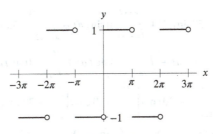

Figure 10.9

17. The signal received on earth is in the form of a periodic function $h(t)$, which can be expanded in a Fourier series

$$h(t) = a_0 + a_1\cos t + a_2\cos 2t + a_3\cos 3t + \cdots$$
$$+ b_1\sin t + b_2\sin 2t + b_3\sin 3t + \cdots$$

If the periodic noise consists of *only* the second and higher harmonics of the Fourier series, then the original signal contributed the fundamental harmonic plus the constant term, i.e.,

$$\underbrace{a_0}_{\text{constant term}} + \underbrace{a_1\cos t + b_1\sin t}_{\text{fundamental harmonic}} = \underbrace{A\cos t}_{\text{original signal}}.$$

In order to find A, we need to find a_0, a_1, and b_1. Looking at the graph of $h(t)$, we see

$$a_0 = \text{average value of } h(t) = \frac{1}{2\pi}(\text{Area above the } x\text{-axis} - \text{Area below the } x\text{-axis})$$
$$= \frac{1}{2\pi}\left[80\left(\frac{\pi}{2}\right) - \left(50\left(\frac{\pi}{4}\right) + 30\left(\frac{\pi}{4}\right) + 30\left(\frac{\pi}{4}\right) + 50\left(\frac{\pi}{4}\right)\right)\right]$$
$$= \frac{1}{2\pi}\left[80\left(\frac{\pi}{2}\right) - 80\left(\frac{\pi}{2}\right)\right] = \frac{1}{2\pi}\cdot 0, = 0$$

$$a_1 = \frac{1}{\pi}\int_{-\pi}^{\pi} h(t)\cos t\,dt$$

$$= \frac{1}{\pi}\left[\int_{-\pi}^{-3\pi/4} -50\cos t\,dt + \int_{-3\pi/4}^{-\pi/2} 0\cos t\,dt + \int_{-\pi/2}^{-\pi/4} -30\cos t\,dt \right.$$

$$\left. + \int_{-\pi/4}^{\pi/4} 80\cos t\,dt + \int_{\pi/4}^{\pi/2} -30\cos t\,dt + \int_{\pi/2}^{3\pi/4} 0\cos t\,dt + \int_{3\pi/4}^{\pi} -50\cos t\,dt\right]$$

$$= \frac{1}{\pi}\left[-50\sin t\Big|_{-\pi}^{-3\pi/4} - 30\sin t\Big|_{-\pi/2}^{-\pi/4}\right.$$

$$\left. +80\sin t\Big|_{-\pi/4}^{\pi/4} - 30\sin t\Big|_{\pi/4}^{\pi/2} - 50\sin t\Big|_{3\pi/4}^{\pi}\right]$$

$$= \frac{1}{\pi}\left[-50\left(-\frac{\sqrt{2}}{2}-0\right) - 30\left(-\frac{\sqrt{2}}{2}-(-1)\right) + 80\left(\frac{\sqrt{2}}{2}-\left(-\frac{\sqrt{2}}{2}\right)\right)\right.$$

$$\left. -30\left(1-\frac{\sqrt{2}}{2}\right) - 50\left(0-\frac{\sqrt{2}}{2}\right)\right]$$

$$= \frac{1}{\pi}[25\sqrt{2}+15\sqrt{2}-30+40\sqrt{2}+40\sqrt{2}-30+15\sqrt{2}+25\sqrt{2}]$$

$$= \frac{1}{\pi}[160\sqrt{2}-60] = 52.93,$$

$$b_1 = \frac{1}{\pi}\int_{-\pi}^{\pi} h(t)\sin t\,dt$$

$$= \frac{1}{\pi}\left[\int_{-\pi}^{-3\pi/4} -50\sin t\,dt + \int_{-3\pi/4}^{-\pi/2} 0\sin t\,dt + \int_{-\pi/2}^{-\pi/4} -30\sin t\,dt\right.$$

$$\left. + \int_{-\pi/4}^{\pi/4} 80\sin t\,dt + \int_{\pi/4}^{\pi/2} -30\sin t\,dt + \int_{\pi/2}^{3\pi/4} 0\sin t\,dt + \int_{3\pi/4}^{\pi} -50\sin t\,dt\right]$$

$$= \frac{1}{\pi}\left[50\cos t\Big|_{-\pi}^{-3\pi/4} + 30\cos t\Big|_{-\pi/2}^{-\pi/4} - 80\cos t\Big|_{-\pi/4}^{\pi/4} + 30\cos t\Big|_{\pi/4}^{\pi/2} + 50\cos t\Big|_{3\pi/4}^{\pi}\right]$$

$$= \frac{1}{\pi}\left[50\left(-\frac{\sqrt{2}}{2}-(-1)\right) + 30\left(\frac{\sqrt{2}}{2}-0\right) - 80\left(\frac{\sqrt{2}}{2}-\frac{\sqrt{2}}{2}\right)\right.$$

$$\left. +30\left(0-\frac{\sqrt{2}}{2}\right) + 50\left(-1-(-\frac{\sqrt{2}}{2})\right)\right]$$

$$= \frac{1}{\pi}\left[-25\sqrt{2}+50+15\sqrt{2}-0-15\sqrt{2}-50+25\sqrt{2}\right] = \frac{1}{\pi}(0) = 0.$$

Also, we could have just noted that $b_1 = \frac{1}{\pi}\int_{-\pi}^{\pi} h(t)\sin t\,dt = 0$ because $h(t)\sin t$ is an odd function. Substituting in, we get

$$a_0 + a_1\cos t + b_1\sin t = 0 + 52.93\cos t + 0 = A\cos t.$$

So $A = 52.93$.

21. The Fourier series for f is

$$f(x) = a_0 + \sum_{k=1}^{\infty} a_k\cos kx + \sum_{k=1}^{\infty} b_k\sin kx.$$

Pick any positive integer m. Then multiply through by $\sin mx$, to get

$$f(x)\sin mx = a_0\sin mx + \sum_{k=1}^{\infty} a_k\cos kx\sin mx + \sum_{k=1}^{\infty} b_k\sin kx\sin mx.$$

Now, integrate term-by-term on the interval $[-\pi, \pi]$ to get

$$\int_{-\pi}^{\pi} f(x)\sin mx\,dx = \int_{-\pi}^{\pi}\left(a_0\sin mx + \sum_{k=1}^{\infty} a_k\cos kx\sin mx + \sum_{k=1}^{\infty} b_k\sin kx\sin mx\right)dx$$

$$= a_0 \int_{-\pi}^{\pi} \sin mx \, dx + \sum_{k=1}^{\infty} \left(a_k \int_{-\pi}^{\pi} \cos kx \sin mx \, dx \right)$$
$$+ \sum_{k=1}^{\infty} \left(b_k \int_{-\pi}^{\pi} \sin kx \sin mx \, dx \right).$$

Since m is a positive integer, we know that the first term of the above expression is zero (because $\int_{-\pi}^{\pi} \sin mx \, dx = 0$). Since $\int_{-\pi}^{\pi} \cos kx \sin mx \, dx = 0$, we know that everything in the first infinite sum is zero. Since $\int_{-\pi}^{\pi} \sin kx \sin mx \, dx = 0$ where $k \neq m$, the second infinite sum reduces down to the case where $k = m$ so

$$\int_{-\pi}^{\pi} f(x) \sin mx \, dx = b_m \int_{-\pi}^{\pi} \sin mx \sin mx \, dx = b_m \pi.$$

Divide by π to get

$$b_m = \frac{1}{\pi} \int_{-\pi}^{\pi} f(x) \sin mx \, dx.$$

25. As c gets closer and closer to 0, the energy of the pulse train will also approach 0, since

$$E = \frac{1}{\pi} \int_{-\pi}^{\pi} (f(x))^2 \, dx = \frac{1}{\pi} \int_{-c/2}^{c/2} 1^2 \, dx = \frac{1}{\pi} \left(\frac{c}{2} - \left(-\frac{c}{2} \right) \right) = \frac{c}{\pi}.$$

The energy spectrum shows the *relative* distribution of the energy of f among its harmonics. The fraction of energy carried by each harmonic gets smaller as c gets closer to 0, as shown by comparing the kth terms of the Fourier series for pulse trains with $c = 2, 1, 0.4$. For instance, notice that the *fraction* or *percentage* of energy carried by the constant term gets smaller as c gets smaller; the same is true for the energy carried by the first harmonic.

　　If each harmonic contributes less energy, then more harmonics are needed to capture a fixed percentage of energy. For example, if $c = 2$, only the constant term and the first two harmonics are needed to capture 90% of the total energy of that pulse train. If $c = 1$, the constant term and the first five harmonics are needed to get 90% of the energy of that pulse train. If $c = 0.4$, the constant term and the first thirteen harmonics are needed to get 90% of the energy of that pulse train. This means that more harmonics, or more terms in the series, are needed to get an accurate approximation. Compare the graphs of the fifth and thirteenth Fourier approximations of f in Problem 23.

29. By formula II-12 of the integral table,

$$\int_{-\pi}^{\pi} \sin kx \cos mx \, dx$$
$$= \frac{1}{m^2 - k^2} \left(m \sin(kx) \sin(mx) + k \cos(kx) \cos(mx) \right) \Big|_{-\pi}^{\pi}$$
$$= \frac{1}{m^2 - k^2} \Big[m \sin(k\pi) \sin(m\pi) + k \cos(k\pi) \cos(m\pi)$$
$$- m \sin(-k\pi) \sin(-m\pi) - k \cos(-k\pi) \cos(-m\pi) \Big].$$

Since k and m are positive integers, $\sin(k\pi) = \sin(m\pi) = \sin(-k\pi) = \sin(-m\pi) = 0$. Also, $\cos(k\pi) = \cos(-k\pi)$ since $\cos x$ is even. Thus this expression reduces to 0. [Note: since $\sin kx \cos mx$ is odd, so $\int_{-\pi}^{\pi} \sin kx \cos mx \, dx$ must be 0.]

Strengthen Your Understanding

33. Unlike Taylor series, Fourier series are good global approximations rather than local ones. Thus, a_0 is the average of $f(x)$ on the interval of approximation.

37. (b). The graph describes an even function, which eliminates (a) and (c). The Fourier series for (d) would have values near π for x close to 0.

Solutions for Chapter 10 Review

Exercises

1. $e^x \approx 1 + e(x-1) + \frac{e}{2}(x-1)^2$

5. $f'(x) = 3x^2 + 14x - 5$, $f''(x) = 6x + 14$, $f'''(x) = 6$. The Taylor polynomial about $x = 1$ is

$$P_3(x) = 4 + \frac{12}{1!}(x-1) + \frac{20}{2!}(x-1)^2 + \frac{6}{3!}(x-1)^3$$
$$= 4 + 12(x-1) + 10(x-1)^2 + (x-1)^3.$$

Notice that if you multiply out and collect terms in $P_3(x)$, you will get $f(x)$ back.

9. The first four nonzero terms of P_7 are given by:

$$i = 1 : \quad \frac{(-1)^{1+1}3^1}{(1-1)!}x^{2 \cdot 1 - 1} = 3x$$

$$i = 2 : \quad \frac{(-1)^{2+1}3^2}{(2-1)!}x^{2 \cdot 2 - 1} = -9x^3$$

$$i = 3 : \quad \frac{(-1)^{3+1}3^3}{(3-1)!}x^{2 \cdot 3 - 1} = \frac{27}{2} \cdot x^5$$

$$i = 4 : \quad \frac{(-1)^{4+1}3^4}{(4-1)!}x^{2 \cdot 4 - 1} = -\frac{27}{2} \cdot x^7.$$

Thus, $P_7 = 3x - 9x^3 + \frac{27}{2} \cdot x^5 - \frac{27}{2} \cdot x^7$.

13.

$$\theta^2 \cos \theta^2 = \theta^2 \left(1 - \frac{(\theta^2)^2}{2!} + \frac{(\theta^2)^4}{4!} - \frac{(\theta^2)^6}{6!} + \cdots\right)$$

$$= \theta^2 - \frac{\theta^6}{2!} + \frac{\theta^{10}}{4!} - \frac{\theta^{14}}{6!} + \cdots$$

17.

$$\frac{1}{\sqrt{4-x}} = \frac{1}{2\sqrt{1 - \frac{x}{2}}} = \frac{1}{2}\left(1 - \frac{x}{2}\right)^{-\frac{1}{2}}$$

$$= \frac{1}{2}\left(1 - \left(-\frac{1}{2}\right)\left(\frac{x}{2}\right) + \frac{1}{2!}\left(-\frac{1}{2}\right)\left(-\frac{3}{2}\right)\left(\frac{x}{2}\right)^2\right.$$

$$\left. - \frac{1}{3!}\left(-\frac{1}{2}\right)\left(-\frac{3}{2}\right)\left(-\frac{5}{2}\right)\left(\frac{x}{2}\right)^3 + \cdots\right)$$

$$= \frac{1}{2} + \frac{1}{8}x + \frac{3}{64}x^2 + \frac{5}{256}x^3 + \cdots$$

21. Using the binomial expansion for $(1+x)^{3/2}$ with $x = y/B$.

$$(B^2 + y^2)^{3/2} = \left(B^2 + B^2\left(\frac{y^2}{B^2}\right)\right)^{3/2} = \left(B^2\left(1 + \left(\frac{y}{B}\right)^2\right)\right)^{3/2} = B^3\left(1 + \left(\frac{y}{B}\right)^2\right)^{3/2}$$

$$= B^3\left(1 + (3/2)\left(\left(\frac{y}{B}\right)^2\right)^1 + \frac{(3/2)(1/2)}{2!}\left(\left(\frac{y}{B}\right)^2\right)^2 + \frac{(3/2)(1/2)(-1/2)}{3!}\left(\left(\frac{y}{B}\right)^2\right)^3 \cdots\right)$$

$$= B^3\left(1 + \frac{3}{2}\left(\frac{y}{B}\right)^2 + \frac{3}{8}\left(\frac{y}{B}\right)^4 - \frac{1}{16}\left(\frac{y}{B}\right)^6 \cdots\right).$$

Problems

25. Infinite geometric series with $a = 1$, $x = -1/3$, so

$$\text{Sum} = \frac{1}{1 - (-1/3)} = \frac{3}{4}.$$

29. (a) Factoring out $7(1.02)^3$ and using the formula for the sum of a finite geometric series with $a = 7(1.02)^3$ and $r = 1/1.02$, we see

$$\text{Sum} = 7(1.02)^3 + 7(1.02)^2 + 7(1.02) + 7 + \frac{7}{(1.02)} + \frac{7}{(1.02)^3} + \cdots + \frac{7}{(1.02)^{100}}$$

$$= 7(1.02)^3 \left(1 + \frac{1}{(1.02)} + \frac{1}{(1.02)^2} + \cdots + \frac{1}{(1.02)^{103}} \right)$$

$$= 7(1.02)^3 \frac{\left(1 - \frac{1}{(1.02)^{104}} \right)}{1 - \frac{1}{1.02}}$$

$$= 7(1.02)^3 \left(\frac{(1.02)^{104} - 1}{(1.02)^{104}} \frac{1.02}{0.02} \right)$$

$$= \frac{7(1.02^{104} - 1)}{0.02(1.02)^{100}}.$$

(b) Using the Taylor expansion for e^x with $x = (0.1)^2$, we see

$$\text{Sum} = 7 + 7(0.1)^2 + \frac{7(0.1)^4}{2!} + \frac{7(0.1)^6}{3!} + \cdots$$

$$= 7 \left(1 + (0.1)^2 + \frac{(0.1)^4}{2!} + \frac{(0.1)^6}{3!} + \cdots \right)$$

$$= 7e^{(0.1)^2}$$

$$= 7e^{0.01}.$$

33. The Taylor series of $\dfrac{1}{1 - 2x}$ around $x = 0$ is

$$\frac{1}{1 - 2x} = 1 + 2x + (2x)^2 + (2x)^3 + \cdots = \sum_{k=0}^{\infty} (2x)^k.$$

To find the radius of convergence, we apply the ratio test with $a_k = (2x)^k$.

$$\lim_{k \to \infty} \frac{|a_{k+1}|}{|a_k|} = \lim_{k \to \infty} \frac{2^{k+1}|x|^{k+1}}{2^k |x|^k} = 2|x|.$$

Hence the radius of convergence is $R = 1/2$.

37. We find the Taylor polynomial for $\cos(x^2)$ by substituting into the series for $\cos x$:

$$\cos \left(x^2 \right) \approx 1 - \frac{1}{2} \left(x^2 \right)^2 + \frac{1}{4!} \left(x^2 \right)^4 = 1 - \frac{x^4}{2} + \frac{x^8}{24}.$$

This means that

$$\int_0^1 \cos \left(x^2 \right) \, dx \approx \int_0^1 \left(1 - \frac{x^4}{2} + \frac{x^8}{24} \right) dx = \left(x - \frac{x^5}{10} + \frac{x^9}{216} \right) \Big|_0^1 = 1 - \frac{1}{10} + \frac{1}{216} = 0.90463.$$

This is a very good estimate; the actual value (found using a computer) is $0.90452 \ldots$.

41. (a) See Figure 10.10. The graph of E_1 looks like a parabola. Since the graph of E_1 is sandwiched between the graph of $y = x^2$ and the x axis, we have

$$|E_1| \leq x^2 \quad \text{for} \quad |x| \leq 0.1.$$

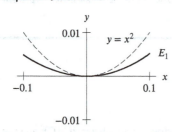

Figure 10.10

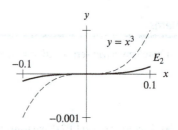

Figure 10.11

(b) See Figure 10.11. The graph of E_2 looks like a cubic, sandwiched between the graph of $y = x^3$ and the x axis, so

$$|E_2| \leq x^3 \quad \text{for} \quad |x| \leq 0.1.$$

(c) Using the Taylor expansion

$$e^x = 1 + x + \frac{x^2}{2!} + \frac{x^3}{3!} + \cdots$$

we see that

$$E_1 = e^x - (1 + x) = \frac{x^2}{2!} + \frac{x^3}{3!} + \frac{x^4}{4!} + \cdots.$$

Thus for small x, the $x^2/2!$ term dominates, so

$$E_1 \approx \frac{x^2}{2!},$$

and so E_1 is approximately a quadratic.

  Similarly

$$E_2 = e^x - (1 + x + \frac{x^2}{2}) = \frac{x^3}{3!} + \frac{x^4}{4!} + \cdots.$$

Thus for small x, the $x^3/3!$ term dominates, so

$$E_2 \approx \frac{x^3}{3!}$$

and so E_2 is approximately a cubic.

45. We have:

$$
\begin{aligned}
P_4(x) &= \sum_{n=1}^{4} \frac{(-n)^{n-1}}{n!} x^n \\
&= \frac{(-1)^{1-1}}{1!} x^1 + \frac{(-2)^{2-1}}{2!} x^2 + \frac{(-3)^{3-1}}{3!} x^3 + \frac{(-4)^{4-1}}{4!} x^4 \\
&= \frac{(-1)^0}{1} x + \frac{(-2)^1}{2} x^2 + \frac{(-3)^2}{6} x^3 + \frac{(-4)^3}{24} x^4 \\
&= x - x^2 + \frac{3}{2} x^3 - \frac{8}{3} x^4.
\end{aligned}
$$

49. (a) $F = \frac{GM}{R^2} + \frac{Gm}{(R+r)^2}$

   (b) $F = \frac{GM}{R^2} + \frac{Gm}{R^2} \frac{1}{(1+\frac{r}{R})^2}$

   Since $\frac{r}{R} < 1$, use the binomial expansion:

$$\frac{1}{(1 + \frac{r}{R})^2} = \left(1 + \frac{r}{R}\right)^{-2} = 1 - 2\left(\frac{r}{R}\right) + (-2)(-3)\frac{(\frac{r}{R})^2}{2!} + \cdots$$

$$F = \frac{GM}{R^2} + \frac{Gm}{R^2}\left[1 - 2\left(\frac{r}{R}\right) + 3\left(\frac{r}{R}\right)^2 - \cdots\right].$$

   (c) Discarding higher power terms, we get

$$
\begin{aligned}
F &\approx \frac{GM}{R^2} + \frac{Gm}{R^2} - \frac{2Gmr}{R^3} \\
&= \frac{G(M + m)}{R^2} - \frac{2Gmr}{R^3}.
\end{aligned}
$$

Looking at the expression, we see that the term $\frac{G(M+m)}{R^2}$ is the field strength at a distance R from a single particle of mass $M + m$. The correction term, $-\frac{2Gmr}{R^3}$, is negative because the field strength exerted by a particle of mass $(M + m)$ at a distance R would clearly be larger than the field strength at P in the question.

53. Expanding $f(y+k)$ and $g(x+h)$ in Taylor series gives

$$f(y+k) = f(y) + f'(y)k + \frac{f''(y)}{2!}k^2 + \cdots,$$

$$g(x+h) = g(x) + g'(x)h + \frac{g''(x)}{2!}h^2 + \cdots.$$

Now let $y = g(x)$ and $y + k = g(x+h)$. Then $k = g(x+h) - g(x)$ so

$$k = g'(x)h + \frac{g''(x)}{2!}h^2 + \cdots.$$

Substituting $g(x+h) = y + k$ and $y = g(x)$ in the series for $f(y+k)$ gives

$$f(g(x+h)) = f(g(x)) + f'(g(x))k + \frac{f''(g(x))}{2!}k^2 + \cdots.$$

Now, substituting for k, we get

$$f(g(x+h)) = f(g(x)) + f'(g(x)) \cdot \left(g'(x)h + \frac{g''(x)}{2!}h^2 + \cdots\right) + \frac{f''(g(x))}{2!}(g'(x)h + \ldots)^2 + \cdots$$

$$= f(g(x)) + (f'(g(x))) \cdot g'(x)h + \text{Terms in } h^2 \text{ and higher powers.}$$

So, substituting for $f(g(x+h))$ and dividing by h, we get

$$\frac{f(g(x+h)) - f(g(x))}{h} = f'(g(x)) \cdot g'(x) + \text{Terms in } h \text{ and higher powers,}$$

and thus, taking the limit as $h \to 0$,

$$\frac{d}{dx}f(g(x)) = \lim_{h \to 0} \frac{f(g(x+h)) - f(g(x))}{h}$$

$$= f'(g(x)) \cdot g'(x).$$

57. (a) Expand $f(x)$ into its Fourier series:

$$f(x) = a_0 + a_1 \cos x + a_2 \cos 2x + a_3 \cos 3x + \cdots + a_k \cos kx + \cdots$$
$$+ b_1 \sin x + b_2 \sin 2x + b_3 \sin 3x + \cdots + b_k \sin kx + \cdots$$

Then differentiate term-by-term:

$$f'(x) = -a_1 \sin x - 2a_2 \sin 2x - 3a_3 \sin 3x - \cdots - ka_k \sin kx - \cdots$$
$$+ b_1 \cos x + 2b_2 \cos 2x + 3b_3 \cos 3x + \cdots + kb_k \cos kx + \cdots$$

Regroup terms:

$$f'(x) = +b_1 \cos x + 2b_2 \cos 2x + 3b_3 \cos 3x + \cdots + kb_k \cos kx + \cdots$$
$$- a_1 \sin x - 2a_2 \sin 2x - 3a_3 \sin 3x - \cdots - ka_k \sin kx - \cdots$$

which forms a Fourier series for the derivative $f'(x)$. The Fourier coefficient of $\cos kx$ is kb_k and the Fourier coefficient of $\sin kx$ is $-ka_k$. Note that there is no constant term as you would expect from the formula ka_k with $k = 0$. Note also that if the kth harmonic f is absent, so is that of f'.

(b) If the amplitude of the kth harmonic of f is

$$A_k = \sqrt{a_k^2 + b_k^2}, \quad k \geq 1,$$

then the amplitude of the kth harmonic of f' is

$$\sqrt{(kb_k)^2 + (-ka_k)^2} = \sqrt{k^2(b_k^2 + a_k^2)} = k\sqrt{a_k^2 + b_k^2} = kA_k.$$

(c) The energy of the kth harmonic of f' is k^2 times the energy of the kth harmonic of f.

CAS Challenge Problems

61. (a) The Taylor polynomials of degree 7 are

$$\text{For } \sin x, \qquad P_7(x) = x - \frac{x^3}{6} + \frac{x^5}{120} - \frac{x^7}{5040}$$

$$\text{For } \sin x \cos x, \qquad Q_7(x) = x - \frac{2x^3}{3} + \frac{2x^5}{15} - \frac{4x^7}{315}$$

(b) The coefficient of x^3 in $Q_7(x)$ is $-2/3$, and the coefficient of x^3 in $P_7(x)$ is $-1/6$, so the ratio is

$$\frac{-2/3}{-1/6} = 4.$$

The corresponding ratios for x^5 and x^7 are

$$\frac{2/15}{1/120} = 16 \quad \text{and} \quad \frac{-4/315}{-1/5040} = 64.$$

(c) It appears that the ratio is always a power of 2. For x^3, it is $4 = 2^2$; for x^5, it is $16 = 2^4$; for x^7, it is $64 = 2^6$. This suggests that in general, for the coefficient of x^n, it is 2^{n-1}.

(d) From the identity $\sin(2x) = 2 \sin x \cos x$, we expect that $P_7(2x) = 2Q_7(x)$. So, if a_n is the coefficient of x^n in $P_7(x)$, and if b_n is the coefficient of x^n in $Q_7(x)$, then, since the x^n terms $P_7(2x)$ and $2Q_7(x)$ must be equal, we have

$$a_n(2x)^n = 2b_n x^n.$$

Dividing both sides by x^n and combining the powers of 2, this gives the pattern we observed. For $a_n \neq 0$,

$$\frac{b_n}{a_n} = 2^{n-1}.$$

CHAPTER ELEVEN

Solutions for Section 11.1

Exercises

1. Since $y = x^3$, we know that $y' = 3x^2$. Substituting $y = x^3$ and $y' = 3x^2$ into the differential equation we get

$$
\begin{aligned}
0 &= xy' - 3y \\
&= x(3x^2) - 3(x^3) \\
&= 3x^3 - 3x^3 \\
&= 0.
\end{aligned}
$$

Since this equation is true for all x, we see that $y = x^3$ is in fact a solution.

5. This is a solution to (VI) since

$$y' = 2x \cdot 0.5e^{x^2} = 2xy.$$

9. (a) We substitute $y = 4x^3$ and its derivative $dy/dx = 12x^2$ into the differential equation:

$$
\begin{aligned}
y\frac{dy}{dx} &= 6x^2 \\
(4x^3) \cdot (12x^2) &= 6x^2? \\
48x^5 &\neq 6x^2.
\end{aligned}
$$

The function $y = 4x^3$ does not satisfy the differential equation so it is not a solution.

(b) We substitute $y = 2x^{3/2}$ and its derivative $dy/dx = 3x^{1/2}$ into the differential equation:

$$
\begin{aligned}
y\frac{dy}{dx} &= 6x^2 \\
(2x^{3/2}) \cdot (3x^{1/2}) &= 6x^2? \\
\text{Yes:} \quad 6x^2 &= 6x^2.
\end{aligned}
$$

The function $y = 2x^{3/2}$ satisfies the differential equation so it is a solution.

(c) We substitute $y = 6x^{3/2}$ and its derivative $dy/dx = 9x^{1/2}$ into the differential equation:

$$
\begin{aligned}
y\frac{dy}{dx} &= 6x^2 \\
(6x^{3/2}) \cdot (9x^{1/2}) &= 6x^2? \\
54x^2 &\neq 6x^2.
\end{aligned}
$$

The function $y = 6x^{3/2}$ does not satisfy the differential equation so it is not a solution.

13. Differentiating $y(x) = Ae^{\lambda x}$ gives

$$y'(x) = A\lambda e^{\lambda x} = \lambda(Ae^{\lambda x}) = \lambda y.$$

Therefore, $y(x)$ is a solution of $y' = \lambda y$ for any value of A.

17. (a) To determine whether Q is increasing or decreasing, we check to see whether dQ/dt is positive or negative. Substituting $Q = 8$ and $t = 2$ into the differential equation, we have:

$$
\begin{aligned}
\frac{dQ}{dt} &= \frac{t}{Q} - 0.5 \\
\frac{dQ}{dt} &= \frac{2}{8} - 0.5 \\
\frac{dQ}{dt} &= -0.25 < 0.
\end{aligned}
$$

Since dQ/dt is negative, we see that Q is decreasing at $t = 2$.

(b) Since $dQ/dt = -0.25$, the rate of change of Q at $t = 2$ is -0.25 per unit of t. If the rate of change stays approximately constant over the interval, then Q changes by approximately -0.25 in going from $t = 2$ to $t = 3$. We have:

$$\text{Value of } Q \text{ at } 3 \approx \text{Value of } Q \text{ at } 2 + \text{ Change in } Q$$
$$= 8 + (-0.25)$$
$$= 7.75.$$

21. Since $P = 5$ when $t = 3$, we have $5 = C/3$; therefore, $C = 15$. So the particular solution is $P = 15/t$.

Problems

25. If $y = \cos \omega t$, then

$$\frac{dy}{dt} = -\omega \sin \omega t, \qquad \frac{d^2y}{dt^2} = -\omega^2 \cos \omega t.$$

Thus, if $\frac{d^2y}{dt^2} + 9y = 0$, then

$$-\omega^2 \cos \omega t + 9 \cos \omega t = 0$$
$$(9 - \omega^2) \cos \omega t = 0.$$

Thus $9 - \omega^2 = 0$, or $\omega^2 = 9$, so $\omega = \pm 3$.

29. If y satisfies the differential equation, then we must have

$$\frac{d\,(5 + 3e^{kx})}{dx} = 10 - 2(5 + 3e^{kx})$$
$$3ke^{kx} = 10 - 10 - 6e^{kx}$$
$$3ke^{kx} = -6e^{kx}$$
$$k = -2.$$

So, if $k = -2$ the formula for y solves the differential equation.

33. (a) (IV) because $dy/dx = k = kx/x = y/x$.
(b) (III) because $dy/dx = ke^{kx} = ky$.
(c) (I) because $dy/dx = kxe^{kx} + e^{kx} = ky + xe^{kx}/x = ky + y/x$.
(d) (II) because $dy/dx = kx^{k-1} = kx^k/x = ky/x$.

37. Since both $f(x)$ and $g(x)$ are solutions of the differential equation, we know that $d(f(x))/dx = x$ and $d(g(x))/dx = x$. Therefore, $f(x) - g(x)$ is a solution to $dy/dx = 0$ since

$$\frac{d}{dx}\left(f(x) - g(x)\right) = \frac{d}{dx}(f(x)) - \frac{d}{dx}(g(x)) = x - x = 0.$$

Strengthen Your Understanding

41. An example is $dy/dx = x/y$ with the condition that $y = 100$ when $x = 0$. Many other examples are possible.

45. An example is $dy/dx = 1/x$. We find the solution by integrating and see that the solution is the function $y = \ln|x| + C$. One solution is the logarithmic function $y = \ln|x|$. Other examples are possible.

49. False. The function $y = t^2$ is a solution to $y'' = 2$.

53. False. If $g(x) > 0$ for all x, then $f(x)$ would have to be increasing for all x so $f(x + p) = f(x)$ would be impossible. For example, let $g(x) = 2 + \cos x$. Then a possibility for f is $f(x) = 2x + \sin x$. Then $g(x)$ is periodic, but $f(x)$ is not.

57. False. The example $f(x) = x^3$ and $g(x) = 3x^2$ shows that you might expect $f(x)$ to be odd. However, the additive constant C can mess things up. For example, still let $g(x) = 3x^2$, but let $f(x) = x^3 + 1$ instead. Then $g(x)$ is still even, but $f(x)$ is not odd (for example, $f(-1) = 0$ but $-f(1) = -2$).

Solutions for Section 11.2

Exercises

1. (a) The slope at any point is given by the derivative, so we find the slope by substituting the x- and y-coordinates into the differential equation $dy/dx = x^2 - y^2$ to find dy/dx.

$$\text{The slope at } (1,0) \text{ is } \frac{dy}{dx} = 1^2 - 0^2 = 1.$$

$$\text{The slope at } (0,1) \text{ is } \frac{dy}{dx} = 0^2 - 1^2 = -1.$$

$$\text{The slope at } (1,1) \text{ is } \frac{dy}{dx} = 1^2 - 1^2 = 0.$$

$$\text{The slope at } (2,1) \text{ is } \frac{dy}{dx} = 2^2 - 1^2 = 3.$$

$$\text{The slope at } (1,2) \text{ is } \frac{dy}{dx} = 1^2 - 2^2 = -3.$$

$$\text{The slope at } (2,2) \text{ is } \frac{dy}{dx} = 2^2 - 2^2 = 0.$$

(b) See Figure 11.1.

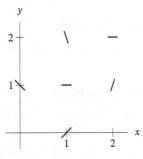

Figure 11.1

5. There are many possible answers. One possibility is shown in Figures 11.2 and 11.3.

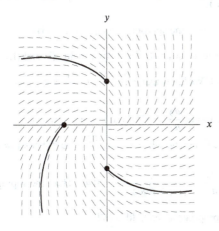

Figure 11.2

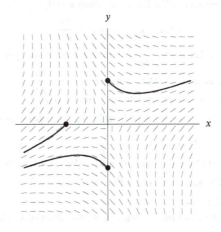

Figure 11.3

9. The solution curve is increasing at $(0,3)$ since $y' > 0$ there:

$$y' = 2y - 3x - 4 = 2 \cdot 3 - 3 \cdot 0 - 4 = 2.$$

Problems

13. (a) Since slope field (A) is constant along horizontal lines, but varies vertically, it represents a differential equation in which y' depends on y but not on t. This is $y' = 0.3y$.

    Slope field (B) is constant along vertical lines, but varies horizontally, so it represents a differential equation in which y' depends on t but not on y. This is $y' = 0.3t$.

    (b) For $y' = 0.3y$, we look at slope field (A).
       (i) Starting at $(0, 1)$, we have $\lim_{t \to \infty} y(t) = \infty$.
       (ii) Starting at $(0, 0)$, we have $\lim_{t \to \infty} y(t) = 0$.

    For $y' = 0.3t$, we look at slope field (B).
       (i) Starting at $(0, 1)$, we have $\lim_{t \to \infty} y(t) = \infty$.
       (ii) Starting at $(0, 0)$, we have $\lim_{t \to \infty} y(t) = \infty$.

17. The second derivative y'' can be used to determine whether the solution is concave up or down, so we first find y'':

$$y' = 2x - 3y - 1$$
$$\text{so} \quad y'' = 2 - 3y'$$
$$= 2 - 3(2x - 3y - 1)$$
$$= 2 - 6x + 9y + 3$$
$$= 5 - 6x + 9y.$$

The solution curve is concave up at $(3, 2)$ since $y'' > 0$ there:

$$y'' = 5 - 6 \cdot 3 + 9 \cdot 2 = 5.$$

21. Notice that $y' = \dfrac{x+y}{x-y}$ is zero when $x = -y$ and is undefined when $x = y$. A solution curve will be horizontal (slope= 0) when passing through a point with $x = -y$, and will be vertical (slope undefined) when passing through a point with $x = y$. The only slope field for which this is true is slope field (b).

25. (a) We know that $\dfrac{dy}{dx} = e^{x^2} > 0$ for all x. As $x \to \infty$, the slopes tend toward infinity. Similarly, as $x \to -\infty$, the slopes tend to infinity. Thus, the appropriate slope field is (IV).

    (b) We know that $\dfrac{dy}{dx} > 0$ for all x. As $x \to \pm\infty$, the slopes tend toward 0. Only slope fields (I) and (III) meet these conditions. We know that when $x = 0$, $\dfrac{dy}{dx} = e^{-2x^2} = e^0 = 1$.
       We know that when $x = 1$, $\dfrac{dy}{dx} = e^{-2x^2} = e^{-2} \approx 0.14$. Thus, the appropriate slope field is (I).

    (c) We know that $\dfrac{dy}{dx} > 0$ for all x. As $x \to \pm\infty$, the slopes tend toward 0. Only slope fields (I) and (III) meet these conditions. We know that when $x = 0$, $\dfrac{dy}{dx} = e^{-x^2/2} = e^0 = 1$. We know that when $x = 1$, $\dfrac{dy}{dx} = e^{-x^2/2} = e^{-1/2} \approx 0.61$. Thus, the appropriate slope field is (III).

    (d) The slope field is both positive and negative. In fact, when $x = 0$, $\dfrac{dy}{dx} = e^{-0.5x} \cos x = e^0 \cos 0 = 1$.
       Also, when $x = 1$, $\dfrac{dy}{dx} = e^{-0.5x} \cos x = e^{-1/2} \cos 1 \approx 0.33$, and when $x = 2$, $\dfrac{dy}{dx} = e^{-0.5x} \cos x = e^{-1} \cos 2 \approx -0.153$. Thus, the appropriate slope field is (V).

    (e) The slope field is positive for all values of x. When $x = 0$, $\dfrac{dy}{dx} = \dfrac{1}{(1 + 0.5 \cos x)^2} \approx 0.44$. Thus, the appropriate slope field is (II).

    (f) The slope field is negative for all values of x. Thus, the appropriate slope field is (VI).

29. Judging from the figure, we see that:
    - The slope depends on x, not on y.
    - The slope is positive for $0 < x < 5$.
    - The slope is zero at $x = 0$ and $x = 5$.

    This corresponds to equation (d): $y' = 0.05x(5 - x)$.

Strengthen Your Understanding

33. Note that $y' = y$ has a slope field with all positive slopes when $y > 0$. The given slope field, however, has negative slopes in the second quadrant.

37. If a derivative dy/dx depends only on y, then the derivative is constant for any fixed value of y. In other words, the slope is the same for all points on any horizontal line. One possibility is the slope field for $dy/dx = y$ in Figure 11.4.

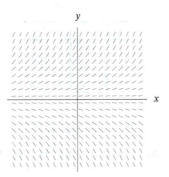

Figure 11.4

41. False. If $y(0) \leq 0$, then $\lim_{x \to \infty} y = -\infty$.

45. False. We have

$$\frac{d^2 y}{dx^2} = \frac{d(x^2 + y^2 + 1)}{dx}$$
$$= 2x + 2y\frac{dy}{dx}$$
$$= 2x + 2y(x^2 + y^2 + 1)$$
$$= 2x + 2y + 2x^2 y + 2y^3.$$

At the point $(x, y) = (-1, 0)$ we have $d^2 y/dx^2 = -2 < 0$. A negative second derivative indicates function concave down. The solution curve of the differential equation that passes through the point $(-1, 0)$ is concave down at $(-1, 0)$.

49. True. Differentiate $dy/dx = 2x - y$, to get:

$$\frac{d^2 y}{dx^2} = \frac{d}{dx}(2x - y) = 2 - \frac{dy}{dx} = 2 - (2x - y).$$

53. False. Suppose that $g(x) = f(x) + C$, where $C \neq 0$. In order to be a solution of $dy/dx = 2x - y$ we would need $g'(x) = 2x - g(x)$. Instead we have:

$$g'(x) = f'(x) = 2x - f(x) = 2x - (g(x) - C) = 2x - g(x) + C.$$

Since $C \neq 0$, this means $g(x)$ is not a solution of $dy/dx = 2x - y$.

Solutions for Section 11.3

Exercises

1. Using Euler's method, we have:

At $(x, y) = (0, 4)$: $y' = (0 - 2)(4 - 3)$ $= -2$

so at $x = 0.1$: $y = 4.0 - 2(0.1)$ $= 3.8$ because $\Delta x = 0.1$

At $(x, y) = (0.1, 3.8)$: $y' = (0.1 - 2)(3.8 - 3)$ $= -1.52$

so at $x = 0.2$: $y = 3.8 - 1.52(0.1)$ $= 3.648$ because $\Delta x = 0.1$

At $(x, y) = (0.2, 3.648)$: $y' = (0.2 - 2)(3.648 - 3) = -1.1664$.

Thus, the completed table is

x	y	y'
0.0	4.0	-2
0.1	3.8	-1.52
0.2	3.648	-1.1664

5. At $(0, 2)$, the slope is given by $y' = 2 \cdot 2 - 3 \cdot 0 - 4 = 0$. This means

$$\Delta y = (\text{slope at } (0, 2)) \cdot \Delta x = 0(0.1) = 0,$$

and therefore,

$$y_1 = (y\text{-value at } (0, 2)) + \Delta y = 2 + 0 = 2.$$

At $(0.1, 2)$, the slope is given by $y' = 2 \cdot 2 - 3 \cdot 0.1 - 4 = -0.3$. This means

$$\Delta y = (\text{slope at } (0.1, 2)) \cdot \Delta x = -0.3(0.1) = -0.03$$

and therefore,

$$y_2 = (y\text{-value at } (0.1, 2)) + \Delta y = 2 + (-0.03) = 1.97.$$

9. (a) See Figure 11.5.

(b) See Table 11.1. At $x = 1$, Euler's method gives $y \approx 0.16$.

(c) Our answer to (a) appears to be an underestimate. This is as we would expect, since the solution curve is concave up.

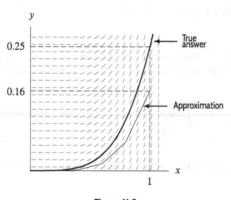

Figure 11.5

Table 11.1

x	y	$\Delta y = (\text{slope})\Delta x$
0	0	0
0.2	0	0.0016
0.4	0.0016	0.0128
0.6	0.0144	0.0432
0.8	0.0576	0.1024
1	0.1600	

Problems

13. Since the error is proportional to one over the number of subintervals, the error using 10 intervals should be roughly half the error obtained using 5 intervals. Since both the estimates are underestimates, if we let A be the actual value we have:

$$\frac{1}{2}(A - 0.667) = A - 0.710$$

$$A - 0.667 = 2A - 1.420$$

$$A = 0.753$$

Therefore, 0.753 should be a better approximation.

17. (a) **(i)** Using one step, $\frac{\Delta B}{\Delta t} = 0.05$, so $\Delta B = \left(\frac{\Delta B}{\Delta t}\right)\Delta t = 50$. Therefore we get an approximation of $B \approx 1050$ after one year.

(ii) With two steps, $\Delta t = 0.5$ and we have

Table 11.2

t	B	$\Delta B = (0.05B)\Delta t$
0	1000	25
0.5	1025	25.63
1.0	1050.63	

(iii) Keeping track to the nearest hundredth with $\Delta t = 0.25$, we have

Table 11.3

t	B	$\Delta B = (0.05B)\Delta t$
0	1000	12.5
0.25	1012.5	12.66
0.5	1025.16	12.81
0.75	1037.97	12.97
1	1050.94	

(b) In part (i), we get our approximation by making a single increment, ΔB, where ΔB is just $0.05B$. If we think in terms of interest, ΔB is just like getting one end of the year interest payment. Since ΔB is 0.05 times the balance B, it is like getting 5% interest at the end of the year.

(c) Part (ii) is equivalent to computing the final amount in an account that begins with $1000 and earns 5% interest compounded twice annually. Each step is like computing the interest after 6 months. When $t = 0.5$, for example, the interest is $\Delta B = (0.05B) \cdot \frac{1}{2}$, and we add this to $1000 to get the new balance.

Similarly, part (iii) is equivalent to the final amount in an account that has an initial balance of $1000 and earns 5% interest compounded quarterly.

Strengthen Your Understanding

21. If $x(0) > 0$, then the statement is true. However, if $x(0) < 0$, then the solution curve is decreasing. Since we are using the value of x at the beginning of a subinterval to estimate the rate of change of x on an interval, we obtain an overestimate for $x(1)$.

25. True. Both lead to
$$y(1) \approx f(0) \cdot 0.2 + f(0.2) \cdot 0.2 + f(0.4) \cdot 0.2 + f(0.6) \cdot 0.2 + f(0.8) \cdot 0.2.$$

29. The length of the interval is 30, so dividing into n equal sized pieces with $\Delta x = 0.1$ would require $n = 300$ pieces.

Solutions for Section 11.4

Exercises

1. (a) Yes   **(b)** No   **(c)** Yes
   (d) No   **(e)** Yes   **(f)** Yes
   (g) No   **(h)** Yes   **(i)** No
   (j) Yes   **(k)** Yes   **(l)** No

5. For this equation, we can separate variables as follows:

$$y' - xy' = y$$
$$y'(1 - x) = y$$
$$\frac{dy}{dx} = \frac{y}{(1 - x)}$$
$$\frac{dy}{y} = \frac{dx}{1 - x}.$$

9. Separating variables gives

$$\int \frac{dQ}{Q} = \int \frac{dt}{5},$$

so

$$\ln|Q| = \frac{1}{5}t + C.$$

So

$$|Q| = e^{\frac{1}{5}t+C} = e^{\frac{1}{5}t}e^C$$

and

$$Q = Ae^{\frac{1}{5}t}, \text{ where } A = \pm e^C.$$

From the initial conditions we know that $Q(0) = 50$, so $Q(0) = Ae^{(\frac{1}{5})\cdot 0} = A = 50$. Thus

$$Q = 50e^{\frac{1}{5}t}.$$

13. Separating variables gives

$$\int \frac{dz}{z} = \int 5\, dt$$
$$\ln|z| = 5t + C.$$

Solving for z, we have

$$z = Ae^{5t}, \text{ where } A = \pm e^C.$$

Using the fact that $z(1) = 5$, we have $z(1) = Ae^5 = 5$, so $A = 5/e^5$. Therefore,

$$z = \frac{5}{e^5}e^{5t} = 5e^{5t-5}.$$

17. Separating variables gives

$$\int \frac{dy}{y} = -\int \frac{1}{3}\, dx$$
$$\ln|y| = -\frac{1}{3}x + C.$$

Solving for y, we have

$$y = Ae^{-\frac{1}{3}x}, \text{ where } A = \pm e^C.$$

Since $y(0) = A = 10$, we have

$$y = 10e^{-\frac{1}{3}x}.$$

21. Factoring and separating variables gives

$$\frac{dQ}{dt} = 0.3(Q - 400)$$
$$\int \frac{dQ}{Q - 400} = \int 0.3\, dt$$
$$\ln|Q - 400| = 0.3t + C$$
$$Q = 400 + Ae^{0.3t}, \quad \text{where } A = \pm e^C.$$

The initial condition, $Q(0) = 50$, gives

$$50 = 400 + A \quad \text{so} \quad A = -350.$$

Thus

$$Q = 400 - 350e^{0.3t}.$$

25. Write

$$\int \frac{1}{y} dy = \int \frac{1}{3+t} dt$$

and so

$$\ln|y| = \ln|3+t| + C$$

or

$$\ln|y| = \ln D|3+t|$$

where $\ln D = C$. Therefore

$$y = D(3+t).$$

The initial condition $y(0) = 1$ gives $D = \frac{1}{3}$, so

$$y = \frac{1}{3}(3+t).$$

29. Separating variables gives

$$\frac{dz}{dt} = z + zt^2 = z(1 + t^2)$$

$$\int \frac{dz}{z} = \int (1 + t^2) dt,$$

so

$$\ln|z| = t + \frac{t^3}{3} + C,$$

giving

$$z = Ae^{t + t^3/3}.$$

We have $z = 5$ when $t = 0$, so $A = 5$ and

$$z = 5e^{t + t^3/3}.$$

Problems

33. (a) We separate variables and integrate:

$$\frac{dy}{dx} = \frac{4x}{y^2}$$

$$y^2 \, dy = 4x \, dx$$

$$\int y^2 \, dy = \int 4x \, dx$$

$$\frac{y^3}{3} = 2x^2 + C$$

$$y = \sqrt[3]{6x^2 + B}.$$

Here, we use B as the arbitrary constant, replacing $3C$ when we multiply through by 3.

(b) When we substitute $y = 1$ when $x = 0$, we have:

$$1 = \sqrt[3]{6(0^2) + B}$$

$$1 = \sqrt[3]{B}$$

$$B = 1^3 = 1.$$

The particular solution satisfying $y(0) = 1$ is $y = \sqrt[3]{6x^2 + 1}$.
When we substitute $y = 2$ when $x = 0$, we have:

$$2 = \sqrt[3]{6(0^2) + B}$$
$$2 = \sqrt[3]{B}$$
$$B = 2^3 = 8.$$

The particular solution satisfying $y(0) = 2$ is $y = \sqrt[3]{6x^2 + 8}$.
When we substitute $y = 3$ when $x = 0$, we have:

$$3 = \sqrt[3]{6(0^2) + B}$$
$$3 = \sqrt[3]{B}$$
$$B = 3^3 = 27.$$

The particular solution satisfying $y(0) = 3$ is $y = \sqrt[3]{6x^2 + 27}$.
The three solutions are shown in Figure 11.6.

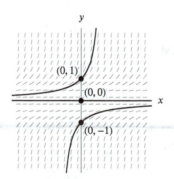

Figure 11.6

37. (a) See Figure 11.7.

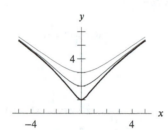

Figure 11.7

(b) It appears that the solution curves in the upper half plane are asymptotic to $y = 0$ as x tends to $-\infty$, and that they are unbounded above as x increases.

It appears that the solution curves in the lower half plane are asymptotic to $y = 0$ as x tends to $+\infty$, and that they are unbounded below as x decreases.

The solution curve through the origin is the x-axis, so it is asymptotic to $y = 0$ on both ends.

(c) We separate variables. From $dy/dx = y^2$ we get, for $y \neq 0$,

$$\int \frac{1}{y^2} dy = \int dx$$
$$-\frac{1}{y} = x + C$$
$$y = \frac{-1}{x + C}.$$

(d) For a given number C, the y-value is not defined for $x = -C$, so the formula gives two solution curves. The first is

$$y = \frac{-1}{x+C} \quad \text{for } x < -C.$$

This curve is in the upper half plane, has a vertical asymptote at $x = -C$ and satisfies $\lim_{x \to -\infty} y = 0$. The second curve is

$$y = \frac{-1}{x+C} \quad \text{for } x > -C,$$

which is in the lower half plane, has a vertical asymptote at $x = -C$, and satisfies $\lim_{x \to +\infty} y = 0$.

41. Separating variables gives

$$\int \frac{dQ}{b-Q} = \int dt.$$

Integrating yields

$$-\ln|b-Q| = t + C,$$

so

$$|b-Q| = e^{-(t+C)} = e^{-t}e^{-C}$$
$$Q = b - Ae^{-t}, \quad \text{where } A = \pm e^{-C} \quad \text{or } A = 0.$$

45. Separating variables and integrating gives

$$\int \frac{1}{y^2} dy = \int k(1+t^2) dt$$

or

$$-\frac{1}{y} = k\left(t + \frac{1}{3}t^3\right) + C.$$

Hence,

$$y = \frac{-1}{k(t + \frac{1}{3}t^3) + C}.$$

49. Separating variables gives

$$\frac{dx}{dt} = \frac{x \ln x}{t},$$

so

$$\int \frac{dx}{x \ln x} = \int \frac{dt}{t},$$

and thus

$$\ln|\ln x| = \ln t + C,$$

so

$$|\ln x| = e^C e^{\ln t} = e^C t.$$

Therefore

$$\ln x = At, \quad \text{where } A = \pm e^C \quad \text{or} \quad A = 0, \quad \text{so} \quad x = e^{At}.$$

53. (a), (b) See Figure 11.8.

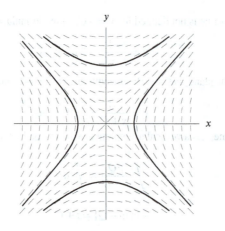

Figure 11.8

(c) Since $dy/dx = x/y$, we have

$$\int y\,dy = \int x\,dx,$$

and thus

$$\frac{y^2}{2} = \frac{x^2}{2} + C,$$

or

$$y^2 - x^2 = 2C.$$

This is the equation of the hyperbolas in part (b).

Strengthen Your Understanding

57. It is impossible to separate variables in the differential equation $dy/dx = x+y$. If we subtract y from both sides, we obtain $dy/dx - y = x$. If we then try to separate the dx, we have $dy - y\,dx = x\,dx$. The variables cannot be separated in this differential equation.

61. An expression such as $f(x) = \cos x$ satisfies the requirement since $dy/dx = \cos x + xy - \cos x = xy$ is a separable differential equation. Other examples are possible.

65. True. We can write $dy/dx = f(x)h(y)$ where $f(x) = 1$ and $h(y) = 1/g(y)$.

69. False. It is true that $y = x^3$ is a solution of the differential equation, since $dy/dx = 3x^2 = 3y^{2/3}$, but it is not the only solution passing through $(0,0)$. Another solution is the constant function $y = 0$. Usually there is only one solution curve to a differential equation passing through a given point, but not always.

Solutions for Section 11.5

Exercises

1. (a) = (III), (b) = (IV), (c) = (I), (d) = (II).

5. For $y = \alpha$, we have $dy/dt = 0$, so the solution curve is horizontal. This is an equilibrium solution.
  For $y > \alpha$, we have $dy/dt < 0$, so the function is decreasing.
  For $y < \alpha$, we have $dy/dt > 0$, so the function is increasing. See Figure 11.9.

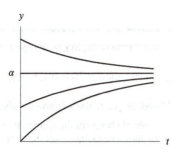

Figure 11.9

9. (a) A very hot cup of coffee cools faster than one near room temperature. The differential equation given says that the rate at which the coffee cools is proportional to the difference between the temperature of the surrounding air and the temperature of the coffee. Since $dH/dt < 0$ (the coffee is cooling) and $H - 20 > 0$ (the coffee is warmer than room temperature), k must be positive.

   (b) Separating variables gives

$$\int \frac{1}{H - 20} dH = \int -k\, dt$$

   and so

$$\ln|H - 20| = -kt + C$$

   and

$$H(t) = 20 + Ae^{-kt}.$$

   If the coffee is initially boiling (100°C), then $A = 80$ and so

$$H(t) = 20 + 80e^{-kt}.$$

   When $t = 2$, the coffee is at 90°C and so $90 = 20 + 80e^{-2k}$ so that $k = \frac{1}{2}\ln\frac{8}{7}$.
   Let the time when the coffee reaches 60°C be H_d, so that

$$60 = 20 + 80e^{-kH_d}$$

$$e^{-kH_d} = \frac{1}{2}.$$

   Therefore,

$$H_d = \frac{1}{k}\ln 2 = \frac{2\ln 2}{\ln\frac{8}{7}} \approx 10 \text{ minutes.}$$

13. Letting $k = 0.10$ and $r = 500$, we have:

$$\frac{dP}{dt} = kP - r$$

$$= 0.10(10{,}000) - 500 = 500.$$

   Thus, at $t = 0$, $dP/dt = 500$. Therefore, the tangent line approximation to P at the point $t = 0$ is

$$P \approx 10{,}000 + 500t.$$

   Using the tangent line approximation at $t = 0.5$, we get $P(0.5) \approx 10{,}000 + 500 \cdot (0.5) = 10{,}250$.

17. Since it takes 6 years to reduce the pollution to 10%, another 6 years would reduce the pollution to 10% of 10%, which is equivalent to 1% of the original. Therefore it takes 12 years for 99% of the pollution to be removed. (Note that the value of Q_0 does not affect this.) Thus the second time is double the first because the fraction remaining, 0.01, in the second instance is the square of the fraction remaining, 0.1, in the first instance.

21. Substituting in the constant function $y = 1$ on the left-hand side of the differential equation gives $d(1)/dx = 0$. The graph of $g(y)$ shows that $g(1) = 0$, so the differential equation is satisfied. Therefore, $y = 1$ is an equilibrium solution to $dy/dx = g(y)$.

Problems

25. (a) We define N to be the amount of nicotine in the body, in mg, at time t, where t represents the number of hours since smoking the cigarette.

(b) We have $dN/dt = -0.347N$ with initial condition $N(0) = 0.4$. Notice that the constant -0.347 is negative since the quantity of nicotine is decreasing.

(c) The general solution is $N = Ce^{-0.347t}$ and the particular solution satisfying the initial condition is $N = 0.4e^{-0.347t}$.

29. (a) Newton's Law of Heating states that the rate of change of the temperature is proportional to the temperature difference. Therefore, if H is the temperature of the milk at time t, we get the differential equation

$$\frac{dH}{dt} = -k(H - 22) \qquad \text{for some constant } k.$$

When $H < 22$, the temperature is increasing, so dH/dt is positive. If k is positive, then $-k(H - 22)$ is also positive.

(b) The general solution to the differential equation is obtained by separating variables

$$\frac{dH}{dt} = -k(H - 22)$$
$$\int \frac{dH}{H - 22} = \int -k\,dt$$
$$\ln|H - 22| = -kt + A$$
$$|H - 22| = e^{-kt}e^{A}$$
$$H = Ce^{-kt} + 22.$$

Since at $t = 0$, $H = 3$, we have $3 = 22 + C$, so $C = -19$. Thus, $H = 22 - 19e^{-kt}$ for some positive constant k.

33. (a) With quantity Q present at time t, we know

$$\frac{dQ}{dt} = -kQ.$$

The constant k is positive, because dQ/dt is negative since Q is decreasing.

(b) To find the half-life, we first solve the differential equation, getting

$$Q = Q_0 e^{-kt}.$$

The half-life, T, is given by solving:

$$\frac{1}{2} = e^{-kT}$$
$$T = -\frac{\ln(1/2)}{k} = \frac{\ln 2}{k}.$$

(c) The half-life is a decreasing function of k, since k is in the denominator and $\ln 2$ is positive. This is as expected as larger k means a faster decay rate and a shorter half-life.

37. (a) If $C' = -kC$, then $C = C_0 e^{-kt}$. Since the half-life is 5730 years,

$$\frac{1}{2}C_0 = C_0 e^{-5730k}.$$

Solving for k, we have

$$-5730k = \ln\left(\frac{1}{2}\right)$$

so

$$k = -\frac{\ln(1/2)}{5730} \approx 0.000121.$$

(b) From the given information, we have $0.91 = e^{-kt}$, where t is the age of the shroud in 1988. Solving for t, we have

$$t = -\frac{\ln 0.91}{k} = 779.4 \text{ years.}$$

41. (a) Since speed is the derivative of distance, Galileo's mistaken conjecture was $\frac{dD}{dt} = kD$.

(b) We know that if Galileo's conjecture were true, then $D(t) = D_0 e^{kt}$, where D_0 would be the initial distance fallen. But if we drop an object, it starts out not having traveled any distance, so $D_0 = 0$. This would lead to $D(t) = 0$ for all t.

Strengthen Your Understanding

45. When $y = 2$, we have $dy/dx = 8 - 8x \neq 0$. An equilibrium solution must have derivative equal to zero for all values of x.

49. Some possible answers are $dQ/dt = Q - 500$ or $dQ/dx = 2(Q - 500)$. Other answers are possible. We must have Q as the dependent variable and the derivative equal to zero when $Q = 500$.

Solutions for Section 11.6

Exercises

1. (a) (III) An island can only sustain the population up to a certain size. The population will grow until it reaches this limiting value.

(b) (V) The ingot will get hot and then cool off, so the temperature will increase and then decrease.

(c) (I) The speed of the car is constant, and then decreases linearly when the breaks are applied uniformly.

(d) (II) Carbon-14 decays exponentially.

(e) (IV) Tree pollen is seasonal, and therefore cyclical.

5. The balance is decreasing at a rate of 0.065 times the current balance and is also decreasing at a rate of 50,000 dollars per year. The differential equation is

$$\frac{dB}{dt} = -0.065B - 50{,}000.$$

9. Since mg is constant and $a = dv/dt$, differentiating $ma = mg - kv$ gives

$$m\frac{da}{dt} = -k\frac{dv}{dt} = -ma.$$

Thus, the differential equation is

$$\frac{da}{dt} = -\frac{k}{m}a.$$

Solving for a gives

$$a = a_0 e^{-kt/m}.$$

At $t = 0$, we have $a = g$, the acceleration due to gravity. Thus, $a_0 = g$, so

$$a = ge^{-kt/m}.$$

Problems

13. Caffeine is leaving the body at a rate of 17% per hour and is entering the body at a rate of 130 mg per hour, so the differential equation is

$$\frac{dA}{dt} = -0.17A + 130.$$

Writing the differential equation as

$$\frac{dA}{dt} = -0.17(A - 764.706),$$

we use separation of variables to see that the general solution is $A = 764.706 + Ce^{-0.17t}$. Since the initial condition is $A_0 = 0$, we have $C = -764.706$ so the particular solution is

$$A = 764.706 - 764.706e^{-0.17t}.$$

We substitute $t = 10$ to find the amount of caffeine at 5 pm:

$$A = 764.706 - 764.706e^{-0.17(10)} = 625.007.$$

At 5 pm, the person has about 625 mg of caffeine in the body.

17. (a) If P = pressure and h = height, $\frac{dP}{dh} = -3.7 \times 10^{-5} P$, so $P = P_0 e^{-3.7 \times 10^{-5} h}$. Now $P_0 = 29.92$, since pressure at sea level (when $h = 0$) is 29.92, so $P = 29.92 e^{-3.7 \times 10^{-5} h}$. At the top of Mt. Whitney, the pressure is

$$P = 29.92 e^{-3.7 \times 10^{-5}(14500)} \approx 17.50 \text{ inches of mercury.}$$

At the top of Mt. Everest, the pressure is

$$P = 29.92 e^{-3.7 \times 10^{-5}(29000)} \approx 10.23 \text{ inches of mercury.}$$

(b) The pressure is 15 inches of mercury when

$$15 = 29.92 e^{-3.7 \times 10^{-5} h}$$

Solving for h gives $h = \frac{-1}{3.7 \times 10^{-5}} \ln(\frac{15}{29.92}) \approx 18{,}661.5$ feet.

21. We are given that the rate of change of pressure with respect to volume, dP/dV is proportional to P/V, so that

$$\frac{dP}{dV} = k \frac{P}{V}.$$

Using separation of variables and integrating gives

$$\int \frac{dP}{P} = k \int \frac{dV}{V}.$$

Evaluating these integral gives

$$\ln P = k \ln V + c$$

or equivalently,

$$P = AV^k.$$

25. (a) We have

$$\frac{dp}{dt} = -k(p - p^*),$$

where k is constant. Notice that $k > 0$, since if $p > p^*$ then dp/dt should be negative, and if $p < p^*$ then dp/dt should be positive.

(b) Separating variables, we have

$$\int \frac{dp}{p - p^*} = \int -k \, dt.$$

Solving, we find $p = p^* + (p_0 - p^*)e^{-kt}$, where p_0 is the initial price.

(c) See Figure 11.10.

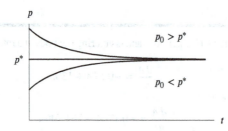

Figure 11.10

(d) As $t \to \infty$, $p \to p^*$. We see this in the solution in part (b), since as $t \to \infty$, $e^{-kt} \to 0$. In other words, as $t \to \infty$, p approaches the equilibrium price p^*.

29. We have $c = 0.05 - 0.05e^{-3\times10^{-5}t}$. We want to solve for t when $c = 0.0002$:

$$0.0002 = 0.05 - 0.05e^{-3\times10^{-5}t}$$

$$-0.0498 = -0.05e^{-3\times10^{-5}t}$$

$$e^{-3\times10^{-5}t} = 0.996$$

$$t = \frac{-\ln(0.996)}{3\times10^{-5}} = 133.601 \text{ min.}$$

33. (a) Since the rod has length a, the ball is initially at $(0, a)$.

(b) In Figure 11.11, the point (x, y) is a typical point on the tractrix. Since the rod is tangent to the tractrix, the slope is given by

$$\frac{dy}{dx} = -\frac{y}{\sqrt{a^2 - y^2}}.$$

(c) Since $x = a(t - \tanh t)$, we have

$$\frac{dx}{dt} = a\left(1 - \frac{1}{\cosh^2 t}\right) = a\frac{\cosh^2 t - 1}{\cosh^2 t} = a\frac{\sinh^2 t}{\cosh^2 t}.$$

Since $y = a/\cosh t$, we have

$$\frac{dy}{dt} = -a\frac{\sinh t}{\cosh^2 t}$$

Thus, the left-hand side of the differential equation is

$$\frac{dy}{dx} = \frac{dy/dt}{dx/dt} = -a\frac{\sinh t}{\cosh^2 t} \cdot \frac{\cosh^2 t}{a \sinh^2 t} = -\frac{1}{\sinh t}.$$

The right-hand side of the differential equation is

$$-\frac{y}{\sqrt{a^2 - y^2}} = -\frac{a/\cosh t}{\sqrt{a^2 - a^2/\cosh^2 t}} = -\frac{a}{\cosh t} \cdot \frac{\cosh t}{a \sinh t} = -\frac{1}{\sinh t}$$

Since the left and right sides of the equation are equal, the differential equation is satisfied.

At $t = 0$, we have $x = a(0 - \tanh 0) = 0$ and $y = a/\cosh 0 = a$, so the initial position of the ball is at $(0, a)$. Thus, the initial condition is also satisfied.

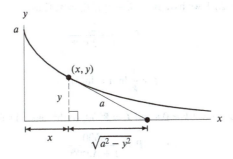

Figure 11.11

Strengthen Your Understanding

37. We use Q for the size of the quantity at time t. The rate due to its growth on its own is proportional to the cube root (such as $0.2\sqrt[3]{Q}$) and the rate due to the external contribution is constant (such as 100). One possible answer is $dQ/dt = 0.2\sqrt[3]{Q} + 100$.

Solutions for Section 11.7

Exercises

1. (a) $P = \frac{1}{1+e^{-t}} = (1 + e^{-t})^{-1}$

    $\frac{dP}{dt} = -(1 + e^{-t})^{-2}(-e^{-t}) = \frac{e^{-t}}{(1+e^{-t})^2}$.

    Then $P(1 - P) = \frac{1}{1+e^{-t}} \left(1 - \frac{1}{1+e^{-t}}\right) = \left(\frac{1}{1+e^{-t}}\right)\left(\frac{e^{-t}}{1+e^{-t}}\right) = \frac{e^{-t}}{(1+e^{-t})^2} = \frac{dP}{dt}$.

    (b) As t tends to ∞, e^{-t} goes to 0. Thus $\lim_{t\to\infty} \frac{1}{1+e^{-t}} = 1$.

5. Since dA/dt is a quadratic function of A with negative leading coefficient, the graph of dA/dt against A is a parabola opening down. Since $dA/dt = 0$ when $A = 0$ and when $A = 1/0.0002 = 5000$, the horizontal intercepts of the parabola are at 0 and 5000. See Figure 11.12. Since we don't know the value of k, we can't put a scale on the vertical axis.

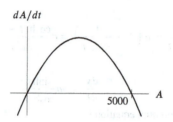

Figure 11.12

9. (a) Equilibrium values are values where $dP/dt = 0$, so the equilibrium values are at $P = 0$ and $P = 2000$.
    (b) At $P = 500$, we see that dP/dt is positive so P is increasing.

13. We see from the differential equation that $k = 0.05$ and $L = 2800$, so the general solution is

$$P = \frac{2800}{1 + Ae^{-0.05t}}.$$

17. We rewrite

$$10P - 5P^2 = 10P\left(1 - \frac{P}{2}\right),$$

so $k = 10$ and $L = 2$. Since $P_0 = L/4$, we have $A = (L - P_0)/P_0 = 3$. Thus

$$P = \frac{2}{1 + 3e^{-10t}}.$$

The time to peak dP/dt is

$$t = \frac{1}{k}\ln A = \ln(3)/10.$$

21. We see from the differential equation that $k = 0.8$ and $L = 8500$, so the general solution is

$$P = \frac{8500}{1 + Ae^{-0.8t}}.$$

We substitute $t = 0$ and $P = 500$ to solve for the constant A:

$$500 = \frac{8500}{1 + Ae^0}$$

$$\frac{1}{500} = \frac{1 + A}{8500}$$

$$17 = 1 + A$$

$$A = 16.$$

The solution to this initial value problem is

$$P = \frac{8500}{1 + 16e^{-0.8t}}.$$

Problems

25. (a) The long run number of cases is 539,226.

(b) The number of new cases in a week is approximated by dN/dt. For some constants k and L, logistic growth follows

$$\frac{dN}{dt} = kN\left(1 - \frac{N}{L}\right).$$

The maximum value of dN/dt occurs at $N = L/2 = 539{,}226/2 = 269{,}613$ for the Dominican Republic. Since $k = 0.35$, rounding to the nearest person, we have

$$\text{Max new cases in a week} = 0.35 \cdot 269{,}613 \left(1 - \frac{269{,}613}{539{,}226}\right) = 47{,}182.$$

29. First we find the time t when the population is 10% of its carrying capacity L. We have

$$P(t) = 0.1L$$

$$\frac{L}{1 + 10e^{-0.2t}} = 0.1L$$

$$1 + 10e^{-0.2t} = \frac{1}{0.1} = 10$$

$$-0.2t = \ln 0.9$$

$$t = \frac{\ln 0.9}{-0.2} = 0.527 \text{ days.}$$

Next we find the time t when the population is 90% of its carrying capacity L. We have

$$P(t) = 0.9L$$

$$\frac{L}{1 + 10e^{-0.2t}} = 0.9L$$

$$1 + 10e^{-0.2t} = \frac{1}{0.9} = 10/9$$

$$-0.2t = \ln(1/90)$$

$$t = \frac{\ln(1/90)}{-0.2} = 22.499 \text{ days.}$$

Therefore, it takes $22.499 - 0.527 = 21.972$ days, or approximately 22 days, for this population to grow from 10% to 90% of its carrying capacity.

33. (a) The equilibrium population occurs when dP/dt is zero. Solving

$$\frac{dP}{dt} = 1 - 0.0004P = 0$$

gives $P = 2500$ fish as the equilibrium population.

(b) The solution of the differential equation is

$$P(t) = \frac{2500}{1 + Ae^{-0.25t}}$$

subject to $P(-10) = 1000$ if $t = 0$ represents the present time. So we have

$$1000 = \frac{2500}{(1 + Ae^{2.5})}$$

from which $A = 0.123127$ and

$$P(0) = \frac{2500}{(1 + 0.123127)} \approx 2230.$$

Therefore, the current population is approximately 2230 fish.

(c) The effect of losing 10% of the fish each year gives the revised differential equation

$$\frac{dP}{dt} = (0.25 - 0.0001P)P - 0.1P$$

or

$$\frac{dP}{dt} = (0.15 - 0.0001P)P.$$

The revised equilibrium population occurs where $dP/dt = 0$, or about 1500 fish.

37. Let r be the relative growth rate, that is $r = (1/N)dN/dt$. Since r is a linear function of N, its graph contains the points $(N, r) = (5, 15\%)$ and $(N, r) = (10, 14.5\%)$, so its slope is

$$m = \frac{\Delta r}{\Delta N} = \frac{0.145 - 0.15}{10 - 5} = \frac{-0.005}{5} = -0.001.$$

Using the point $(5, 0.15)$, we have

$$r = 0.15 - 0.001(N - 5)$$
$$r = 0.155 - 0.001N.$$

This gives us the differential equation

$$\frac{1}{N}\frac{dN}{dt} = 0.155 - 0.001N$$
$$\frac{dN}{dt} = N(0.155 - 0.001N).$$

We see that $dN/dt = 0$ where $N = 0$ or where

$$0.155 - 0.001N = 0$$
$$0.001N = 0.155$$
$$N = 155.$$

Our model predicts the spread of pigweed will halt when 155 million acres are afflicted.

41. (a) Since $t = 0$ is in 1993, for 2014, we substitute $t = 21$ into the logistic function

$$P = \frac{2695}{1 + 2.72e^{-0.0415t}} \quad \text{so} \quad P = \frac{2695}{1 + 2.72e^{-0.0415 \cdot 21}} = 1261 \text{ bn barrels.}$$

(b) The derivative satisfies the differential equation

$$\frac{dP}{dt} = 0.0415P\left(1 - \frac{P}{2695}\right).$$

We can estimate the quantity of oil produced during 2014 using this differential equation at $t = 21$, or $P = 1261$ as found in part (a). We get

$$\left.\frac{dP}{dt}\right|_{t=21} = 27.9 \text{ bn barrels.}$$

(c) Since the original reserves were estimated to be $L = 2695$ bn barrels, the oil projected to remain is $2695 - 1261 = 1434$ billion barrels.

(d) The difference between the projection using the derivative and the actual value, is $28.4 - 27.9 = 0.5$ billion barrels, or about 500 million barrels in 2014.

More precisely, the estimate is $0.4/28.4 = 0.018 \approx$ less than 2% too low—very close.

Strengthen Your Understanding

45. The curve has the shape we expect of a logistic curve. However, since the graph is leveling off at a carrying capacity of 100, we expect the inflection point to be at a height of 50, but in the graph the inflection point is above 60.

49. Since P is growing logistically, the graph of dP/dt against P is a parabola opening down which passes through the origin. Since P increases when $0 < P < 20$, we know that dP/dt is positive when $0 < P < 20$. Since P decreases when $P < 0$ or $P > 20$, we know that dP/dt is negative for those values of P. See Figure 11.13.

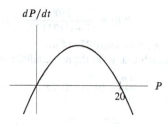

Figure 11.13

Additional Problems (online only)

53. (a) Figure 11.14 shows that the yeast population seems to stabilize at about 13, so we take this to be the limiting value, L.

(b) Solving $\dfrac{dP}{dt} = kP(1 - \dfrac{P}{L})$ for k, we get:

$$k = \frac{dP/dt}{P \cdot (1 - P/L)}.$$

We now find dP/dt from the first two data points.

$$\frac{dP}{dt} \approx \frac{\Delta P}{\Delta t} = \frac{P(10) - P(0)}{10 - 0} = \frac{8.87 - 0.37}{10} = 0.85.$$

Putting in our values for dP/dt, L, and $P(10)$, we get:

$$k \approx \frac{dP/dt}{P(10) \cdot (1 - P(10)/L)} = \frac{0.85}{(8.87)(1 - 8.87/13)} = 0.30.$$

(c) For $k = 0.3$ and $L = 13$,

$$A = \frac{L - P_0}{P_0} = \frac{13 - 0.37}{0.37} = 34.1.$$

Putting this into the equation for P we get:

$$P = \frac{13}{1 + Ae^{-kt}} = \frac{13}{1 + 34.1e^{-0.3t}},$$

which is plotted in Figure 11.15.

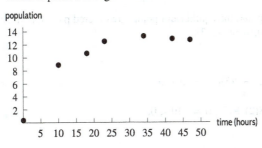

population

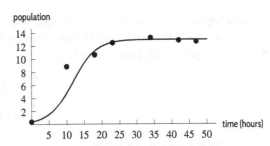

population

Figure 11.14

Figure 11.15: $P = 13 / \left(1 + 34.1e^{-0.3t}\right)$

57. (a) At equilibrium $dP/dt = 0$, so

$$\frac{dP}{dt} = kP\left(1 - \frac{P}{L}\right) - cP = 0,$$

Thus if $P \neq 0$,

$$kL - kP - cL = 0$$
$$P = \frac{k - c}{k}L.$$

At equilibrium, the annual harvest is

$$H = cP = \frac{c(k - c)}{k}L.$$

(b) Since k and L are constant and the annual harvest is $H = c(k - c)L/k$, at the maximum where $dH/dc = 0$, we have

$$\frac{dH}{dc} = \frac{(k - 2c)}{k}L = 0$$
$$c = \frac{k}{2}.$$

The maximum value of H is then

$$H = \left(\frac{k}{2}\right)\frac{(k - k/2)}{k}L = \frac{k}{4}L.$$

(c) The equilibrium population is represented by the P-value (horizontal coordinate) at the point of intersection of the line and parabola. The annual harvest at the equilibrium, cP, is represented by the vertical coordinate at the point of intersection.

As c increases toward k, the slope of the line $dP/dt = cP$ gets steeper and the intersection point between the line and parabola moves closer to the origin. For $c > k/2$, the equilibrium population, $P = (k-c)L/k$, and the annual harvest, $H = c(k - c)L/k$, get smaller and smaller. At $c = k$, the population becomes extinct.

Solutions for Section 11.8

Exercises

1. Here x and y both increase at about the same rate.

5. We set each derivative equal to zero and solve:

$$\frac{dx}{dt} = 0$$
$$-3x + xy = 0$$
$$x(-3 + y) = 0$$
$$x = 0 \text{ or } y = 3.$$

Also,

$$\frac{dy}{dt} = 0$$
$$5y - xy = 0$$
$$y(5 - x) = 0$$
$$y = 0 \text{ or } x = 5.$$

Since both derivatives must be zero at an equilibrium point, the equilibrium points are ordered pairs for which $x = 0$ or $y = 3$ *and* $y = 0$ or $x = 5$. The equilibrium points are $(0, 0)$ and $(5, 3)$.

9. (a) At the point $x = 3$ and $y = 2$, we have

$$\frac{dx}{dt} = 5(3) - 3(3)(2) = -3 < 0$$
$$\frac{dy}{dt} = -8(2) + (3)(2) = -10 < 0,$$

so both x and y are decreasing.

(b) At the point $x = 5$ and $y = 1$, we have

$$\frac{dx}{dt} = 5(5) - 3(5)(1) = 10 > 0$$
$$\frac{dy}{dt} = -8(1) + (5)(1) = -3 < 0,$$

so x is increasing and y is decreasing.

13. We set each derivative equal to zero and solve:

$$\frac{dx}{dt} = 0$$
$$x(3 - x + y) = 0$$
$$x = 0 \text{ or } y = x - 3.$$

Also,

$$\frac{dy}{dt} = 0$$
$$y(6 - x - y) = 0$$
$$y = 0 \text{ or } y = 6 - x.$$

Since both derivatives must be zero at an equilibrium point, the equilibrium points are ordered pairs for which $x = 0$ or $y = x - 3$ *and* $y = 0$ or $y = 6 - x$. If $x = 0$, we get the points $(0, 0)$ and $(0, 6)$, and if $y = 0$ we get the additional point $(3, 0)$. The last equilibrium point $(4.5, 1.5)$ is found by solving the system of equations when both $y = x - 3$ and $y = 6 - x$.

17. When $w > 0$, the term $0.003w$ contributes a positive term to the relative growth of v, so it is helpful.

Problems

21. (a) The human population shrinks as humans are turned into zombies, so parameter a is negative. The interaction term for zombies is positive, since an interaction between a human and a zombie increases the zombie population, so c is positive. The zombie population shrinks by a certain percentage each time interval because some zombies will starve to death, so b is negative.

(b) The terms aHZ and cHZ both indicate the rate of human to zombie conversions. Since each loss of one human is directly a gain of one zombie, $aHZ = -cHZ$. Therefore, the parameter a is exactly the negative of the parameter c. Thus, $a = -c$.

25. If there are no worms, then $w = 0$, and $\frac{dr}{dt} = -r$ giving $r = r_0 e^{-t}$, where r_0 is the initial robin population. If there are no robins, then $r = 0$, and $\frac{dw}{dt} = w$ giving $w = w_0 e^{t}$, where w_0 is the initial worm population.

29.

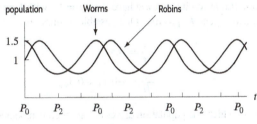

Figure 11.16

33. (a) Symbiosis, because both populations decrease while alone but are helped by the presence of the other.

(b)

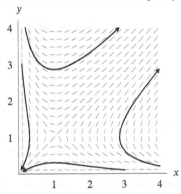

Both populations tend to infinity or both tend to zero.

37. We set each derivative equal to zero and solve:

$$\frac{dp}{dt} = 0$$
$$p(0.01q - 0.3) = 0$$
$$p = 0 \text{ or } q = 30.$$

Also,

$$\frac{dq}{dt} = 0$$
$$q(0.02p - 0.2) = 0$$
$$q = 0 \text{ or } p = 10.$$

Since $p, q \neq 0$, the only equilibrium is $(10, 30)$.

41. (a) Lanchester's square law for the battle of Iwo Jima is

$$0.05y^2 - 0.01x^2 = C.$$

If we measure x and y in thousands, $x_0 = 54$ and $y_0 = 21.5$, so $0.05(21.5)^2 - 0.01(54)^2 = C$ giving $C = -6.0475$. Thus the equation of the trajectory is

$$0.05y^2 - 0.01x^2 = -6.0475$$

giving

$$x^2 - 5y^2 = 604.75.$$

(b) Assuming that the battle did not end until all the Japanese were dead or wounded, that is, $y = 0$, then the number of US soldiers remaining is given by $x^2 - 5(0)^2 = 604.75$. This gives $x = 24.59$, or about 25,000 troops. This is approximately what happened.

Strengthen Your Understanding

45. Since X is indifferent to Y and thrives on its own, we have $dx/dt = kx$ with k positive. Since Y needs X to survive, we have $dy/dt = -k_1 y + k_2 xy$, with k_1 and k_2 positive. One possible example is

$$\frac{dx}{dt} = 0.5x$$
$$\frac{dy}{dt} = -0.1y + 0.3xy.$$

49. False. Competitive exclusion, in which one population drives out another, is modeled by a system of differential equations.

Additional Problems (online only)

53. (a) At an equilibrium point both w and r are constant, so

$$\frac{dw}{dt} = 0 \text{ and } \frac{dr}{dt} = 0.$$

Therefore, we need to solve

$$aw - cwr = 0 \text{ and } -br + kwr = 0.$$

Rearranging gives

$$w(a - cr) = 0 \text{ and } -r(b - kw) = 0$$

so the only equilibrium points are $w = 0, r = 0$ and

$$w = \frac{b}{k} \text{ and } r = \frac{a}{c}.$$

(b) If the insecticide causes a decline in the number of worms then the model becomes

$$\frac{dw}{dt} = aw - cwr - pw \text{ and } \frac{dr}{dt} = -br + kwr$$

where p is a positive constant. Solving as before,

$$w(a - p - cr) = 0 \text{ and } -r(b - kw) = 0,$$

so the equilibrium points are $w = 0, r = 0$ and

$$w = \frac{b}{k} \text{ and } r = \frac{a - p}{c}.$$

So, the equilibrium worm population is unchanged but the equilibrium robin population falls.

Solutions for Section 11.9

Exercises

1. Equilibrium points occur where both derivatives dx/dt and dy/dt are zero. Thus, equilibrium are located at the intersection of a nullcline with vertical line segments and a nullcline with horizontal line segments. We see in Figure 11.17 there is only one equilibrium point located at the point $(4, 10)$.

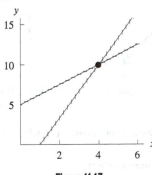

Figure 11.17

5. We can see in Figure 11.91 that any trajectory eventually tends toward the equilibrium point $(4, 10)$.

9. At the point $(10, 4)$, we have $dx/dt < 0$ and $dy/dt < 0$. The trajectory goes down and to the left, crossing the top nullcline with $dy/dt = 0$. The trajectory then goes up and to the left since dy/dt has changed from negative to positive. The trajectory eventually tends toward the equilibrium point $(0, 6)$. See Figure 11.18.

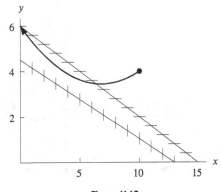

Figure 11.18

Problems

13. We first find the nullclines. Again, we assume $x, y \geq 0$.
    Vertical nullclines occur where $dx/dt = 0$, which happens when $\frac{dx}{dt} = x(2 - x - y) = 0$,
    i.e. when $x = 0$ or $x + y = 2$.
    Horizontal nullclines occur where $dy/dt = 0$, which happens when $\frac{dy}{dt} = y(1 - x - y) = 0$, i.e. when $y = 0$ or $x + y = 1$.
    These nullclines are shown in Figure 11.19.

    Equilibrium points (also shown in Figure 11.19) occur where both dy/dt and dx/dt are 0, i.e. at the intersections of vertical and horizontal nullclines. There are three such points for these equations: $(0, 0)$, $(0, 1)$, and $(2, 0)$.

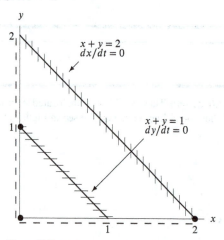

Figure 11.19: Nullclines and equilibrium points (dots)

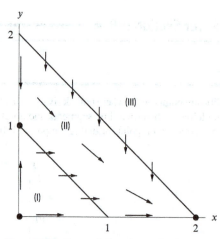

Figure 11.20: General directions of trajectories and equilibrium points (dots)

Looking at sectors in Figure 11.20, we see that no matter in what sector the initial point lies, the trajectory will head toward the equilibrium point $(2, 0)$.

17. We first find the nullclines. Again, we assume $x, y \geq 0$.

$\frac{dx}{dt} = x(1 - x - \frac{y}{3}) = 0$ when $x = 0$ or $x + y/3 = 1$.

$\frac{dy}{dt} = y(1 - y - \frac{x}{2}) = 0$ when $y = 0$ or $y + x/2 = 1$.

These nullclines are shown in Figure 11.21. There are four equilibrium points for these equations. Three of them are the points, $(0, 0)$, $(0, 1)$, and $(1, 0)$. The fourth is the intersection of the two lines $x + y/3 = 1$ and $y + x/2 = 1$. This point is $(\frac{4}{5}, \frac{3}{5})$.

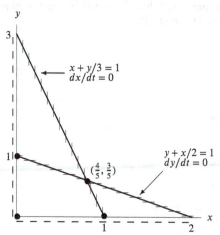

Figure 11.21: Nullclines and equilibrium points (dots)

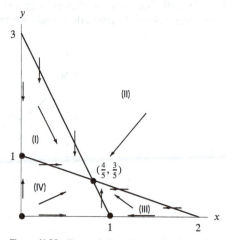

Figure 11.22: General directions of trajectories and equilibrium points (dots)

Looking at sectors in Figure 11.22, we see that no matter in what sector the initial point lies, the trajectory will head toward the equilibrium point $(\frac{4}{5}, \frac{3}{5})$. Only if the initial point lies on the x- or y-axis, will the trajectory head toward the equilibrium points at $(1, 0)$, $(0, 1)$, or $(0, 0)$. In fact, the trajectory will go to $(0, 0)$ only if it starts there, in which case $x(t) = y(t) = 0$ for all t. From direction of the trajectories in Figure 11.22, it appears that if the initial point is in sectors (I) or (III), then it will remain in that sector as it heads toward the equilibrium.

21. (a) If B were not present, then we'd have $A' = 2A$, so company A's net worth would grow exponentially. Similarly, if A were not present, B would grow exponentially. The two companies restrain each other's growth, probably by competing for the market.

(b) To find equilibrium points, find the solutions of the pair of equations

$$A' = 2A - AB = 0$$
$$B' = B - AB = 0$$

The first equation has solutions $A = 0$ or $B = 2$. The second has solutions $B = 0$ or $A = 1$. Thus the equilibrium points are (0,0) and (1,2).

(c) In the long run, one of the companies will go out of business. Two of the trajectories in the figure below go toward the A axis; they represent A surviving and B going out of business. The trajectories going toward the B axis represent A going out of business. Notice both the equilibrium points are unstable.

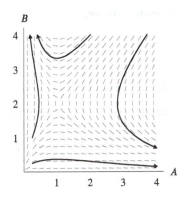

Strengthen Your Understanding

25. Equilibrium occurs when both $dx/dt = 0$ and $dy/dt = 0$. Both nullclines have horizontal line segments indicating points that $dy/dt = 0$. There are no nullclines that indicate where $dx/dt = 0$. Thus, the point (6, 6) has $dy/dt = 0$ but $dx/dt \neq 0$ and is not an equilibrium point.

29. False. For example, the system $dx/dt = x(y - x^2), dy/dt = y$ has a nullcline of $y = x^2$.

Solutions for Section 11.10

Exercises

1. If we write $y = 3 \sin 2t + 4 \cos 2t$ in the form $y(t) = A \sin(2t + \phi)$, then $A = \sqrt{3^2 + 4^2} = 5$.

5. (a)

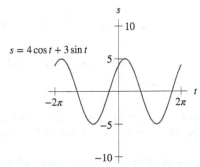

(b) Trace along the curve to the highest point; which has coordinates of about (0.66, 5), so $A \approx 5$. If $s = 5 \sin(t + \phi)$, then the maximum occurs where $t \approx 0.66$ and $t + \phi = \pi/2$, that is $0.66 + \phi \approx 1.57$, giving $\phi \approx 0.91$.

(c) Analytically

$$A = \sqrt{4^2 + 3^2} = 5$$

and

$$\tan \phi = \frac{4}{3} \quad \text{so} \quad \phi = \arctan\left(\frac{4}{3}\right) = 0.93.$$

9. If $y(t) = A \sin(2t) + B \cos(2t)$ then

$$y' = 2A \cos(2t) - 2B \sin(2t)$$
$$y'' = -4A \sin(2t) - 4B \cos(2t)$$

therefore

$$y'' + 4y = -4A \sin(2t) - 4B \cos(2t) + 4(A \sin(2t) + B \cos(2t)) = 0$$

for all values of A and B, so the given function is a solution.

13. The characteristic equation of $9z'' + z = 0$ is

$$9r^2 + 1 = 0$$

If we write this in the form $r^2 + br + c = 0$, we have that $r^2 + 1/9 = 0$ and

$$b^2 - 4c = 0 - (4)(1/9) = -4/9 < 0$$

This indicates underdamped motion and since the roots of the characteristic equation are $r = \pm\frac{1}{3}i$, the general equation is

$$y(t) = C_1 \cos\left(\frac{1}{3}t\right) + C_2 \sin\left(\frac{1}{3}t\right)$$

Problems

17. At $t = 0$, we find that $y = 0$. Since $-1 \le \sin 3t \le 1$, y ranges from -0.5 to 0.5, so at $t = 0$ it is starting in the middle. Since $y' = -1.5 \cos 3t$, we see $y' = -1.5$ when $t = 0$, so the mass is moving downward.

21. (a) General solution

$$x(t) = A \cos 4t + B \sin 4t.$$

Thus,

$$5 = A \cos 0 + B \sin 0 \quad \text{so } A = 5.$$

Since $x'(0) = 0$, we have

$$0 = -4A \sin 0 + 4B \cos 0 \quad \text{so } B = 0.$$

Thus,

$$x(t) = 5 \cos 4t$$

so amplitude $= 5$, period $= \frac{2\pi}{4} = \frac{\pi}{2}$.

(b) General solution

$$x(t) = A \cos\left(\frac{t}{5}\right) + B \sin\left(\frac{t}{5}\right).$$

Since $x(0) = -1$, we have $A = -1$.
Since $x'(0) = 2$, we have

$$2 = -\frac{A}{5} \sin 0 + \frac{B}{5} \cos 0 \quad \text{so } B = 10.$$

Thus,

$$x(t) = -\cos\left(\frac{t}{5}\right) + 10 \sin\left(\frac{t}{5}\right).$$

So, amplitude $= \sqrt{(-1)^2 + 10^2} = \sqrt{101}$, period $= \frac{2\pi}{1/5} = 10\pi$.

25. (a) If x_0 is increased, the amplitude of the function x is increased, but the period remains the same. In other words, the pendulum will start higher, but the time to swing back and forth will stay the same.

(b) If l is increased, the period of the function x is increased. (Remember, the period of $x_0 \cos\sqrt{\frac{g}{l}}t$ is $\frac{2\pi}{\sqrt{g/l}} = 2\pi\sqrt{l/g}$.)
In other words, it will take longer for the pendulum to swing back and forth.

29. The equation we have for the charge tells us that:

$$\frac{d^2Q}{dt^2} = -\frac{Q}{LC},$$

where L and C are positive.
If we let $\omega = \sqrt{\frac{1}{LC}}$, we know the solution is of the form:

$$Q = C_1 \cos \omega t + C_2 \sin \omega t.$$

Since $Q(0) = 0$, we find that $C_1 = 0$, so $Q = C_2 \sin \omega t$.

Since $Q'(0) = 4$, and $Q' = \omega C_2 \cos \omega t$, we have $C_2 = \dfrac{4}{\omega}$, so $Q = \dfrac{4}{\omega} \sin \omega t$.

But we want the maximum charge, meaning the amplitude of Q, to be $2\sqrt{2}$ coulombs. Thus, we have $\dfrac{4}{\omega} = 2\sqrt{2}$, which

gives us $\omega = \sqrt{2}$.

So we now have: $\sqrt{2} = \dfrac{1}{\sqrt{LC}} = \dfrac{1}{\sqrt{10C}}$. Thus, $C = \dfrac{1}{20}$ farads.

Strengthen Your Understanding

33. A second-order differential equation is a differential equation involving the second derivative, so we find the second derivative of $y = e^{2x}$. We have $y' = 2e^{2x}$ and $y'' = 4e^{2x}$. Thus $y'' = 4y$ is a second order differential equation which has $y = e^{2x}$ as a solution. Other answers are possible.

Solutions for Section 11.11

Exercises

1. The characteristic equation is $r^2 + 4r + 3 = 0$, so $r = -1$ or -3.
Therefore $y(t) = C_1 e^{-t} + C_2 e^{-3t}$.

5. The characteristic equation is $r^2 + 7 = 0$, so $r = \pm\sqrt{7}i$.
Therefore $s(t) = C_1 \cos \sqrt{7}t + C_2 \sin \sqrt{7}t$.

9. The characteristic equation is $r^2 + r + 1 = 0$, so $r = -\dfrac{1}{2} \pm \dfrac{\sqrt{3}}{2}i$.
Therefore $p(t) = C_1 e^{-t/2} \cos \dfrac{\sqrt{3}}{2}t + C_2 e^{-t/2} \sin \dfrac{\sqrt{3}}{2}t$.

13. The characteristic equation of $9z'' - z = 0$ is
$$9r^2 - 1 = 0.$$
If this is written in the form $r^2 + br + c = 0$, we have that $r^2 - 1/9 = 0$ and
$$b^2 - 4c = 0 - (4)(-1/9) = 4/9 > 0$$
This indicates overdamped motion and since the roots of the characteristic equation are $r = \pm 1/3$, the general solution is
$$y(t) = C_1 e^{\frac{1}{3}t} + C_2 e^{-\frac{1}{3}t}.$$

17. The characteristic equation is
$$r^2 + 5r + 6 = 0$$
which has the solutions $r = -2$ and $r = -3$ so that
$$y(t) = Ae^{-3t} + Be^{-2t}$$
The initial condition $y(0) = 1$ gives
$$A + B = 1$$
and $y'(0) = 0$ gives
$$-3A - 2B = 0$$
so that $A = -2$ and $B = 3$ and
$$y(t) = -2e^{-3t} + 3e^{-2t}$$

21. The characteristic equation is $r^2 + 6r + 5 = 0$, so $r = -1$ or -5.
Therefore $y(t) = C_1 e^{-t} + C_2 e^{-5t}$.
$y'(t) = -C_1 e^{-t} - 5C_2 e^{-5t}$
$y'(0) = 0 = -C_1 - 5C_2$
$y(0) = 1 = C_1 + C_2$
Therefore $C_2 = -1/4$, $C_1 = 5/4$ and $y(t) = \dfrac{5}{4}e^{-t} - \dfrac{1}{4}e^{-5t}$.

25. The characteristic equation is

$$r^2 + 5r + 6 = 0$$

which has the solutions $r = -2$ and $r = -3$ so that

$$y(t) = Ae^{-2t} + Be^{-3t}$$

The initial condition $y(0) = 1$ gives

$$A + B = 1$$

and $y(1) = 0$ gives

$$Ae^{-2} + Be^{-3} = 0$$

so that $A = \dfrac{1}{1-e}$ and $B = -\dfrac{e}{1-e}$ and

$$y(t) = \frac{1}{1-e}e^{-2t} + \frac{-e}{1-e}e^{-3t}$$

Problems

29. (a) $x'' + 4x = 0$ represents an undamped oscillator, and so goes with (IV).

(b) $x'' - 4x = 0$ has characteristic equation $r^2 - 4 = 0$ and so $r = \pm 2$. The solution is $C_1 e^{-2t} + C_2 e^{2t}$. This represents non-oscillating motion, so it goes with (II).

(c) $x'' - 0.2x' + 1.01x = 0$ has characteristic equation $r^2 - 0.2 + 1.01 = 0$ so $b^2 - 4ac = 0.04 - 4.04 = -4$, and $r = 0.1 \pm i$. So the solution is

$$C_1 e^{(0.1+i)t} + C_2 e^{(0.1-i)t} = e^{0.1t}(A \sin t + B \cos t).$$

The negative coefficient in the x' term represents an amplifying force. This is reflected in the solution by $e^{0.1t}$, which increases as t increases, so this goes with (I).

(d) $x'' + 0.2x' + 1.01x$ has characteristic equation $r^2 + 0.2r + 1.01 = 0$ so $b^2 - 4ac = -4$. This represents a damped oscillator. We have $r = -0.1 \pm i$ and so the solution is $x = e^{-0.1t}(A \sin t + B \cos t)$, which goes with (III).

33. The frictional force is $F_{\text{drag}} = -c\frac{ds}{dt}$. Thus spring (iv) has the smallest frictional force.

37. Recall that $s'' + bs' + cs = 0$ is overdamped if the discriminant $b^2 - 4c > 0$, critically damped if $b^2 - 4c = 0$, and underdamped if $b^2 - 4c < 0$. Since $b^2 - 4c = 8 - 4c$, the solution is overdamped if $c < 2$, critically damped if $c = 2$, and underdamped if $c > 2$.

41. (a) We have $k/m = 10$ so $k = 10m = 100$. We have $a/m = 7$ so $a = 7m = 70$.

(b) Since $b^2 - 4c = 7^2 - 4(10) = 9 > 0$, the motion is overdamped. The solutions of the characteristic equation $r^2 + 7r + 10 = 0$ are $r_1 = -2$ and $r_2 = -5$. The general solution is

$$s(t) = C_1 e^{-2t} + C_2 e^{-5t}.$$

To find the constants C_1 and C_2, we use the initial conditions:

$$s(0) = C_1 e^0 + C_2 e^0 = C_1 + C_2 = -1.$$

Since $s'(t) = -2C_1 e^{-2t} - 5C_2 e^{-5t}$, we have

$$s'(0) = -2C_1 e^0 - 5C_2 e^0 = -2C_1 - 5C_2 = -7.$$

Solving the simultaneous equations $C_1 + C_2 = -1$ and $-2C_1 - 5C_2 = -7$, we have $C_1 = -4$ and $C_2 = 3$. The solution satisfying the given initial conditions is

$$s(t) = -4e^{-2t} + 3e^{-5t}.$$

(c) We see on a graph of the solution that the lowest point is at approximately $s = -1.58$. The spring is overdamped and never goes above equilibrium, so the highest point is at equilibrium $s = 0$. See Figure 11.23.

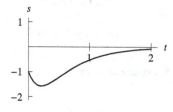

Figure 11.23

(d) We use the graph of the solution to see that the mass is within 0.1 of equilibrium for $t \geq 1.843$.

45. In the underdamped case, $b^2 - 4c < 0$ so $4c - b^2 > 0$. Since the roots of the characteristic equation are

$$\alpha \pm i\beta = \frac{-b \pm \sqrt{b^2 - 4c}}{2} = \frac{-b \pm i\sqrt{4c - b^2}}{2}$$

we have $\alpha = -b/2$ and $\beta = (\sqrt{4c - b^2})/2$ or $\beta = -(\sqrt{4c - b^2})/2$. Since the general solution is

$$y = C_1 e^{\alpha t} \cos \beta t + C_2 e^{\alpha t} \sin \beta t$$

and since α is negative, $y \to 0$ as $t \to \infty$.

49. (a) $\dfrac{d^2 y}{dt^2} = -\dfrac{dx}{dt} = y$ so $\dfrac{d^2 y}{dt^2} - y = 0$.

   (b) Characteristic equation $r^2 - 1 = 0$, so $r = \pm 1$.
   The general solution for y is $y = C_1 e^t + C_2 e^{-t}$, so $x = C_2 e^{-t} - C_1 e^t$.

53. In this case, the differential equation describing the charge is $Q'' + Q' + \frac{1}{4}Q = 0$, so the characteristic equation is $r^2 + r + \frac{1}{4} = 0$. This equation has one root, $r = -\frac{1}{2}$, so the equation for charge is

$$Q(t) = (C_1 + C_2 t)e^{-\frac{1}{2}t},$$
$$Q'(t) = -\frac{1}{2}(C_1 + C_2 t)e^{-\frac{1}{2}t} + C_2 e^{-\frac{1}{2}t}$$
$$= \left(C_2 - \frac{C_1}{2} - \frac{C_2 t}{2} \right) e^{-\frac{1}{2}t}.$$

   (a) We have

$$Q(0) = C_1 = 0,$$
$$Q'(0) = C_2 - \frac{C_1}{2} = 2.$$

   Thus, $C_1 = 0$, $C_2 = 2$, and

$$Q(t) = 2te^{-\frac{1}{2}t}.$$

   (b) We have

$$Q(0) = C_1 = 2,$$
$$Q'(0) = C_2 - \frac{C_1}{2} = 0.$$

   Thus, $C_1 = 2$, $C_2 = 1$, and

$$Q(t) = (2 + t)e^{-\frac{1}{2}t}.$$

   (c) The resistance was decreased by exactly the amount to switch the circuit from the overdamped case to the critically damped case. Comparing the solutions of parts (a) and (b) in Problems 52, we find that in the critically damped case the net charge goes to 0 much faster as $t \to \infty$.

Strengthen Your Understanding

57. The curve is decreasing at $t = 0$, so $s'(0)$ must be negative and cannot be 2.

Solutions for Chapter 11 Review

Exercises

1. For (I): $y = xe^x$, $y' = e^x + xe^x$, and $y'' = 2e^x + xe^x$.
   For (II): $y = xe^{-x}$, $y' = e^{-x} - xe^{-x}$, and $y'' = -2e^{-x} + xe^{-x}$.
   Thus (I) satisfies equation (d).
   (II) satisfies equation (c).

5. This equation is separable, so we integrate, giving

$$\int \frac{1}{0.2y - 8} \, dy = \int dx$$

so

$$\frac{1}{0.2} \ln |0.2y - 8| = x + C.$$

Thus,

$$y(x) = 40 + Ae^{0.2x}.$$

9. This equation is separable, so we integrate, using the table of integrals or partial fractions, to get:

$$\int \frac{250}{100P - P^2} \, dP = \int dt$$

$$\frac{250}{100} \left(\int \frac{1}{P} \, dP + \int \frac{1}{100 - P} \, dP \right) = \int dt$$

so

$$2.5(\ln |P| - \ln |100 - P|) = t + C$$

$$2.5 \ln \left| \frac{P}{100 - P} \right| = t + C$$

$$\frac{P}{100 - P} = Ae^{0.4t}$$

$$P = \frac{100Ae^{0.4t}}{1 + Ae^{0.4t}}$$

13. We have $2 \sin x - y^2 \frac{dy}{dx} = 0$, giving $2 \sin x = y^2 \frac{dy}{dx}$. Then $\int 2 \sin x \, dx = \int y^2 \, dy$ so $-2 \cos x = \frac{y^3}{3} + C$. Since $y(0) = 3$ we have $-2 = 9 + C$, so $C = -11$. Thus, $-2 \cos x = \frac{y^3}{3} - 11$ giving $y = \sqrt[3]{33 - 6 \cos x}$.

17. This equation is separable and so we write it as

$$\frac{1}{z(z - 1)} \frac{dz}{dt} = 1.$$

We integrate with respect to t, giving

$$\int \frac{1}{z(z - 1)} \, dz = \int dt$$

$$\int \frac{1}{z - 1} \, dz - \int \frac{1}{z} \, dz = \int dt$$

$$\ln |z - 1| - \ln |z| = t + C$$

$$\ln \left| \frac{z - 1}{z} \right| = t + C,$$

so that

$$\frac{z - 1}{z} = e^{t+C} = ke^t.$$

Solving for z gives

$$z(t) = \frac{1}{1 - ke^t}.$$

The initial condition $z(0) = 10$ gives

$$\frac{1}{1 - k} = 10$$

or $k = 0.9$. The solution is therefore

$$z(t) = \frac{1}{1 - 0.9e^t}.$$

21. We have $\frac{dy}{dx} = e^{x-y}$, giving $\int e^y \, dy = \int e^x \, dx$, so $e^y = e^x + C$. Since $y(0) = 1$, we have $e^1 = e^0 + C$ so $C = e - 1$. Thus, $e^y = e^x + e - 1$, so $y = \ln(e^x + e - 1)$.
[Note: $e^x + e - 1 > 0$ always.]

25. We have

$$\frac{dy}{dt} = 2^y \sin^3 t,$$

so

$$\int 2^{-y}\, dy = \int \sin^3 t\, dt.$$

Using Integral Table Formula 17, gives

$$-\frac{1}{\ln 2} 2^{-y} = -\frac{1}{3}\sin^2 t \cos t - \frac{2}{3}\cos t + C.$$

According to the initial conditions: $y(0) = 0$ so

$$-\frac{1}{\ln 2} = -\frac{2}{3} + C, \quad \text{and} \quad C = \frac{2}{3} - \frac{1}{\ln 2}.$$

Thus,

$$-\frac{1}{\ln 2} 2^{-y} = -\frac{1}{3}\sin^2 t \cos t - \frac{2}{3}\cos t + \frac{2}{3} - \frac{1}{\ln 2}.$$

Solving for y gives

$$2^{-y} = \frac{\ln 2}{3}\sin^2 t \cos t + \frac{2\ln 2}{3}\cos t - \frac{2\ln 2}{3} + 1.$$

It can be shown that the right side is always > 0, so we can take natural logs.

$$y\ln 2 = -\ln\left(\frac{\ln 2}{3}\sin^2 t \cos t + \frac{2\ln 2}{3}\cos t - \frac{2\ln 2}{3} + 1\right),$$

so

$$y = \frac{-\ln\left(\frac{\ln 2}{3}\sin^2 t \cos t + \frac{2\ln 2}{3}\cos t - \frac{2\ln 2}{3} + 1\right)}{\ln 2}.$$

Problems

29. Since $y = f(x)$ is a solution to $y' = xy - y$, we know that

$$f'(x) = xf(x) - f(x).$$

Letting $y = 2 + f(x)$, we want to find out if the left-hand side of this differential equation, y', equals the right-hand side, $xy - y$. First we consider the left-hand side:

$$\begin{aligned} y' &= (f(x) + 2)' \\ &= f'(x) \\ &= xf(x) - f(x) \quad \text{since } f'(x) = xf(x) - f(x). \end{aligned}$$

Turning to the right-hand side, we see that:

$$\begin{aligned} xy - y &= x(2 + f(x)) - (2 + f(x)) \quad \text{because } y = 2 + f(x) \\ &= xf(x) - f(x) + 2x - 2. \end{aligned}$$

We see that the left-hand side, $xf(x) - f(x)$, is not the same as the right-hand side, $xf(x) - f(x) + 2x - 2$, so $y = 2 + f(x)$ is not a solution to the equation.

33. (a) Assuming that the world's population grows exponentially, satisfying $dP/dt = cP$, and that the land in use for crops is proportional to the population, we expect A to satisfy $dA/dt = kA$.

(b) We have $A(t) = A_0 e^{kt} = 4.55 \cdot 10^9 e^{kt}$, where t is the number of years after 1966. Since 30 years later the amount of land in use is 4.93 billion hectares, we have

$$4.93 \cdot 10^9 = (4.55 \cdot 10^9)e^{k(30)},$$

so

$$e^{30k} = \frac{4.93}{4.55}.$$

Solving for k gives

$$k = \frac{\ln(4.93/4.55)}{30} = 0.00267.$$

Thus,

$$A = (4.55 \cdot 10^9)e^{0.00267t}.$$

We want to find t such that

$$6 \cdot 10^9 = A(t) = (4.55 \cdot 10^9)e^{0.00267t}.$$

Taking logarithms gives

$$t = \frac{\ln(6/4.55)}{0.00267} = 103.608 \text{ years.}$$

This model predicts land will have run out 104 years after 1966, that is by the year 2070.

37. Using (Rate balance changes) = (Rate interest is added)− (Rate payments are made), when the interest rate is i, we have

$$\frac{dB}{dt} = iB - 500.$$

Solving this equation, we find:

$$\frac{dB}{dt} = i\left(B - \frac{500}{i}\right)$$

$$\int \frac{dB}{B - \frac{500}{i}} = \int i\,dt$$

$$\ln\left|B - \frac{500}{i}\right| = it + C$$

$$B - \frac{500}{i} = Ae^{it}, \text{ where } A = \pm e^C.$$

At time $t = 0$ we start with a balance of \$25,000. Thus,
$25000 - \frac{500}{i} = Ae^0$, so $A = 25000 - \frac{500}{i}$.
Thus, $B = \frac{500}{i} + (25000 - \frac{500}{i})e^{it}$.
When $i = 0.01$, $B = 50000 - 25000e^{0.01t}$.
When $i = 0.02$, $B = 25000$.
When $i = 0.03$, $B = 16,666.67 + 8333.33e^{0.03t}$.
We now look at the graph in Figure 11.24 when $i = 0.01, i = 0.02$, and $i = 0.03$.

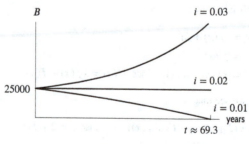

Figure 11.24

41. Let $C(t)$ be the current flowing in the circuit at time t, then

$$\frac{dC}{dt} = -\alpha C$$

where $\alpha > 0$ is the constant of proportionality between the rate at which the current decays and the current itself.

The general solution of this differential equation is $C(t) = Ae^{-\alpha t}$ but since $C(0) = 30$, we have that $A = 30$, and so we get the particular solution $C(t) = 30e^{-\alpha t}$.

When $t = 0.01$, the current has decayed to 11 amps so that $11 = 30e^{-\alpha 0.01}$ which gives $\alpha = -100\ln(11/30) = 100.33$, so that

$$C(t) = 30e^{-100.33t}.$$

45. (a) Quantity of A present at time t equals $(a - x)$.

Quantity of B present at time t equals $(b - x)$.

So

$$\text{Rate of formation of } C = k(\text{Quantity of } A)(\text{Quantity of } B)$$

gives

$$\frac{dx}{dt} = k(a - x)(b - x)$$

(b) Separating gives

$$\int \frac{dx}{(a - x)(b - x)} = \int k \, dt.$$

Rewriting the denominator as $(a - x)(b - x) = (x - a)(x - b)$ enables us to use Formula 26 in the Table of Integrals, since $a \neq b$. For some constant K, this gives

$$\frac{1}{a - b}(\ln|x - a| - \ln|x - b|) = kt + K.$$

Thus,

$$\ln\left|\frac{x - a}{x - b}\right| = (a - b)kt + K(a - b)$$

$$\left|\frac{x - a}{x - b}\right| = e^{K(a-b)}e^{(a-b)kt}$$

$$\frac{x - a}{x - b} = Me^{(a-b)kt} \quad \text{where } M = \pm e^{K(a-b)}.$$

Since $x = 0$ when $t = 0$, we have $M = \frac{a}{b}$. Thus,

$$\frac{x - a}{x - b} = \frac{a}{b}e^{(a-b)kt}.$$

Solving for x, we have

$$bx - ba = ae^{(a-b)kt}(x - b)$$

$$x(b - ae^{(a-b)kt}) = ab - abe^{(a-b)kt}$$

$$x = \frac{ab(1 - e^{(a-b)kt})}{b - ae^{(a-b)kt}} = \frac{ab(e^{bkt} - e^{akt})}{be^{bkt} - ae^{akt}}.$$

49. (a) The x population is unaffected by the y population—it grows exponentially no matter what the y population is, even if $y = 0$. If alone, the y population decreases to zero exponentially, because its equation becomes $dy/dt = -0.1y$.

(b) Here, interaction between the two populations helps the y population but does not effect the x population. This is not a predator-prey relationship; instead, this is a one-way relationship, where the y population is helped by the existence of x's. These equations could, for instance, model the interaction of rhinoceroses (x) and dung beetles (y).

53. (a) The insects grow exponentially with no birds around (the equation becomes $dx/dt = 3x$); the birds die out exponentially with no insects to feed on $dy/dt = -10y$. The interaction increases the birds' growth rate (the $+0.001xy$ term is positive), but decreases the insects' growth rate (the $-0.02xy$ term is negative). This is as we would expect: having the insects around helps the birds; having birds around hurts the insects.

(b) Equilibrium solutions occur where both derivatives are zero:

$$(3 - 0.02y)x = 0$$

$$-(10 - 0.001x)y = 0.$$

We see that the solutions are $(0, 0)$ and $(10,000, 150)$

(c) The chain rule gives an equation for dy/dx:

$$\frac{dy}{dx} = \frac{dy/dt}{dx/dt} = \frac{y(-10 + 0.001x)}{x(3 - 0.02y)}.$$

Separation of variables gives

$$\int \frac{-10 + 0.001x}{x} dx = \int \frac{3 - 0.02y}{y} dy,$$

which yields $3 \ln y - 0.02y = -10 \ln x + 0.001x + C$.

Using the initial point $A = (10,000, 160)$, we have

$$3 \ln 160 - 0.02(160) = -10 \ln 10,000 + 0.001(10,000) + C.$$

Thus, $C \approx 94.13$ and the solution is $3 \ln y - 0.02y = -10 \ln x + 0.001x + 94.1$

(d) We can check that the equation is satisfied by points B, C, D by substituting the coordinates into the equation $3 \ln y - 0.02y = -10 \ln x + 0.001x + 94.1$.

(e) The phase plane and the trajectory is in Figure 11.25.

(f) Consider point A. We have

$$\frac{dy}{dt} = 0, \quad \text{and} \quad \frac{dx}{dt} = 3(10,000) - 0.02(10,000)(160) = -2000 < 0.$$

Thus, x is decreasing at point A. Hence the rotation is counterclockwise in the phase plane, and the order of traversal is $A \longrightarrow B \longrightarrow C \longrightarrow D$.

(g) The graphs of x and y versus t are in Figure 11.26.

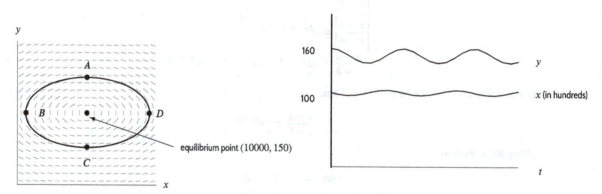

Figure 11.25   **Figure 11.26**

(h) At points A and C, we have $dy/dx = 0$, and at B and D, we have $dx/dy = 0$, so these points are extrema: y is maximized at A, minimized at C; x is maximized at D, minimized at B.

57. If $P > 2L$, then $P/L > 1$ and $P/(2L) > 1$:

$$\frac{dP}{dt} = - \underbrace{kP}_{+} \underbrace{\left(1 - \frac{P}{L}\right)}_{-} \underbrace{\left(1 - \frac{P}{2L}\right)}_{-} < 0.$$

Thus, if initially there are more than $2L$ animals, $dP/dt < 0$ and the population will decrease. Since $P = 2L$ is an equilibrium solution, the population will decrease towards $P = 2L$.